AF736065

The Geometric and Arithmetic Volume of Shimura Varieties of Orthogonal Type

Volume 35

CRM
MONOGRAPH
SERIES

Centre de Recherches Mathématiques
Montréal

The Geometric and Arithmetic Volume of Shimura Varieties of Orthogonal Type

Fritz Hörmann

The Centre de Recherches Mathématiques (CRM) was created in 1968 to promote research in pure and applied mathematics and related disciplines. Among its activities are special theme years, summer schools, workshops, postdoctoral programs, and publishing. The CRM receives funding from the Natural Sciences and Engineering Research Council (Canada), the FRQNT (Québec), the NSF (USA), and its partner universities (Université de Montréal, McGill, UQAM, Concordia, Université Laval, Université de Sherbrooke and University of Ottawa). It is affiliated with the Institut des Sciences Mathématiques (ISM). For more information visit www.crm.math.ca.

American Mathematical Society
Providence, Rhode Island USA

2010 *Mathematics Subject Classification.* Primary 11F27, 11F23, 11G18, 14G35, 14G40.

For additional information and updates on this book, visit
www.ams.org/bookpages/crmm-35

Library of Congress Cataloging-in-Publication Data

Hörmann, Fritz, 1978– author.
The geometric and arithmetic volume of Shimura varieties of orthogonal type / Fritz Hörmann.
pages cm. – (CRM monograph series ; 35)
Includes bibliographical references and index.
ISBN 978-1-4704-1912-7 (alk. paper)
1. Shimura varieties. 2. Arithmetical algebraic geometry. I. Title.
QA242.5.H67 2014
516.3′5–dc23

2014027530

Copying and reprinting. Individual readers of this publication, and nonprofit libraries acting for them, are permitted to make fair use of the material, such as to copy select pages for use in teaching or research. Permission is granted to quote brief passages from this publication in reviews, provided the customary acknowledgment of the source is given.

Republication, systematic copying, or multiple reproduction of any material in this publication is permitted only under license from the American Mathematical Society. Permissions to reuse portions of AMS publication content are handled by Copyright Clearance Center's RightsLink® service. For more information, please visit: `http://www.ams.org/rightslink`.

Send requests for translation rights and licensed reprints to `reprint-permission@ams.org`.

Excluded from these provisions is material for which the author holds copyright. In such cases, requests for permission to reuse or reprint material should be addressed directly to the author(s). Copyright ownership is indicated on the copyright page, or on the lower right-hand corner of the first page of each article within proceedings volumes.

© 2014 by the American Mathematical Society. All rights reserved.
The American Mathematical Society retains all rights
except those granted to the United States Government.
Printed in the United States of America.

∞ The paper used in this book is acid-free and falls within the guidelines
established to ensure permanence and durability.
Visit the AMS home page at `http://www.ams.org/`

10 9 8 7 6 5 4 3 2 1 19 18 17 16 15 14

Contents

CHAPTER 1

Overview

1.1. Introduction

A Shimura variety of orthogonal type arises from the Shimura datum consisting of the orthogonal group $\mathrm{SO}(L_{\mathbb{Q}})$ of a quadratic space $L_{\mathbb{Q}}$ of signature $(l-2, 2)$ and the set

$$\mathbb{D} := \{N \subseteq L_{\mathbb{R}} \mid N \text{ oriented negative definite plane}\}$$

which has the structure of an Hermitian symmetric domain and can be interpreted as a conjugacy class of morphisms $\mathbb{S} = \mathrm{Res}_{\mathbb{R}}^{\mathbb{C}} \mathbb{G}_{\mathrm{m}} \to \mathrm{SO}(L_{\mathbb{R}})$. For any compact open subgroup $K \subseteq \mathrm{SO}(L_{\mathbb{A}^{(\infty)}})$ we can form the Shimura variety

$$[\mathrm{SO}(L_{\mathbb{Q}}) \backslash \mathbb{D} \times \mathrm{SO}(L_{\mathbb{A}^{(\infty)}})/K]$$

in which we always consider the quotient as an orbifold. It is a smooth manifold for sufficiently small K, and it always has an algebraic model M^K, in general a smooth Deligne–Mumford stack.[1]

For simplicity we assume for the moment that there is an integral lattice $L_{\mathbb{Z}} \subset L_{\mathbb{Q}}$ with cyclic discriminant group $L_{\mathbb{Z}}^*/L_{\mathbb{Z}}$ of square-free order ε. Let K be the group of those integral isometries that induce the identity of the discriminant group. It goes back to Siegel [60] that in this case

$$(1.1) \quad \mathrm{vol}(\mathrm{M}^K) = 2^{-\lfloor (l-4)/2 \rfloor} \left(\prod_{i=1}^{\lfloor (l-1)/2 \rfloor} \zeta(1-2i) \right) \cdot \begin{cases} L\left(1 - \frac{l}{2}, \chi\right) & 2 \mid l, \\ (-1)^{(l^2-1)/8} \prod_{p \mid \epsilon/2} \left(p^{(l-1)/2} + \left(\frac{-1}{p}\right)^{(l-1)/2} \Phi_p \right) & 2 \nmid l, \end{cases}$$

where χ is the associated quadratic character (possibly trivial), Φ_p is the Hasse invariant at p, and where the volume is understood w.r.t. to a natural volume form (highest power of the Chern class of a canonical ample automorphic line bundle). This volume is in fact an intersection number and should therefore be a rational number. Inserting the well-known formulas

$$\zeta(1-2i) = -\frac{B_{2i}}{2i}, \quad L\left(1 - \frac{l}{2}, \chi\right) = -\frac{B_{l/2,\chi}}{l/2},$$

we see that the quantity (1.1) is indeed a rational number. It is also (up to a factor 2) the proportionality factor that occurs in the famous proportionality principle of Hirzebruch and Mumford [55] (cf. 2.6.22) applied to this case. For $l = 2$,

[1] defined over $\mathbb{Q}$ for $l > 2$

equation (1.1) is nothing but the classical analytic class number formula for definite binary quadratic forms. There exist formulas of similar shape for all locally symmetric spaces (see, e.g., [47] for the case of Chevalley groups). In the special case considered here M^K is, in fact, canonically defined over[2] $\mathbb{Z}[\frac{1}{2}]$. Therefore, using Arakelov theory, one obtains an analogous arithmetic 'volume element' (the highest power of the *arithmetic* Chern class of the same automorphic line bundle). To compute its Arakelov degree, which naturally is called the *arithmetic volume* of M^K, is the main objective of this book. The result is (in the special case considered here):

$$\widehat{\mathrm{vol}}(\mathrm{M}^K) \equiv \mathrm{vol}(\mathrm{M}^K) \tag{1.2}$$
$$\cdot \left(\frac{l-1}{2}C + \sum_{i=1}^{\lfloor (l-1)/2 \rfloor} \left(-2\frac{\zeta'(1-2i)}{\zeta(1-2i)} - N_{2i} \right) \right.$$
$$\left. + \begin{cases} -\dfrac{L'(1-l/2,\chi)}{L(1-l/2,\chi)} - \dfrac{1}{2}N_{l/2} - \dfrac{1}{2}\log\left|\dfrac{\varepsilon}{2}\right| & 2 \mid l, \\ \dfrac{1}{2}\displaystyle\sum_{p|\epsilon/2} \dfrac{p^{(l-1)/2} - \left(\frac{-1}{p}\right)^{(l-1)/2}\Phi_p}{p^{(l-1)/2} + \left(\frac{-1}{p}\right)^{(l-1)/2}\Phi_p} \log(p) + \dfrac{1}{2}\log(2) & 2 \nmid l, \end{cases} \right)$$

modulo rational multiples of $\log(2)$ (see B.1.3 for missing definitions).

We observe, of course, an apparent similarity between the formulas (1.1) and (1.2), and in fact, we have:

$$\mathrm{vol}(\mathrm{M}^K) = 4\tilde{\lambda}^{-1}(L_{\mathbb{Z}};0) \quad \widehat{\mathrm{vol}}(\mathrm{M}^K) \equiv 4\frac{\mathrm{d}}{\mathrm{d}s}\tilde{\lambda}^{-1}(L_{\mathbb{Z}};s)\bigg|_{s=0}$$

for the function

$$\tilde{\lambda}^{-1}(L_{\mathbb{Z}};s) = 2^{-\lfloor l/2 \rfloor}(2\pi)^{s(l-1)/2}|\varepsilon|^{-s/2} \prod_{i=1}^{\lfloor (l-1)/2 \rfloor} \frac{\Gamma(s+2i)\zeta(1-2s-2i)}{\Gamma(2s+2i)\cos(\pi(s+i))}$$
$$\cdot \begin{cases} 2^{s/2}\dfrac{\Gamma(s/2+l/2)L(1-s-l/2,\chi)}{\Gamma(s+l/2)\cos(\pi(s/2+\lfloor l/4 \rfloor))} & \text{if } l \text{ is even}, \\ 2^s \displaystyle\prod_{p|\epsilon/2} \left(p^{(l-1)/2+s} + \left(\frac{-1}{p}\right)^{(l-1)/2} \Phi_p \right) & \text{if } l \text{ is odd}, \end{cases}$$

The function $\lambda^{-1}(L_{\mathbb{Z}};s)$ has, however, an intrinsic and much more general definition in terms of representation densities. Its study was the main subject of the paper [26] by the author. Formula (1.2) had been known only in some cases for $l \leq 4$, and was conjectured in some cases for $l = 5$.

Our proof uses only information from the "Archimedean fibre," that is, we do not need explicit computations of local intersection numbers. We generalize work of Bruinier, Burgos, and Kühn [10] which dealt with the case of Hilbert modular surfaces. Borcherds' construction of orthogonal modular forms [2], and a computation of the integral of their norm [11,35] are used.

---

[2] and conjecturally over $\mathbb{Z}$

The orthogonal Shimura varieties are interesting in particular because they contain special algebraic cycles of arbitrary codimension whose arithmetic, respectively geometric volumes are conjectured to be encoded as a special value, respectively derivative of Fourier coefficients of Eisenstein series. This relation, in turn, has deep arithmetic consequences, including for example the formula of Gross and Zagier [21] and vast generalizations of it. It is also the key to the calculation of geometric and arithmetic volumes.

More precisely, these cycles are constructed as follows: For an isometry $x\colon M \hookrightarrow L$, where M is positive definite consider the subset

$$\mathbb{D}_x := \{N \in \mathbb{D} \mid N \perp x(M)\}$$

of $\mathbb{D}$. If (K-stable) lattices $L_\mathbb{Z}$ and $M_\mathbb{Z}$ or more generally a K-invariant Schwartz function $\varphi \in S(M^*_{\mathbb{A}^{(\infty)}} \otimes L_{\mathbb{A}^{(\infty)}})$ is chosen, we define cycles $\mathrm{Z}(L, M, \varphi; K)$ on the Shimura variety by taking the quotient of the union of the $\mathbb{D}_x$ over all integral isometries (respectively all isometries in the support of φ in a weighted way). See Section 3.1 for the precise definition. For singular lattices M analogous cycles can be defined. Consider a model $\mathrm{M}({}^K_\Delta\mathbf{O}(L))$ of a toroidal compactification of the Shimura variety (the notation will be explained below). We consider the generating series

$$\Theta_m(L, \varphi; \tau) = \sum_{Q \in \mathrm{Sym}^2((\mathbb{Z}^m)^*)} [\mathrm{Z}(L, \langle Q \rangle, \varphi; K)] \exp(2\pi i Q\tau)$$

with values in its algebraic Chow group $\mathrm{CH}^m\left(\mathrm{M}({}^K_\Delta\mathbf{O}(L))_\mathbb{C}\right) \otimes \mathbb{C}$. Assume now that $\mathrm{M}({}^K_\Delta\mathbf{O}(L))$ is even defined over $\mathbb{Z}$ in a "reasonably canonical" way. Kudla proposes a definition of arithmetic cycles $\hat{\mathrm{Z}}(L, M, \varphi; K, \nu)$, depending on the imaginary part ν of τ, too, and also for singular and for indefinite M, such that

$$\hat{\Theta}_m(L, \varphi; \tau) = \sum_{Q \in \mathrm{Sym}^2((\mathbb{Z}^m)^*)} [\hat{\mathrm{Z}}(L, \langle Q \rangle, \varphi; K, \nu)] \exp(2\pi i Q\tau)$$

should have values in a suitable Arakelov Chow group $\widehat{\mathrm{CH}}^m\left(\mathrm{M}({}^K_\Delta\mathbf{O}(L))\right) \otimes \mathbb{C}$. He proposes specific Green's functions which have singularities at the boundary.

The orthogonal Shimura varieties come equipped with a natural Hermitian automorphic line bundle $\Xi^*\overline{\mathcal{E}}$ whose metric also has singularities along the boundary. This provides us (via multiplication with a suitable power of its first Chern class and taking pushforward) with geometric (respectively arithmetic) degree maps $\deg\colon \mathrm{CH}^p(\cdots) \to \mathbb{Z}$ (respectively $\widehat{\deg}\colon \widehat{\mathrm{CH}}^p(\cdots) \to \mathbb{R}$). Assuming that an Arakelov theory can be set up to deal with all different occurring singularities, Kudla conjectures (for the geometric part this goes back to Siegel, Hirzebruch, Zagier, Kudla–Millson, Borcherds, etc.)[3]:

(K1) Θ_m and $\hat{\Theta}_m$ are (holomorphic, respectively nonholomorphic) Siegel modular forms of weight $l/2$ and genus m.

(K2) $\deg(\Theta_m)$ and $\widehat{\deg}(\hat{\Theta}_m)$ are equal to a special value (respectively the special derivative at the same point) of a *normalization* of the standard Eisenstein series of weight $l/2$ associated with the Weil representation of L [26, Section 4].

[3]If $l - r \leq m + 1$, the statement has to be modified. Here r is the Witt-rank of L.

(K3) $\Theta_{m_1}(L, \varphi_1; \tau_1) \cdot \Theta_{m_2}(L, \varphi_2; \tau_2) = \Theta_{m_1+m_2}\big(L, \varphi_1 \oplus \varphi_2, \left(\begin{smallmatrix} \tau_1 & \\ & \tau_2 \end{smallmatrix}\right)\big)$ and similarly for $\widehat{\Theta}$.

(K4) $\widehat{\Theta}_{l-1}$ can be defined with coefficients being zero-cycles on the arithmetic model, and it satisfies the properties above.

Kudla shows (see [36] for an overview) that this implies almost formally vast generalizations of the formula of Gross–Zagier [21]. In full generality the conjectures are known only for Shimura curves [44]. Section 1.3 contains a brief overview on what is known in other special cases.

(K2) is closedly related to the calculation of geometric and arithmetic volumes because the cycles in question consist themselves (for sufficiently good reduction) of models of orthogonal Shimura varieties of smaller dimension.

We are thus able to obtain partial results towards the *arithmetic part* of (K2) for *all* Shimura varieties of orthogonal type. More precisely, we prove the following:

3.5.5. Main Theorem. *Let $L_\mathbb{Z} \subset L_\mathbb{Q}$ be an integral lattice in a quadratic space of signature $(l-2, 2)$. Let K be its discriminant kernel. Let D' be the product of the primes p such that $p^2 \mid D$, where D is the discriminant of $L_\mathbb{Z}$. We have*

(1) $\mathrm{vol}_\mathcal{E}\Big(\mathrm{M}(^K_\Delta\mathbf{O}(L_\mathbb{Z}))\Big) = 4\tilde{\lambda}^{-1}(L_\mathbb{Z}; 0)$,

(2) $\widehat{\mathrm{vol}}_{\overline{\mathcal{E}}}\Big(\mathrm{M}(^K_\Delta\mathbf{O}(L_\mathbb{Z}))\Big) \equiv \frac{\mathrm{d}}{\mathrm{d}s}4\tilde{\lambda}^{-1}(L_\mathbb{Z}; s)|_{s=0}$ *in* $\mathbb{R}_{2D'}$.

Let $M_\mathbb{Z}$ be a lattice of dimension m with positive definite $Q_M \in \mathrm{Sym}^2(M^*_\mathbb{Q})$. Let D'' be the product of the primes p such that $M_{\mathbb{Z}_p} \not\subseteq M^*_{\mathbb{Z}_p}$ or $M^*_{\mathbb{Z}_p}/M_{\mathbb{Z}_p}$ is not cyclic. Assume

- $l - m \geq 4$, or
- $l = 4$, $m = 1$, *and $L_\mathbb{Q}$ has Witt rank* 1.

Then we have for each $\kappa \in \mathbb{Z}[(L^*_\mathbb{Z}/L_\mathbb{Z}) \otimes M^*_\mathbb{Z}]$:

(3) $\mathrm{vol}_\mathcal{E}\big(\mathrm{Z}(L_\mathbb{Z}, M_\mathbb{Z}, \kappa; K)\big) = 4\tilde{\lambda}^{-1}(L_\mathbb{Z}; s)\widetilde{\mu}(L_\mathbb{Z}, M_\mathbb{Z}, \kappa; 0)$,

(4) $\widehat{\mathrm{vol}}_{\overline{\mathcal{E}}}\big(\mathrm{Z}(L_\mathbb{Z}, M_\mathbb{Z}, \kappa; K)\big) \equiv \frac{\mathrm{d}}{\mathrm{d}s}4\big(\tilde{\lambda}^{-1}(L_\mathbb{Z}; s)\widetilde{\mu}(L_\mathbb{Z}, M_\mathbb{Z}, \kappa; s)\big)\big|_{s=0}$ *in* $\mathbb{R}_{2DD''}$.

Here $\mathbb{R}_N$ is $\mathbb{R}$ modulo rational multiples of $\log(p)$ for $p \mid N$, and the $\tilde{\lambda}$ and $\widetilde{\mu}$ are functions in $s \in \mathbb{C}$, given by certain Euler products (3.2.12) associated with representation densities of $L_\mathbb{Z}$ and $M_\mathbb{Z}$. The function $\widetilde{\mu}$ appears as the "holomorphic part" of a Fourier coefficient of the standard Eisenstein series associated with the Weil representation of L. Moreover K is the discriminant kernel and $\mathrm{M}(^K_\Delta\mathbf{O}(L_\mathbb{Z}))$ is any toroidal compactification of the orthogonal Shimura variety (see below). Finally $\overline{\mathcal{E}}$ is the integral tautological bundle on the compact dual equipped with an equivariant metric on the restriction of its complex fibre to $\mathbb{D}$; the geometric/arithmetic volumes are computed w.r.t. the associated arithmetic automorphic line bundle (Definition 2.6.4) $\Xi^*\overline{\mathcal{E}}$ on $\mathrm{M}(^K_\Delta\mathbf{O}(L_\mathbb{Z}))$.

In view of the Main Theorem it seems plausible that, if $L^*_\mathbb{Z}/L_\mathbb{Z}$ is cyclic, the function $4\tilde{\lambda}^{-1}(L; s)$ is always the correct normalizing factor in (K2).[4] This is in accordance with the observations of Kudla–Rapoport–Yang [42] in the case of Shimura curves.

[4] Of course this is only a statement about its first 2 Taylor coefficients.

For a more detailed introduction to the method of proof, we refer the reader to Section 1.4. An overview on known results in the direction of the conjectures will be given in Section 1.3.

In Chapter 2 up to Section 2.5 we recall the functorial theory of

- canonical integral models of toroidal compactifications of mixed Shimura varieties of Abelian type,
- their arithmetic automorphic vector bundles,

developed in the thesis of the author [25]. This theory, for general Hodge or even Abelian type, relies on an assumption regarding the stratification of the compactification (2.5.6) which was recently proven by Madapusi [49] and previously for the orthogonal Shimura varieties—spin-version—for $l \leq 5$ by Lan [46], (the Shimura varieties are of P.E.L. type in that case). The theory is crucial even for the precise formulation of our main result mentioned above. The reader is assumed to be familiar with the theory of rational Shimura varieties as developed by Deligne [16, 17] and to some extent with Pink's thesis [58] which extends the theory to the mixed case and contains the construction of rational canonical models of toroidal compactifications.

Our models are constructed locally (i.e. over an extension of $\mathbb{Z}_{(p)}$). Input data for the theory are p-integral mixed Shimura data (abbreviated p-MSD) $\mathbf{X} = (P_\mathbf{X}, \mathbb{D}_\mathbf{X}, h_\mathbf{X})$ consisting of an affine group scheme $P_\mathbf{X}$ over $\operatorname{Spec}(\mathbb{Z}_{(p)})$ of a certain rigid type (P) (see 2.1.6) and a set $\mathbb{D}_\mathbf{X}$ which comes equipped with a finite covering $h_\mathbf{X}\colon \mathbb{D}_\mathbf{X} \to \operatorname{Hom}(\mathbb{S}_\mathbb{C}, P_{\mathbf{X},\mathbb{C}})$ onto a certain conjugacy class, subject to some axioms, which are roughly Pink's mixed extension [58] of Deligne's axioms for a pure Shimura datum. Consider a compact open subgroup $K \subseteq P_\mathbf{X}(\mathbb{A}^{(\infty)})$ of the form $K^{(p)} \times P_\mathbf{X}(\mathbb{Z}_p)$ for a compact open subgroup $K^{(p)} \subseteq P_\mathbf{X}(\mathbb{A}^{(\infty,p)})$ (we call those admissible). For the toroidal compactification a certain rational polyhedral cone decomposition Δ of a natural conical complex $C_\mathbf{X}$ associated with $\mathbf{X}$ is needed. We call the collection ${}^K\mathbf{X}$ (respectively ${}^K_\Delta\mathbf{X}$) extended (compactified) p-integral mixed Shimura data (abbreviated p-EMSD, respectively p-ECMSD). These form categories where the morphisms ${}^K_\Delta\mathbf{X} \to {}^{K'}_{\Delta'}\mathbf{Y}$ are pairs (α, ρ) of a morphism α of Shimura data and $\rho \in P_\mathbf{Y}(\mathbb{A}^{(\infty,p)})$ satisfying compatibility conditions with the K's and Δ's. The construction of the models defines a functor M from p-ECMSD to the category of Deligne–Mumford stacks over reflex rings (above $\mathbb{Z}_{(p)}$). The functor, base-changed to $\mathbb{C}$ and restricted to p-EMSD, becomes naturally isomorphic to the one given by the construction of the analytic mixed Shimura variety. It is characterized uniquely by: Deligne's canonical model condition; Milne's extension property (integral canonicity); a stratification of the boundary into mixed Shimura varieties, together with boundary isomorphisms of the formal completions along these strata with similar completions of other (more mixed) Shimura varieties. These boundary isomorphisms, for the case of the symplectic Shimura varieties, are given by Mumford's construction [19, Appendix]. There is also a functor $\mathrm{M}^\vee$ ('compact' dual) from p-MSD to the category of schemes over reflex rings. The duals come equipped with an action of the group scheme $P_\mathbf{X}$, and we have morphisms of Artin stacks

$$\Xi_\mathbf{X}\colon \mathrm{M}({}^K_\Delta\mathbf{X}) \to [\mathrm{M}^\vee(\mathbf{X})/P_{\mathbf{X},\mathcal{O}_\mathbf{X}}].$$

Those constitute a pseudonatural transformation of functors, are a model of the usual construction over $\mathbb{C}$ if Δ is trivial, and are compatible with the boundary isomorphisms. This is the theory of integral automorphic vector bundles. For more information on the 'philosophy' of these objects in terms of motives see Section 1.5.

In particular this enables us to do the following construction: Let $\mathbf{X}$ be, for simplicity, a pure Shimura datum with reflex field $\mathbb{Q}$, and let $\mathcal{E}$ be a $P_{\mathbf{X}}$-equivariant vector bundle on $\mathrm{M}^\vee(\mathbf{X})$ equipped with a $P_{\mathbf{X},\mathbb{R}}$-equivariant Hermitian metric $h_{\mathcal{E}}$ on its restriction to the image of the Borel embedding $h_{\mathbf{X}}(\mathbb{D}_{\mathbf{X}}) \hookrightarrow \mathrm{M}^\vee(\mathbf{X})(\mathbb{C})$. The above morphism of stacks and its 2-isomorphism over $\mathbb{C}$ with the analytic period construction allows us to obtain a *well-defined* Arakelov vector bundle $(\Xi^*\mathcal{E}, \Xi^*h_{\mathcal{E}})$. The metric $\Xi^*h_{\mathcal{E}}$, however, has (rather mild) singularities along the boundaries of toroidal compactifications (see below). Without the existence of this canonical pullback, it would not even make sense to speak of arithmetic volumes.

The remaining sections of Chapter 2 are completely revised with respect to [25]: In Section 2.6 we define integral Hermitian automorphic vector bundles and the notions of arithmetic and geometric volume. Furthermore we set up an Arakelov theory which has enough properties to deal with singularities of the natural Hermitian metrics on "fully decomposed" automorphic vector bundles. This uses work of Burgos, Kramer, and Kühn [14, 15] but the arithmetic Chow groups are defined using a technically simpler method.

In Sections 2.7 and 2.8 a precise general q-expansion principle is derived from the abstract properties of Section 2.5.

Chapter 3 is concerned entirely with the theory of orthogonal Shimura varieties: In Section 3.1 the structure of the models of orthogonal Shimura varieties is investigated and special cycles are defined.

In Section 3.2 we use the general q-expansion principle to prove that Borcherds products with their natural norm yield *integral* sections of an appropriate integral Hermitian line bundle. Among other things the product expansions of Borcherds' are adelized and their Galois properties investigated.

In Section 3.3 we prove that the bundle of vector valued modular forms for the Weil representation (which appear as input forms in the construction of Borcherds products) has a rational structure. Then we use this to construct input forms with special properties which will be needed later.

In Section 3.4 we relate Borcherds' theory and Arakelov geometry on the orthogonal Shimura varieties. Its main result is an (averaged) arithmetic Siegel–Weil formula which will be crucial for the proof of the Main Theorem.

In Section 3.5 the Main Theorem is proven. An overview on the proof may be found in Section 1.4.

In Appendix A additional material on quadratic forms, on "lacunarity of modular forms," and on semilinear representations is provided which will be needed in the proofs in Section 3.5. Appendix B contains a continuation of the calculation in [26] of the function $\lambda(L_{\mathbb{Z}}; s)$ for the special case of lattices with square-free discriminant.

Finally, it is a pleasure to thank the Department of Mathematics and Statistics at McGill University and especially Henri Darmon, Jayce Getz and Eyal Goren, for providing a very inspiring working environment during the preparation of this book.

1.2. A brief introduction to Siegel–Weil theory

1.2.1. Consider two lattices $L_{\mathbb{Z}} \cong \mathbb{Z}^l$, and $M_{\mathbb{Z}} \cong \mathbb{Z}^m$ with integral and positive definite quadratic forms Q_L, and Q_M. It is a classical problem, to which already Gauss, Euler and in particular Siegel devoted themselves, to determine the representation number, that is, the number of elements in the set of isometric embeddings

$$\mathrm{I}(L_{\mathbb{Z}}, M_{\mathbb{Z}}) = \{\alpha \colon M_{\mathbb{Z}} \hookrightarrow L_{\mathbb{Z}} \mid \alpha \text{ is an isometry}\}.$$

It includes (for $m = 1$) questions like: "In how many ways can an integer be represented as a sum of l squares?".

If $M_{\mathbb{Z}} = \mathbb{Z}^m$ then Q_M is given by an element in $\mathrm{Sym}^2((\mathbb{Z}^m)^*)$. We write $\langle Q \rangle$ for the lattice $\mathbb{Z}^m$ with quadratic form given by Q. The *generating series*, the *theta series* of $L_{\mathbb{Z}}$,

$$\Theta_m(L_{\mathbb{Z}}; \tau) = \sum_{Q \in \mathrm{Sym}^2((\mathbb{Z}^m)^*)} \#\mathrm{I}(L_{\mathbb{Z}}, \langle Q \rangle) \exp(2\pi i Q \cdot \tau), \tag{1.3}$$

(here τ is an element in Siegel's upper half space $\mathbb{H}_m \subset (\mathbb{C}^m \otimes \mathbb{C}^m)^s$, the subset of elements with positive definite imaginary part) is a *Siegel modular form* of weight $l/2$ for a certain congruence subgroup of Sp_{2m}' (the symplectic or metaplectic group, according to the parity of l). For example $\Theta_1(\langle 1 \rangle; \tau)$ is just the classical theta function.

Under some conditions on the dimensions, a certain weighted sum of these theta functions over all *classes* $L_{\mathbb{Z}}^{(i)}$ in the *genus* $L_{\hat{\mathbb{Z}}}$ is an Eisenstein series (cf. [26, Section 4] for details):

1.2.2. Theorem (Siegel–Weil). *If $l > 2m + 2$, we have*

$$\sum_i c_i \Theta_m(L_{\mathbb{Z}}^{(i)}; \tau) = E_m(\Phi; \tau, s_0).$$

The additional parameter s_0 indicates that this Eisenstein series is in fact the (holomorphic) special value of a nonholomorphic Eisenstein series $E_m(\Phi; \tau, s)$ at $s = s_0 := (l - m + 1)/2$. The Fourier coefficients of the series are given by a product formula

$$\mu(L_{\mathbb{Z}}, \langle Q \rangle; s, y) = \mu_\infty(L, \langle Q \rangle; s, y) \prod_p \mu_p(L_{\mathbb{Z}_p}, \langle Q \rangle; s). \tag{1.4}$$

Here y is the imaginary part of τ—its appearance indicates that this series is nonholomorphic for general s. At $s = 0$ the μ_p are just the p-adic volumes of the 'spheres' $I(L, \langle Q \rangle)(\mathbb{Z}_p)$, classically called *representation densities*. For almost all p, the functions μ_p are very simple polynomials in p^{-s} (see, e.g., [26, Theorem 8.1]). Otherwise they may be computed by determining sufficiently many representation numbers of the congruences modulo p^n.

Essentially, the Siegel–Weil formula (1.2.2) is valid, if and only if $l > m + 1$, but if $l \leq 2m + 2$, the value of the Eisenstein series has to be defined via analytic continuation in s and the theta function has sometimes to be complemented by indefinite coefficients.

The mere fact that the representation numbers (in an average over classes) should be given by a product over local volumes or densities can be explained easily in the adelic language:

1.2.3. Let $L_{\mathbb{Z}} \subset L_{\mathbb{Q}}$ now be an integral lattice in an arbitrary quadratic space (not necessarily definite) and $M_{\mathbb{Q}}$ a positive definite quadratic space. Assume $l - m \geq 3$, for simplicity, for the rest of the discussion. On the adelic points $\mathrm{SO}(L_{\mathbb{A}})$ of the special orthogonal group of $L_{\mathbb{Q}}$, there is a canonical measure μ. It is a product over local measures μ_ν on the various $\mathrm{SO}(L_{\mathbb{Q}_\nu})$, constructed by any algebraic volume form defined over $\mathbb{Q}$ [65]. The product μ is independent of the choice of this form. The volume of $\mathrm{SO}(L_{\mathbb{Q}})\backslash \mathrm{SO}(L_{\mathbb{A}})$ which turns out to be finite, is called the **Tamagawa number**, and we have

1.2.4. Theorem ([65]). *For $l \geq 3$*

$$\mathrm{vol}\big(\mathrm{SO}(L_{\mathbb{Q}})\backslash \mathrm{SO}(L_{\mathbb{A}})\big) = 2.$$

From this the fact that the representation numbers (in an average over classes) should be given by a product over local volumes already follows, as we will explain now (in a slightly broader context):

1.2.5. Let $\varphi \in S(L_{\mathbb{A}^{(\infty)}} \otimes M^*_{\mathbb{A}^{(\infty)}})$ be a Schwartz-Bruhat function (i.e., locally constant with compact support). Let $K = \prod_p K_p$ be a compact open subgroup of $\mathrm{SO}(L_{\mathbb{A}^{(\infty)}})$ which stabilizes φ. For example, K could be the stabilizer of the lattice $L_{\widehat{\mathbb{Z}}}$ and φ the characteristic function of $L_{\widehat{\mathbb{Z}}}$. Let K_∞ be a maximal compact subgroup of $\mathrm{SO}(L_{\mathbb{R}})$. (If L is definite, this will be equal to $\mathrm{SO}(L_{\mathbb{R}})$.)

From 1.2.4 we may infer that the volume of the real analytic orbifold

$$[\mathrm{SO}(L_{\mathbb{Q}})\backslash(\mathrm{SO}(L_{\mathbb{A}})/K_\infty K)],$$

induced by the quotient of μ_∞ and some measure on K_∞, is:

$$2\prod_{\nu} \mathrm{vol}_\nu^{-1}(K_\nu). \tag{1.5}$$

We have a finite disjoint decomposition

$$\mathrm{I}(L, M)(\mathbb{A}^{(\infty)}) \cap \mathrm{supp}(\varphi) = \bigcup_i K\alpha_i$$

If this set is nonempty, we have by Hasse's principle an $\alpha' \in \mathrm{I}(L, M)(\mathbb{Q})$ and hence $g_i \in \mathrm{SO}(L_{\mathbb{A}^{(\infty)}})$ with $g_i^{-1}\alpha' = \alpha_i$. There is a lattice $L_{\mathbb{Z}}^{(i)} \subset L_{\mathbb{Q}}$ satisfying $L_{\widehat{\mathbb{Z}}}^{(i)} = g_i^{-1}L_{\widehat{\mathbb{Z}}}$. We denote by abuse of notation by $\alpha_i^\perp$ the lattice $\mathrm{im}(\alpha')^\perp \cap L_{\mathbb{Z}}^{(i)}$. We have $\alpha_i^\perp \otimes \widehat{\mathbb{Z}} \cong \mathrm{im}(\alpha_i)^\perp$. However, only the genus of $\alpha_i^\perp$ is well-defined, but we will use the notation only for objects which depend only on this genus.

Consider the symmetric space[5]

$$\mathbb{D}(L) = \{\text{maximal negative definite subspaces of } L_{\mathbb{R}}\} = \mathrm{SO}(L_{\mathbb{R}})/K_\infty.$$

We have an embedding $\mathbb{D}(\alpha_i^\perp) \times \mathrm{SO}\big((\alpha_i^\perp)_{\mathbb{A}^{(\infty)}}\big) \hookrightarrow \mathbb{D}(L) \times \mathrm{SO}(L_{\mathbb{A}^{(\infty)}})$, given by the natural inclusion of $\mathbb{D}(\alpha_i^\perp) \hookrightarrow \mathbb{D}(L)$ and multiplication of the adelic part by g_i from the right.

We form the *special cycle* $\mathrm{Z}(L, M, \varphi; K)$, the following formal sum (with real coefficients):

$$\sum_i \varphi(\alpha_i)\Big[\mathrm{SO}((\alpha_i^\perp)_{\mathbb{Q}})\backslash\mathbb{D}(\alpha_i^\perp) \times \mathrm{SO}\big((\alpha_i^\perp)_{\mathbb{A}^{(\infty)}}\big)\Big/\Big({}^{g_i}K \cap \mathrm{SO}\big((\alpha_i^\perp)_{\mathbb{A}^{(\infty)}}\big)\Big)\Big]$$

---

[5] *Caution*: This definition differs from the later definition of $\mathbb{D}_{\mathbf{O}(L)}$ in case that L has signature $(l-2, 2)$

which we consider, by means of the embeddings above, as a formal sum of real analytic suborbifolds of $[\mathrm{SO}(L_{\mathbb{Q}})\backslash\mathbb{D}(L) \times (\mathrm{SO}(L_{\mathbb{A}^{(\infty)}})/K)]$. It does not depend on the choices made above.

The *canonical* measures [26, 2.4] on $\mathrm{SO}(L_{\mathbb{Q}})$, $\mathrm{SO}(\alpha_i^{\perp})$ and $\mathrm{I}(L,M)$ over any $\mathbb{Q}_\nu$ are related by an orbit equation [26, 5.6]—an equation of the shape:

$$\text{`volume of space'} = \sum_{\text{orbits}} \frac{\text{`volume of group'}}{\text{`volume of stabilizer'}},$$

similar to the corresponding formula for actions of finite groups on sets.

From this and (1.5) above

$$\frac{\mathrm{vol}(Z(L,M,\varphi;K))}{\mathrm{vol}(\mathrm{SO}(L_{\mathbb{Q}})\backslash\mathbb{D}(L) \times \mathrm{SO}(L_{\mathbb{A}^{(\infty)}})/K)} = \frac{\mathrm{vol}(K_\infty)}{\mathrm{vol}(K'_\infty)} \int_{\mathrm{I}(L,M)(\mathbb{A}^{(\infty)})} \varphi(\alpha)\mu(\alpha) \tag{1.6}$$

follows immediately. K'_∞ is any maximal compact subgroup of any of the $\mathrm{SO}(\alpha_{i,\mathbb{R}}^{\perp})$. We define $\mu_\infty(L,M)$ to be the quantity $\mathrm{vol}(K_\infty)/\mathrm{vol}(K'_\infty)$ (computed w.r.t. the canonical measures). If L is definite, it is equal to:

$$\mathrm{vol}(\mathrm{I}(L,M)(\mathbb{R})) = \prod_{k=l-m+1}^{l} 2\frac{\pi^{k/2}}{\Gamma(k/2)}.$$

Observe that

$$[\mathrm{SO}(L_{\mathbb{Q}})\backslash\mathbb{D}(L) \times (\mathrm{SO}(L_{\mathbb{A}^{(\infty)}})/K)] = \bigcup_j [(\mathrm{SO}(L_{\mathbb{Q}}) \cap {}^{g_j}K)\backslash\mathbb{D}(L)],$$

with respect to a set $\{g_j\}_j$ of representatives of $\mathrm{SO}(L_{\mathbb{Q}})\backslash \mathrm{SO}(L_{\mathbb{A}^{(\infty)}})/K$, i.e., of the *classes of* $\mathrm{SO}(L_{\mathbb{Q}})$ with respect to the compact open group K. (If K is the stabilizer of a lattice $L_{\widehat{\mathbb{Z}}}$, this coincides with the classical notion of classes in the genus $L_{\widehat{\mathbb{Z}}}$.) Similarly, we have

$$\mathrm{Z}(L,M,\varphi;K) = \sum_{i,k} \varphi(\alpha_{ik})\big[\big(\mathrm{SO}\big((\alpha_i^{\perp})_{\mathbb{Q}}\big) \cap K^{g_{ik}}\big)\backslash\mathbb{D}(\alpha_i^{\perp})\big], \tag{1.7}$$

where $\{g_{ik}\}_k$ is a set of representatives of the classes of $\mathrm{SO}\big((\alpha_i^{\perp})_{\mathbb{Q}}\big)$ w.r.t. ${}^{g_i}K \cap \mathrm{SO}\big((\alpha_i^{\perp})_{\mathbb{A}^{(\infty)}}\big)$.

Let now K be the stabilizer of $L_{\widehat{\mathbb{Z}}}$ and φ the characteristic function. We have the following easy

1.2.6. Lemma. *There is a bijection*

$$\left\{\begin{array}{l}\textit{class } L_{\mathbb{Z}}^{(j)} \textit{ in the genus } L_{\widehat{\mathbb{Z}}},\\ \mathrm{SO}(L_{\mathbb{Z}}^{(j)})\textit{-orbit } \mathrm{SO}(L_{\mathbb{Z}}^{(j)})\alpha \textit{ in } \mathrm{I}(L^{(j)},M)(\mathbb{Z})\end{array}\right\}$$
$$\xrightarrow{\sim} \left\{\begin{array}{l}\mathrm{SO}(L_{\widehat{\mathbb{Z}}})\textit{-orbit } \mathrm{SO}(L_{\widehat{\mathbb{Z}}})\alpha \textit{ in } \mathrm{I}(L,M)(\widehat{\mathbb{Z}}),\\ \textit{class in } \mathrm{SO}(\alpha_{\mathbb{Q}}^{\perp})\backslash \mathrm{SO}(\alpha_{\mathbb{A}^{(\infty)}}^{\perp})/K \cap \mathrm{SO}(\alpha_{\mathbb{A}^{(\infty)}}^{\perp})\end{array}\right\}.$$

We have, of course, a similar statement for any K.

We denote the cycle in this case by $\mathrm{Z}(L_{\mathbb{Z}}, M_{\mathbb{Z}})$ and it is, according to the lemma and (1.7), equal to:

$$\mathrm{Z}(L_{\mathbb{Z}}, M_{\mathbb{Z}}) = \sum_j \sum_{\mathrm{SO}(L_{\mathbb{Z}}^{(j)})\alpha \subset \mathrm{I}(L^{(j)},M)(\mathbb{Z})} \big[\big(\mathrm{SO}(\alpha_{\mathbb{Z}}^{\perp}) \cap \mathrm{SO}(L_{\mathbb{Z}}^{(j)})\big)\backslash\mathbb{D}(L)\big].$$

1.2.7. Now, if the form Q_L is *positive definite*, the quotient of volumes (1.6) has an interpretation as a global representation number. For this observe that in this case

$$\mathrm{vol}\big(\mathrm{SO}(L_{\mathbb{Z}})\backslash\mathbb{D}(L)\big) = \frac{1}{\#\,\mathrm{SO}(L_{\mathbb{Z}})}$$

and similarly

$$\mathrm{vol}\big((\mathrm{SO}(\alpha_{\mathbb{Z}}^{\perp}) \cap \mathrm{SO}(L_{\mathbb{Z}}^{(j)}))\backslash\mathbb{D}(\alpha^{\perp})\big) = \frac{1}{\#\big(\mathrm{SO}(\alpha_{\mathbb{Z}}^{\perp}) \cap \mathrm{SO}(L_{\mathbb{Z}}^{(j)})\big)}.$$

Furthermore, we have by the set theoretical orbit equation,

$$\frac{\#\,\mathrm{I}(L^{(j)}, M)(\mathbb{Z})}{\#\,\mathrm{SO}(L_{\mathbb{Z}}^{(j)})} = \sum_{\mathrm{SO}(L_{\mathbb{Z}}^{(j)})\alpha \in \mathrm{I}(L^{(j)}, M)(\mathbb{Z})} \frac{1}{\#\big(\mathrm{SO}(\alpha_{\mathbb{Z}}^{\perp}) \cap \mathrm{SO}(L_{\mathbb{Z}}^{(j)})\big)}.$$

Hence we get

$$\frac{\mathrm{vol}\big(\mathrm{Z}(L_{\mathbb{Z}}, M_{\mathbb{Z}})\big)}{\mathrm{vol}(\mathrm{SO}(L_{\mathbb{Q}})\backslash\mathbb{D}(L) \times \mathrm{SO}(L_{\mathbb{A}^{(\infty)}})/K)} = \frac{\sum_j \#\,\mathrm{I}(L_{\mathbb{Z}}^{(j)}, M)(\mathbb{Z})/\#\,\mathrm{SO}(L_{\mathbb{Z}}^{(j)})}{\sum_j 1/\#\,\mathrm{SO}(L_{\mathbb{Z}}^{(j)})}$$

which is precisely a weighted sum over the representation numbers. Combined with (1.6), we get *Siegel's formula*. The deep part, of course, is hidden in Theorem 1.2.4.

1.2.8. If the quadratic form on L is *indefinite*, say of signature (p, q), then these representation numbers do not make sense because there are always infinitely many isometries. However, equation (1.6) tells us, what the correct analogue in the indefinite case is: the quotient of volumes

$$\frac{\mathrm{vol}\big(\mathrm{Z}(L, M, \varphi; K)\big)}{\mathrm{vol}([\mathrm{SO}(L_{\mathbb{Q}})\backslash\mathbb{D}(L) \times (\mathrm{SO}(L_{\mathbb{A}^{(\infty)}})/K)])}.$$

For every cohomology theory H (in a very broad sense) one might in addition consider the *classes* $[Z(L, \langle Q\rangle, \varphi; K)]^H$ of these cycles and define their generating theta series:

$$\Theta_m^H(L, \varphi; \tau) = \sum_{Q \in \mathrm{Sym}^2((\mathbb{Z}^m)^*)} [\mathrm{Z}(L, \langle Q\rangle, \varphi; K)]^H \cup e_q^{m-r(Q)} \exp(2\pi i Q \cdot \tau),$$

where e_q is a certain Euler class, and $r(Q)$ is the rank of Q. One always expects modularity of this function and a relation to Eisenstein series.

Kudla and Millson [37, 38] have shown (generalizing work of Hirzebruch and Zagier [24]) that the generating series

$$\Theta_m^{\mathrm{B}}(L, \varphi; \tau) = \sum_{Q \in \mathrm{Sym}^2((\mathbb{Z}^m)^*)} [\mathrm{Z}(L, \langle Q\rangle, \varphi; K)]^{\mathrm{B}} \cup e_q^{m-r(Q)} \exp(2\pi i Q \cdot \tau),$$

with values in the Betti cohomology groups

$$H^{(p-m)q}([\mathrm{SO}(L_{\mathbb{Q}})\backslash\mathbb{D}(L) \times (\mathrm{SO}(L_{\mathbb{A}^{(\infty)}})/K)], \mathbb{C})$$

is a modular form. Under certain conditions on l, m and the Witt rank of L, its 'degree' is the special value of an Eisenstein series:

$$\langle \Theta_m^{\mathrm{B}}(L, \varphi; \tau), e_q^{l-m}\rangle = \mathrm{vol}_{e_q}([\mathrm{SO}(L_{\mathbb{Q}})\backslash\mathbb{D}(L) \times (\mathrm{SO}(L_{\mathbb{A}^{(\infty)}})/K)]) E_m(\Phi; \tau, s_0).$$

The latter equation follows essentially again from the Siegel–Weil formula (in its full generality) or the Tamagawa number result, respectively. If $L_{\mathbb{Q}}$ is anisotropic, the locally symmetric space is compact and the pairing on the left is the degree of the product in cohomology (Poincaré duality pairing). If $L_{\mathbb{Q}}$ is isotropic, the locally symmetric space is noncompact but the expression still makes sense, because the natural forms defining e_q^{l-m} are integrable on the special cycles. For details see [36].

The $\Theta_m^H(L, \varphi; \tau)$'s are also always expected to satisfy a product relation like:

$$\Theta_{m_1}^H(L, \varphi_1; \tau_1) \cup \Theta_{m_2}^H(L, \varphi_2; \tau_2) = \Theta_{m_1+m_2}^H\left(L, \varphi_1 \otimes \varphi_2; \begin{pmatrix} \tau_1 & \\ & \tau_2 \end{pmatrix}\right). \tag{1.8}$$

1.2.9. The above theory is particularly interesting if the signature is $(l-2, 2)$. As mentioned in the beginning, the locally symmetric orbifold

$$[\mathrm{SO}(L_{\mathbb{Q}})\backslash \mathbb{D}(L) \times (\mathrm{SO}(L_{\mathbb{A}^{(\infty)}})/K)]$$

has, in this case, a complex structure and is associated with an algebraic Deligne–Mumford stack $\mathrm{M}({}^K\mathbf{O}(L))$, a Shimura variety of orthogonal type. In particular, the special cycles may be considered as *algebraic* cycles on $\mathrm{M}({}^K\mathbf{O}(L))$ and we have the following theta function:

$$\Theta_m^{\mathrm{CH}}(L, \varphi; \tau) = \sum_{Q \in \mathrm{Sym}^2((\mathbb{Z}^m)^*)} [\mathrm{Z}(L, \langle Q \rangle, \varphi; K)]^{\mathrm{CH}} \cup \mathrm{c}_1(\Xi^* \mathcal{E})^{m-r(Q)} \exp(2\pi i Q \cdot \tau),$$

with values in $\mathrm{CH}^m(\mathrm{M}({}^K_\Delta\mathbf{O}(L))_{\mathbb{C}}) \otimes \mathbb{C}$, where $\mathrm{M}({}^K_\Delta\mathbf{O}(L))$ is a toroidal compactification of $\mathrm{M}({}^K\mathbf{O}(L))$.

$\Xi^*\mathcal{E}$ is a certain ample (automorphic) line bundle on $\mathrm{M}({}^K_\Delta\mathbf{O}(L))$. It is equipped with a Hermitian metric $\Xi^* h_{\mathcal{E}}$ (singular along the boundary of the compactification), whose associated Chern form is (roughly) e_2 above (see 3.2.5). The series is therefore a 'lift' of Θ_m^{B} with respect to the cycle class map.

The only known fact, however, in the direction of modularity *in arbitrary dimensions* is the following theorem:

1.2.10. Theorem. *For* $m = 1, 2$, $\Theta_m^{\mathrm{CH}}(L, \varphi; \tau)$ *is a modular form of weight* $l/2$.

This was proven by Borcherds [3] for $m = 1$, and for $m = 2$ independently by Martin Raum and by Bruinier [8] (using results of Wei Zhang).

The theta functions Θ_m^{B}, in this case, do satisfy the relation (1.8) [32]. An analogue of this for Θ_m^{CH} is not known in general.

1.3. Known results

1.3.1. In this section, some of the recent developments in the direction of Kudla's conjectures will be presented with no aim whatsoever towards completeness.

For lattices of small dimensions the associated Shimura varieties are of P.E.L. type and were already subject to a variety of classical work of Gross, Heegner, Hilbert, Hirzebruch, Riemann, Shimura, Siegel, Zagier and many others. The cases are listed in Table 1.1.

Modularity of $\widehat{\Theta_m^{\mathrm{CH}}}$ is widely unknown, especially for higher-dimensional varieties with nonempty boundary which requires the use of extended Arakelov theories like, e.g., defined in Section 2.6. Modularity was obtained so far only for the cases II and III above—for II, by work of Kudla, Rapoport and Yang [42, 43] culminating in their recent book *Modular forms and special cycles on Shimura curves* [44]. They obtained modularity of $\Theta_1^{\widehat{\mathrm{CH}}}$ and $\Theta_2^{\widehat{\mathrm{CH}}}$, as well as their relations to the corresponding special derivatives of Eisenstein series of genus 1, respectively 2 and of weight $\frac{3}{2}$. Also the formula (K3) was established, yielding inner product formulas. This completed earlier work started by Kudla in the 90s [31, 33, 34] and [40]. We mention also related work of Liu for higher-dimensional unitary Shimura varieties [48].

To obtain results (at least) about the equality of heights with the 'nonholomorphic' part of the special derivatives of Eisenstein series, there are in principle two approaches:

(1) The *first* approach is by comparison of direct calculations of the finite intersection numbers of the cycles $\widehat{\mathrm{Z}}(L, M, \varphi; K, y)$ and of the special derivative of the Eisenstein series, respectively. These lines have been followed predominantly in the above mentioned work. In these cases, the equality of the 'nonholomorphic part' of the special derivative with the integral of the corresponding Kudla-Millson Green's functions has also been verified.

Evidence in higher dimensions had been provided so far only by work of Kudla and Rapoport, [39] for Hilbert–Blumenthal varieties (V), and [41] for Siegel modular varieties (VII, VIII). These approaches rely heavily on explicit use of the underlying moduli problem. In particular, the special cycles are defined algebraically via a submoduli problem involving additional special endomorphisms.

(2) The *second* approach which is used in the main part of this book, is an inductive method, generalized from Burgos, Bruinier and Kühn [10], who investigated a special case of (V) above. However, for cycles of codimension $n > 1$, it seems to be restricted to the case of indices Q_M, where Q_M has 'good shape' at all primes p considered, at least such that $M_{\mathbb{Z}_p}^*/M_{\mathbb{Z}_p}$ is at most cyclic (i.e., essentially the codimension one case). Otherwise it seems to require at least as much knowledge about bad reduction as a direct computation of finite intersection numbers in the first approach requires.

This method has, however, the advantage of giving results in arbitrary dimension, even for non-P.E.L. type Shimura varieties, and with boundary, too—cases which only recently seem to come within reach of the first method. It uses modular forms on these Shimura varieties, constructed by Borcherds [2] by purely analytic means, using ideas from physics. They have a divisor consisting precisely of the codimension one cycles $\mathrm{Z}(L, \langle q \rangle, \varphi; K)$ and have integral Fourier coefficients. A computation of the integral of their norm is needed. This was accomplished before independently by Kudla [35] and by Bruinier and Kühn [11].

TABLE 1.1

	signature	**Witt rk.**	**classical name**
I	(0,2)	0	Heegner points (compact)
II	(1,2)	0	Shimura curves (compact)
III	(1,2)	1	modular curve (moduli space of elliptic curves)
IV	(2,2)	0	Shimura surfaces (compact)
V	(2,2)	1	Hilbert–Blumenthal varieties
VI	(2,2)	2	product of modular curves
VII	(3,2)	1	twisted Siegel modular threefolds
VIII	(3,2)	2	Siegel modular threefold (moduli space of Abelian surfaces)

In this book, using the second approach, we compute the respective arithmetic volumes for *all* Shimura varieties of orthogonal type and *all* cycles $\mathrm{Z}(L, M, \varphi; K)$ on them, but only up to contributions (multiples of $\log(p)$) from primes p, where the above requirement of 'good shape' is violated and up to contributions from $p = 2$, due to the still incomplete theory of good reduction of integral models of Shimura varieties of non-P.E.L. type. In case $m = 1$ (codimension 1) the heights of the special cycles can be computed for *all* $M = \langle q \rangle$, $q > 0$ only up to contributions from bad reduction of the surrounding Shimura variety.

For the modular curve, Yang [66] verified the modularity of $\Theta_1^{\widehat{\mathrm{CH}}}$ and the identity of $\langle \Theta_1^{\widehat{\mathrm{CH}}}, \hat{c}_1(\Xi^*\overline{\mathcal{E}}) \rangle$ with the special derivative of an Eisenstein series, using Chow groups of an extended Arakelov theory as in [14, 15] which, however, for the case needed here (arithmetic surfaces) had already been constructed long before by Kühn [45] and by Bost [4], independently. It should be mentioned that the equality of $\deg(\Theta_1^{\mathrm{B}})$ with the special *value* of the same Eisenstein series in this case is more difficult because the modular curve is, in a sense, an extremal case. One has to introduce also negative, nonholomorphic Fourier coefficients. The positive ones here are given by the class numbers of binary quadratic forms (the $\mathrm{Z}(L_{\mathbb{Z}}, \langle q \rangle_{\mathbb{Z}})$ consist of special *points* in this case). This special value of the Eisenstein series which is accordingly also nonholomorphic, is Zagier's famous Eisenstein series [67] of weight $\frac{3}{2}$. The other conjectures have not been verified so far in this special case, but Bruinier and Yang succeeded in obtaining the formula of Gross and Zagier and generalizations directly, also using Borcherds products [12].

1.4. An overview on the proof

1.4.1. Let $L_{\mathbb{Z}} \subset L_{\mathbb{Q}}$ be a lattice of signature $(l-2, 2)$ and let D be the discriminant of $L_{\mathbb{Z}}$. Let $K \subseteq \mathrm{SO}(L_{\mathbb{A}^{(\infty)}})$ be its discriminant kernel. Let $\mathrm{M}({}^K\mathbf{O}(L_{\mathbb{Z}}))$ be a global canonical model of the Shimura variety

$$[\mathrm{SO}(L_{\mathbb{Q}})\backslash \mathbb{D}_{\mathbf{O}(L)} \times \mathrm{SO}(\mathbb{A}^{(\infty)})/K].$$

We want to prove the assertion

$$\widehat{\mathrm{vol}}\Big(\mathrm{M}({}^K\mathbf{O}(L_{\mathbb{Z}}))\Big) \equiv 4\frac{\mathrm{d}}{\mathrm{d}s}\tilde{\lambda}^{-1}(L_{\mathbb{Z}}; s)\Big|_{s=0} \tag{1.9}$$

(up to some contributions of the form $\mathbb{Q}\log(p)$, but with control over these primes!). To illuminate the main ideas we ignore here completely the issue of compactification. This is legitimate insofar as the arithmetic volume will be shown to be independent of the choice of toroidal compsactification. The idea is to prove (1.9) first up to rational multiples of $\log(p)$, where p is either 2 or such that the Shimura variety has bad reduction there, i.e., for $p \mid 2D$. This will be done by induction on l, the dimension of L, using *codimension 1* cycles. We are immediately reduced (using (1.11) below) to show for each prime $p \nmid 2D$ the statement

$$\widehat{\mathrm{vol}}\Big(\mathrm{M}({}^K\mathbf{O}(L_{\mathbb{Z}}))\Big) \equiv \mathrm{vol}\Big(\mathrm{M}({}^K\mathbf{O}(L_{\mathbb{Z}}))\Big) \cdot \frac{\mathrm{d}\log\tilde{\lambda}^{-1}(L_{\mathbb{Z}}; s)}{\mathrm{d}s}\Big|_{s=0} \tag{1.10}$$

up to rational multiples of $\log(q)$ for all primes $q \neq p$. Here we gain the flexibility of taking any K of the form $\mathrm{SO}(L_{\mathbb{Z}_p})K^{(p)}$, where $K^{(p)} \subset \mathrm{SO}(L_{\mathbb{A}^{(\infty,p)}})$ is arbitrary, and to work with the local canonical model of the Shimura variety over $\mathbb{Z}_{(p)}$. In certain cases (e.g., if $L_{\mathbb{Z}}^*/L_{\mathbb{Z}}$ is cyclic), we will be able to show (1.9) a posteriori merely up to $\mathbb{Q}\log(2)$, where $\mathrm{M}\big({}^{K}\mathbf{O}(L_{\mathbb{Z}})\big)$ has to be interpreted as a model induced by an embedding into a nonsingular $\mathrm{M}\big({}^{K'}\mathbf{O}(L'_{\mathbb{Z}})\big)$ for bigger L'. It seems that, at least if $L_{\mathbb{Z}}^*/L_{\mathbb{Z}}$ is cyclic *of square-free order*, such a model is still *canonical* at all $p \neq 2$ in a way similiar to our notion of canonical for p of good reduction.

We will reprove along the way also the statement:

$$\mathrm{vol}\Big(\mathrm{M}\big({}^{K}\mathbf{O}(L_{\mathbb{Z}})\big)\Big) = 4\tilde{\lambda}^{-1}(L_{\mathbb{Z}};0) \tag{1.11}$$

where K is the discriminant kernel.

1.4.2. Let $q \in \mathbb{Q}$, $q > 0$ be given such that $|q|_p = 1$. Let κ be (the characteristic function of) a coset in $L_{\hat{\mathbb{Z}}}^*/L_{\hat{\mathbb{Z}}}$ and assume that K fixes κ, e.g., let it be contained in the discriminant kernel. First we observe that, by definition, the arithmetic/geometric volume of the special cycle $\mathrm{Z}(L, \langle q\rangle, \kappa; K)$ (here understood to be the Zariski closure of those defined in the introduction, see 3.1.7 for the precise definition) is

$$\mathrm{vol}\big(\mathrm{Z}(L, \langle q\rangle, \kappa; K)\big) = \sum_{K\alpha \subset \mathrm{I}(L,\langle q\rangle)(\mathbb{A}^{(\infty)})\cap\kappa} \mathrm{vol}\Big(\mathrm{M}\big({}^{K_\alpha}\mathbf{O}(\alpha_{\mathbb{Z}}^{\perp})\big)\Big) \tag{1.12}$$

$$\widehat{\mathrm{vol}}\big(\mathrm{Z}(L, \langle q\rangle, \kappa; K)\big) = \sum_{K\alpha \subset \mathrm{I}(L,\langle q\rangle)(\mathbb{A}^{(\infty)})\cap\kappa} \widehat{\mathrm{vol}}\Big(\mathrm{M}\big({}^{K_\alpha}\mathbf{O}(\alpha_{\mathbb{Z}}^{\perp})\big)\Big) \tag{1.13}$$

ignoring the problem of compactifications here for simplicity. The sums run over K-orbits in the adelic sphere $\mathrm{I}(L, \langle q\rangle)(\mathbb{A}^{(\infty)})$ of radius q, more precisely, over those lying in κ. In other words, the decomposition of the special cycles into sub-Shimura varieties is governed by these orbits. We have furthermore:

$$\lambda^{-1}(L_{\mathbb{Z}}; s)\mu(L_{\mathbb{Z}}, \langle q\rangle, \kappa; s) = \sum_{K\alpha \subset \mathrm{I}(L,\langle q\rangle)(\mathbb{A}^{(\infty)})\cap\kappa} \lambda^{-1}(\alpha^{\perp}; s) \tag{1.14}$$

by [26, Theorem 5.10]. Here $\mu(L_{\mathbb{Z}}, \langle q\rangle, \kappa; s)$ is (part of) the qth Fourier coefficient of the Eisenstein series of genus 1 associated with the Weil representation of L, cf. [26, 10.2, Equation (18)].

1.4.3. Taking the *value* of (1.14) at $s = 0$ and inserting (1.11), we get for K, respectively K_α being the respective discriminant kernels:

$$\begin{aligned} \mathrm{vol}\Big(\mathrm{M}\big({}^{K}\mathbf{O}(L_{\mathbb{Z}})\big)\Big)\widetilde{\mu}(L_{\mathbb{Z}}, \langle q\rangle, \kappa; 0) &= \sum_{\alpha K \subset \mathrm{I}(L,\langle q\rangle)(\mathbb{A}^{(\infty)})\cap\kappa} \mathrm{vol}\Big(\mathrm{M}\big({}^{K_\alpha}\mathbf{O}(\alpha_{\mathbb{Z}}^{\perp})\big)\Big) \\ &= \mathrm{vol}\big(\mathrm{Z}(L, \langle q\rangle, \kappa; K)\big). \end{aligned} \tag{1.15}$$

This is just a version of the Siegel–Weil formula for the indefinite case and can be seen *directly* in several ways. For example, using Borcherds theory, see 1.4.5 below, or using the original version of the Siegel–Weil formula and Kudla–Millson theory. Hence equation (1.11) is true for $\mathrm{M}({}^{K}\mathbf{O}(L_{\mathbb{Z}}))$ if and only if it is true for all $\mathrm{M}({}^{K_\alpha}\mathbf{O}(\alpha_{\mathbb{Z}}^{\perp}))$. The result follows by induction.

1.4.4. Taking the *derivative* of (1.14)

$$\begin{aligned} &\frac{\mathrm{d}}{\mathrm{d}s}\lambda^{-1}(L_{\mathbb{Z}}; s)\mu(L_{\mathbb{Z}}, \langle q\rangle, \kappa; s) + \lambda^{-1}(L_{\mathbb{Z}}; s)\frac{\mathrm{d}}{\mathrm{d}s}\mu(L_{\mathbb{Z}}, \langle q\rangle, \kappa; s) \\ &\qquad = \sum_{\alpha K \subset \mathrm{I}(L,\langle q\rangle)(\mathbb{A}^{(\infty)})\cap\kappa} \frac{\mathrm{d}}{\mathrm{d}s}\lambda^{-1}(\alpha_{\mathbb{Z}}^{\perp}; s) \end{aligned}$$

at $s = 0$ (which remains true [26, Theorem 10.5] up to $\mathbb{Q}\log(q)$ for primes $q \neq p$, if we replace λ by $\tilde{\lambda}$ and μ by $\widetilde{\mu}$) and inserting (1.10), we get:

$$\begin{aligned}(1.16)\quad &\widehat{\mathrm{vol}}\Big(\mathrm{M}({}^K\mathbf{O}(L_{\mathbb{Z}}))\Big)\widetilde{\mu}(L_{\mathbb{Z}}, \langle q\rangle, \kappa; 0)\\ &\qquad + \mathrm{vol}\Big(\mathrm{M}({}^K\mathbf{O}(L_{\mathbb{Z}}))\Big)\frac{\mathrm{d}}{\mathrm{d}s}\widetilde{\mu}(L_{\mathbb{Z}}, \langle q\rangle, \kappa; s)\Big|_{s=0}\\ &= \sum_{K\alpha \subset \mathrm{I}(L,\langle q\rangle)(\mathbb{A}^{(\infty)})\cap\kappa} \widehat{\mathrm{vol}}\Big(\mathrm{M}({}^{K_\alpha}\mathbf{O}(\alpha_{\mathbb{Z}}^{\perp}))\Big) = \widehat{\mathrm{vol}}(\mathrm{Z}(L, \langle q\rangle, \kappa; K)).\end{aligned}$$

As soon as we are able to show this relation *directly*, too, we get that (1.10) holds for $\mathrm{M}({}^K\mathbf{O}(L_{\mathbb{Z}}))$ if and only if it holds for all $\mathrm{M}({}^{K_\alpha}\mathbf{O}(\alpha_{\mathbb{Z}}^{\perp}))$. Again, the result follows by induction.

To show (1.16) is the main step of the proof. In fact a certain average of it (cf. Theorem 3.4.2) will be sufficient. In its proof Borcherds' theory enters.

1.4.5. Recall from [2] that a Borcherds product is a meromorphic modular form F on $\mathrm{M}({}^K\mathbf{O}(L_{\mathbb{Z}}))_{\mathbb{C}}$ having singularities precisely at the (codimension $m = 1$) special cycles. It is a multiplicative lift of an integral vector valued modular form f of weight $1 - l/2$, holomorphic in $\mathbb{H}$, and meromorphic in the cusp $i\infty$ for the Weil representation of $\mathrm{Sp}_2'(\mathbb{Z})$ restricted to $\mathbb{C}[L_{\mathbb{Z}}^*/L_{\mathbb{Z}}]$. Here $\mathrm{Sp}_2'(\mathbb{Z})$ is $\mathrm{Sp}_2(\mathbb{Z}) = \mathrm{SL}_2(\mathbb{Z})$ if l is even and $\mathrm{Mp}_2(\mathbb{Z})$ if l is odd. Such an f has a Fourier expansion

$$f = \sum_{k \in \mathbb{Q}_{\gg -\infty}} a_k q^k,$$

where only a_k's with k having bounded denominator actually occur and $a_k \in \mathbb{Z}[L_{\mathbb{Z}}^*/L_{\mathbb{Z}}]$. The divisor of F (on the uncompactified Shimura variety) is given by $\sum_{k<0} \mathrm{Z}(L_{\mathbb{Z}}, \langle -k\rangle, a_k; K)$ and its weight is equal to $a_0(0)$. Assume for the moment (for simplicity) that an f can be found such that $a_k = 0$ for $k < 0$ except $a_{-q} = \kappa$ for some $q > 0$.

Using the corresponding meromorphic function F, Theorem 3.4.2 precisely gives relation (1.15), by an argument involving Serre duality already used by Borcherds [3] to show modularity of the Chow-group-valued theta function. However, we obtain relation (1.16), too, by a calculation in Arakelov theory, using a calculation of the integral of the Hermitian norm of F. This however requires the solution of additional problems:

(1) Need to prove that F is an *integral and 'monic'* modular form. This follows from an application of the q-expansion principle to Borcherds' famous product expansion of F. This implies that also its *arithmetic* divisor (away from the boundary) is precisely $\sum_{k<0} \mathrm{Z}(L_{\mathbb{Z}}, \langle -k\rangle, a_k; K)$, i.e., that no vertical components occur.

(2) Need to show that the boundary divisor of F does not contribute. This requires rather hard estimates on the occurring Green's functions.

Another issue concerns the induction process itself. Since many of the low-dimensional orthogonal Shimura varieties are compact, the here described method (q-expansion principle) cannot be used for them. Hence we start with the modular curve (the variety associated with a lattice of dimension 3 and Witt rank 1) and then for lattices of dimension ≥ 4 ensure that all components in the divisor of F have good reduction at p and are not compact. This yields the result for all non-compact orthogonal Shimura varieties. The compact case is obtained by exploiting the opposite implication of 1.4.4. Consequently, we need to design modular forms f prescribing their negative Fourier coefficients to some extent. This involves showing that the space of input forms f has a rational structure, and Serre duality like arguments to reduce to showing that modular forms with sparsely occurring Fourier coefficients (lacunary modular forms) vanish. This idea of designing appropriate f appears in principle already in [10].

In the same way, exploiting the opposite implication of 1.4.4, we finally also get equation (1.9) only up to contributions of the form $\mathbb{Q}\log(2)$ for an ample class of orthogonal Shimura varieties.

1.5. The philosophy of Shimura varieties

1.5.1. Even to define an intrinsic notion of arithmetic volume for an orthogonal Shimura varieties in higher dimensions, *canonical* integral models are necessary. To apply Arakelov theoretical methods in a reasonable way, the models have to be compactified, too. The arithmetic special cycles $\widehat{\mathrm{Z}}$ (good reduction case) on them are built from models of this kind themselves. Furthermore the Hermitian line bundle $\Xi^*(\mathcal{E}, h)$, involved in the definition of the arithmetic theta function and used to compute the 'degree' of this function, has to be defined as a 'canonical integral model' of an automorphic line bundle. In addition, to be able to work with Borcherds products as sections of them, one needs a 'q-expansion principle' to establish its integrality. The best and broadest context for all of these considerations is a fully functorial theory of canonical integral models of toroidal compactifications of mixed Shimura varieties, of the standard principal bundle on them, and of their 'compact' dual.

Consider a p-integral mixed Shimura datum $\mathbf{X} = (P_{\mathbf{X}}, \mathbb{D}_{\mathbf{X}}, h_{\mathbf{X}})$, where: $P_{\mathbf{X}}$ is a group scheme $P_{\mathbf{X}}$ over $\mathbb{Z}_{(p)}$ of a certain rigid type which we will call type (P) (see Section 2.1); $\mathbb{D}_{\mathbf{X}}$ is a generalized Hermitian symmetric space (a principal $P_{\mathbf{X}}(\mathbb{R})U_{\mathbf{X}}(\mathbb{C})$-space, where $U_{\mathbf{X}}$ is a certain subgroup of the unipotent radical of $P_{\mathbf{X}}$); $h_{\mathbf{X}} : \mathbb{D} \to \mathrm{Hom}(\mathbb{S}_{\mathbb{C}}, P_{\mathbf{X},\mathbb{C}})$ is an equivariant morphism. The data is subject to the condition roughly that $(P_{\mathbf{X},\mathbb{Q}}, \mathbb{D}_{\mathbf{X}}, h_{\mathbf{X}})$ satisfy Pink's axioms for a mixed Shimura datum [58].

To understand why analytic locally symmetric varieties (or orbifolds) of the form

$$[P_{\mathbf{X}}(\mathbb{Q})\backslash\mathbb{D}_{\mathbf{X}} \times (P_{\mathbf{X}}(\mathbb{A}^{(\infty)})/K)]$$

should have *canonical* algebraic models defined over number fields (or even rings of integers) at all, and where this structure is supposed to come from, one should bear in mind the following philosophy:

Let $L_{\mathbb{Z}_{(p)}}$ be free $\mathbb{Z}_{(p)}$-module of finite rank with a faithful representation (closed embedding) ρ: $P_{\mathbf{X}} \to \mathrm{GL}(L_{\mathbb{Z}_{(p)}})$, fixing some weight filtration $\{W_i\}$, $W_i \subset L_{\mathbb{Z}_{(p)}}$ and polarization. There is always a finite set of tensors $v_1, \dots, v_n$, $v_i \in L_{\mathbb{Z}_{(p)}}^{\otimes}$ such that the image of ρ (in the stabilizer of the weight filtration in the similitude group of the polarization form) is precisely the stabilizer of these tensors. The complex manifold $\mathbb{D}_{\mathbf{X}}$ can be seen as an *open* $P_{\mathbf{X}}(\mathbb{R})U_{\mathbf{X}}(\mathbb{C})$-orbit in the parameter space of (polarized) mixed Hodge structures (w.r.t. the filtration $\{W_i\}$) on $L_{\mathbb{C}}$, having the property that all v_i lie in $(L^{\otimes})^{(0,0)}$. Furthermore there is a category (groupoid) of families of mixed Hodge structures on arbitrary local systems over a base analytic space B. It is convenient to take local systems of $\mathbb{Q}$-vector spaces and equip the families with a K-level structure (for a compact open $K \subset P_{\mathbf{X}}(\mathbb{A}^{(\infty)})$). This groupoid is denoted by

$$\left[\boldsymbol{B}\text{-}{}^{\boldsymbol{K}}\mathbf{X}\text{-}\boldsymbol{L}\text{-}\mathbf{loc}\text{-}\mathbf{mhs}\right].$$

In fact, these groupoids (with varying B) form a category fibered in groupoids which is an analytic orbifold represented by the quotient

$$[P_{\mathbf{X}}(\mathbb{Q})\backslash\mathbb{D}_{\mathbf{X}} \times (P_{\mathbf{X}}(\mathbb{A}^{(\infty)})/K)]$$

i.e., the analytic mixed Shimura variety associated with $\mathbf{X}$.

If the group $P_{\mathbf{X}}$ and $\mathbb{D}_{\mathbf{X}}$ form a p-integral Shimura datum (2.2.3) one expects that over any base scheme S over $\mathcal{O}$ (a reflex ring of $\mathbf{X}$) there is a category (groupoid) of mixed motives

$$\left[\boldsymbol{S}\text{-}\mathbf{X}\text{-}\boldsymbol{L}\text{-}\mathbf{mot}\right]$$

which should (very roughly) be seen as the category of those polarized mixed motives M of fixed weight filtration type with morphisms $v_i'\colon \mathbb{Z}(0) \to M^{\otimes}$ which have the following property: etale locally, there is an isomophism (respecting weight filtration and polarization) of some realization (H^{et} or H^{dR}, say) with L that maps $H(v_i')$ to v_i for every i. This will be made precise for certain Shimura data and certain associated standard representations—corresponding to 1-motives—in Section 2.7. It can be made precise for all 'P.E.L. situations' (pure weight 1 and all v_i endomorphisms) and we refer to [30] or [46] for this. For Hodge-type Shimura data, the truth of the Hodge conjecture would as well allow to pose a moduli problem involving the existence of certain algebraic cycles.

Furthermore, one expects (functorial) maps

$$\left[\boldsymbol{S}\text{-}{}^{\boldsymbol{K}}\mathbf{X}\text{-}\boldsymbol{L}\text{-}\mathbf{mot}\right] \to \left[\boldsymbol{S}^{\mathbf{an}}\text{-}{}^{\boldsymbol{K}}\mathbf{X}\text{-}\boldsymbol{L}\text{-}\mathbf{loc}\text{-}\mathbf{mhs}\right], \tag{1.17}$$

if S is of finite type over $\mathbb{C}$, which are *equivalences* for $S = \operatorname{Spec}(\mathbb{C})$. Here, on the left hand side, we consider now motives up to $\mathbb{Z}_{(p)}$-isogeny with a $K^{(p)}$-level structure (on the etale realization in $\mathbb{A}^{(\infty,p)}$-vector spaces), too. (Assume that K is admissible, i.e., of the form $P_{\mathbf{X}}(\mathbb{Z}_p) \times K^{(p)}$, in particular, hyperspecial.)

$\left[\boldsymbol{S}\text{-}{}^{\boldsymbol{K}}\mathbf{X}\text{-}\boldsymbol{L}\text{-}\mathbf{mot}\right]$ should be (represented by) an *algebraic* smooth Deligne–Mumford stack $\mathrm{M}({}^{K}\mathbf{X})$ over $\operatorname{Spec}(\mathcal{O})$, which would be a model of the analytic Shimura variety because (1.17) is an equivalence for $S = \operatorname{Spec}(\mathbb{C})$.

It is also important to look at the categories of motives, like above, equipped with a trivialization of H^{et} (with values in $\mathbb{A}^{(\infty,p)}$ vector spaces, say), H^{dR}, and, in the analytic setting, of H_{B}—in each case *respecting* the $P_{\mathbf{X}}$-structure (given by the tensors, polarization and weight filtration). These groupoids should be represented by ...

- in the etale case:

$$\mathrm{M}^p(\mathbf{X}) := \varprojlim_{K \subset P_{\mathbf{X}}(\mathbb{A}^{(\infty)}) \text{ admissible}} \mathrm{M}({}^{K}\mathbf{X}), \tag{1.18}$$

- in the de Rham case:

$$\mathrm{P}({}^{K(1)}\mathbf{X}), \tag{1.19}$$

which is a right $P_{\mathbf{X}}$-torsor on $\mathrm{M}({}^{K(1)}\mathbf{X})$, called the standard principal bundle,

- in the Betti case:

(1.20) $$\mathbb{D}_{\mathbf{X}}$$

itself, as mentioned above.

Analytic comparison isomorphisms should give embeddings $\mathbb{D}_{\mathbf{X}} \hookrightarrow (\mathrm{M}^p(\mathbf{X})_{\mathbb{C}})^{\mathrm{an}}$ and $\mathbb{D}_{\mathbf{X}} \hookrightarrow (\mathrm{P}({}^{K(1)}\mathbf{X})_{\mathbb{C}})^{\mathrm{an}}$. An equivariant embedding $\mathbb{D}_{\mathbf{X}} \hookrightarrow (\mathrm{P}({}^{K(1)}\mathbf{X})_{\mathbb{C}})^{\mathrm{an}}$ trivializes $\mathrm{P}({}^{K(1)}\mathbf{X})_{\mathbb{C}}^{\mathrm{an}}$ locally up to translation by $P_{\mathbf{X}}(\mathbb{Z})$. The image of a point under this trivialization followed by ρ would be precisely a *period matrix* of the corresponding motive. The standard principal bundle therefore is sometimes also called 'period torsor' because it encodes (when the above philosophy can be realized) relations between periods.

The main point, which makes it possible to approach the theory of these models without having an appropriate theory of mixed motives, is that all objects $\mathrm{M}({}^K\mathbf{X})$, $\mathrm{P}({}^K\mathbf{X})$, $\mathbb{D}_{\mathbf{X}}$, etc. should be *independent of the representation* ρ. Moreover, it *is* possible to characterize models intrinsically, which we call *canonical*. These should always represent the corresponding moduli problem, if an appropriate one in terms of motives can be posed. The intrinsic characterization of the models is as follows:

(1) $\mathbb{D}_{\mathbf{X}}$ is seen as a certain conjugacy class of morphisms in $\mathrm{Hom}(\mathbb{S}_{\mathbb{C}}, P_{\mathbf{X},\mathbb{C}})$ (defined over $\mathbb{R}$ modulo $U_{\mathbf{X}}(\mathbb{C})$, a part of the unipotent radical). If a representation ρ is chosen, composition with it yields morphisms $\mathbb{S}_{\mathbb{C}} \to \mathrm{GL}(L_{\mathbb{C}})$ which are splittings for the corresponding mixed Hodge structures. In particular, this determines already an intrinsic complex analytic structure (via Borel embedding) of the Shimura variety.

(2) The characterization of the projective limit of the rational models $\mathrm{M}({}^K\mathbf{X})_E$ is reduced, requiring functoriality in Shimura data, to the case where $P_{\mathbf{X}}$ is a torus and accordingly the analytic Shimura variety is 0-dimensional. The characterization in that case is in terms of class field theory and is motivated by the theory of complex multiplication of Abelian varieties. This marvelous idea is due to Deligne [16, 17] and was extended to the mixed case by Pink [58]. The characterization of the integral model $\mathrm{M}({}^K\mathbf{X})$ itself is as follows. One requires the limit $\mathrm{M}^p(\mathbf{X})$ to satisfy an extension property very similar to the Neron property[6]. This idea appears first in [50, 51].

(3) The characterization of $\mathrm{P}({}^K\mathbf{X})$ can be reduced via functoriality, at least for a wide class of (mixed) Shimura data, to the case of the symplectic Shimura data, where a moduli problem in terms of 1-motives is available. It is then possible to show independence of an embedding into a symplectic Shimura datum directly. The author does not know of a better characterization which works in the integral case, too.

If a faithful representation $\rho\colon P_{\mathbf{X}} \hookrightarrow \mathrm{GL}(L_{\mathbb{Z}_{(p)}})$ is given, the objects (1.18)–(1.20) yield an l-adic sheaf (for every $l \neq p$), a vector bundle with connection, and a local system (in the analytic case), respectively, on the Shimura variety. Whenever it is possible to precise the moduli problem determined by this representation, these objects should be equal to the corresponding realizations of the universal mixed motive.

However, it should be possible to recover the filtration steps of the de Rham bundle and tensor constructions of them, too. This is seen as follows: If a moduli problem exists and $\mathrm{P}({}^K\mathbf{X})$ represents motives together with a trivialization of the

[6]in fact this *is* the Neron property for the first step in an unipotent extension

de Rham realization, the filtration on the realization yields a filtration on L_S, compatible with the $P_{\mathbf{X}}$-structure (determined by the tensors, polarization and weight filtration). Filtrations of this type on L_S with varying S are represented by a quasi-projective (projective, if $\mathbf{X}$ is pure) variety $\mathrm{M}^\vee(\mathbf{X})$, called the 'compact' dual. It is defined over $\mathcal{O}$ and independent of ρ, too. Hence we get a $P_{\mathbf{X}}$-equivariant morphism $\mathrm{P}({}^K\mathbf{X}) \to \mathrm{M}^\vee(\mathbf{X})$, or, in other words, a morphism of Artin stacks

$$\Xi\colon \mathrm{M}({}^K\mathbf{X}) \to [\mathrm{M}^\vee(\mathbf{X})/P_{\mathbf{X}}]. \tag{1.21}$$

This allows to associate with *every* $P_{\mathbf{X}}$-bundle $\mathcal{E}$ on $\mathrm{M}^\vee(\mathbf{X})$ a bundle $\Xi^*\mathcal{E}$ on $\mathrm{M}({}^K\mathbf{X})$ called an (integral) automorphic vector bundle. (In particular this holds for the bundles in the universal filtration associated with ρ.) The integral structure, however, is of course not pinned down by considering $\Xi^*\mathcal{E}$ as an abstract sheaf. The analytic comparison isomorphism, however, allows to compare this map with the Borel embedding

$$\mathbb{D}_{\mathbf{X}} \hookrightarrow \mathrm{M}^\vee(\mathbf{X})(\mathbb{C}).$$

Therefore, if $\mathcal{E}_{\mathbb{C}}|_{\mathbb{D}_{\mathbf{X}}}$ is equipped with a $P_{\mathbf{X}}(\mathbb{R})U_{\mathbf{X}}(\mathbb{C})$-invariant Hermitian metric h, we may define $\Xi^*(\mathcal{E}, h)$ (by slight abuse of notation). It is a well-defined Hermitian arithmetic vector bundle on $\mathrm{M}({}^K\mathbf{X})$.

For many purposes, in particular for the main purpose of this book, this is not sufficient because $\mathrm{M}({}^K\mathbf{X})$ is not proper. Desirable are toroidal compactifications $\mathrm{M}({}^K_\Delta\mathbf{X})$, depending on a rational polyhedral cone decomposition Δ of the conical complex $C_{\mathbf{X}}$ associated with $\mathbf{X}$. Furthermore, an extension $\mathrm{P}({}^K_\Delta\mathbf{X})$ of $\mathrm{P}({}^K\mathbf{X})$, or equivalently of the morphism (1.21), is needed to extend automorphic vector bundles. This would yield proper varieties and Hermitian automorphic vector bundles $\Xi^*(\mathcal{E}, h)$ on them. It turns out that there is only one meaningful way to extend $\mathrm{P}({}^K\mathbf{X})$, forced by the structure of Abelian unipotent extension as a torus torsor. In fact, this structure trivializes the standard principal bundle along this unipotent fibre and since the compactification along the unipotent fibre is defined by a torus embedding of the corresponding torus, the trivialization defines a 'trivial' extension of the bundle. This pins down the extensions in general, if one requires functoriality with respect to boundary maps (which are, in the algebraic setting, maps between formal completions). This functoriality also yields a 'q-expansion principle' for integral automorphic forms.

The extension of $\mathrm{P}({}^K\mathbf{X})$ determines canonical extensions of all automorphic vector bundles. In the rational case these are the same as described before by Mumford [55] (for fully decomposed bundles) and Deligne (for local systems).

The corresponding theory, developed in Part I of the thesis of the author [25], is presented in Sections 2.1–2.5.

CHAPTER 2

Integral Models of Toroidal Compactifications of Mixed Shimura Varieties

2.1. Group schemes of type (P)

Our definition of integral mixed Shimura datum involves a special kind of group schemes over $\mathrm{Spec}(\mathbb{Z}_{(p)})$ which are—in a sense—as rigid as reductive group schemes. We call them group schemes of type (P). They are a slight generalization of those of type (R) and (RR) considered by Grothendieck in [22, XXII, 5.1] which occur as parabolics of reductive group schemes. We use extensively the notation and results of [loc. cit.]. Proofs of the theorems in this section can be found in [25, Section 1].

2.1.1. In this section, S denotes a base scheme. If s is a point of S, we denote by $\bar{s}$ a geometric point above s. For a group scheme X over S we consider the functor $D(X) := \underline{\mathrm{Hom}}(X, \mathbb{G}_{\mathrm{m}})$. For an ordinary Abelian group M, we denote by M_S the associated constant group scheme over S.

2.1.2. For a coherent sheaf E on S, we denote by $\mathbb{W}(E)$ its *additive group scheme*, see [22, I, 4.6]. This is a group functor from the category of S-schemes to the category of abelian groups, and it is defined as follows:

$$\mathbb{W}(E)(S') = (E \otimes_{\mathcal{O}_S} \mathcal{O}_{S'})(S'),$$

where the right-hand side is equipped with its additive group structure. This is a representable and additive functor. We have $\mathbb{W}(\mathcal{O}_S) = \mathbb{G}_{\mathrm{a}}$, and if E is locally free, then $\mathbb{W}(E)$ is a smooth group scheme over S. Moreover, there is a natural isomorphism $\mathrm{Lie} \circ \mathbb{W} \cong \mathrm{id}$ where Lie is the functor associating to a group scheme its Lie algebra, see [22, II, 3].

2.1.3. Let G be a smooth group scheme over S with connected fibers and T a maximal torus. Then T acts via conjugation on G and the Lie algebra of each geometric fiber

$$\mathrm{Lie}(G_{\bar{s}}) = \mathrm{Lie}(G) \otimes_{\mathcal{O}_S} \mathcal{O}_{\bar{s}}$$

decomposes as

$$\mathrm{Lie}(G_{\bar{s}}) = \mathrm{Lie}(T_{\bar{s}}) \oplus \bigoplus_{r \in D(T)_{\bar{s}}} \mathrm{Lie}(G_{\bar{s}})^r .$$

With this notations, and following [22, XIX, 3.2, 3.6], we define systems of roots:

2.1.4. Definition (Root). We call $r \in D(T)(S)$ a **root** of G with respect to T, if $r_{\bar{s}}$ occurs for each s as a nontrivial root in the decomposition of $\mathrm{Lie}(G_{\bar{s}})$, i.e., $r_{\bar{s}} \neq 0$ and $\mathrm{Lie}(G_{\bar{s}})^{r_{\bar{s}}} \neq 0$.

2.1.5. Definition (System of roots). A subset $R \subset D(T)(S)$ is called a **system of roots** for G with respect to T if it consists of roots and the following equivalent conditions are satisfied:

(1) for each s, $r \mapsto r_{\bar{s}}$ is a bijection of R with the set of roots of $G_{\bar{s}}$;
(2) for each s, the elements of $\{r_{\bar{s}}\}_{r\in R}$ are distinct and

$$\dim(G_{\bar{s}}) - \dim(T_{\bar{s}}) = \# R;$$

(3) We have

$$\mathrm{Lie}(G) = \mathrm{Lie}(T) \oplus \bigoplus_{r\in R} \mathrm{Lie}(G)^r.$$

We now introduce the notion of group scheme of type (P):

2.1.6. Definition (Group scheme of type (P)). Let P be a group scheme of finite type over S. We call P of **type (P)** if there is a closed unipotent normal subgroup scheme W and S is covered by etale neighborhoods $S' \to S$ for which the following exist:

(1) a closed reductive subgroup scheme G of $P_{S'}$ such that $P_{S'}$ is isomorphic to the semidirect product of $W_{S'}$ with G;
(2) a split maximal torus T of G isomorphic to $D(M_{S'})$ for some lattice M;
(3) a system of roots $R = R_G \dot{\cup} R_W \subset M$, satisfying:

$$\mathrm{Lie}(G) = \mathrm{Lie}(T) \oplus \bigoplus_{r\in R_G} \mathrm{Lie}(G)^r$$

$$\mathrm{Lie}(W_{S'}) = \bigoplus_{r\in R_W} \mathrm{Lie}(W_{S'})^r$$

where W is the unipotent radical of P, and with the property that any two different $r_1 \in R$, $r_2 \in R_W$ are linearly independent;
(4) closed embeddings

$$\exp_r \colon \mathbb{W}(\mathrm{Lie}(W_{S'})^r) \to W_{S'}$$

for all $r \in R_W$, inducing via Lie the inclusion and satisfying

$$\mathrm{int}(t)\exp_r(\xi) = \exp_r(r(t)\xi)$$

for all $S'' \to S'$, for every $t \in T(S'')$ and for every $\xi \in \mathrm{Lie}(W_{S''})^r(S'')$.

The group scheme W of the previous definition is called the *unipotent radical* of P.

2.1.7. Definition (Split of type (P)). In the previous definition, if the objects required in conditions (1)–(4) are already defined over S we call (P, G, T, M, R) a **split group of type (P)**.

2.1.8. Remark. The property of being of type (P) is obviously stable under base change. Moreover, condition (1) of the definition implies that P is smooth and affine over S, and there exist maximal split tori locally in the etale topology, see [22, XII, 1.7].

2.1.9. Remark. If (P, G, T, M, R) is split of type (P) then we have

$$\mathrm{Lie}(P) = \mathrm{Lie}(G) \oplus \mathrm{Lie}(W)$$

because P is the semidirect product of G and W. Moreover, the action of T on $\mathrm{Lie}(P)$ respects this decomposition. Consequently, we have

$$\mathrm{Lie}(P) = \mathrm{Lie}(T) \bigoplus_{r \in R} \mathrm{Lie}(P)^r.$$

Group schemes of type (P) have a nice theory of (quasi-)parabolics. We first recall the definition of parabolic for group schemes of finite type, following [22, XV, 6.1]:

2.1.10. Definition (Parabolic). Let P be a group scheme of finite type over S. A **parabolic** subgroup scheme of P is a smooth subgroup scheme Q of P, such that for each $s \in S$, $Q_{\bar{s}}$ is a parabolic subgroup scheme of $P_{\bar{s}}$, (i.e., such that $P_{\bar{s}}/Q_{\bar{s}}$ is proper).

2.1.11. Proposition. *Let P be a group scheme of type* (P) *over S and W its unipotent radical. Let $\widetilde{G} = P/W$. The parabolics of P are in* 1:1 *correspondence with the parabolics of $\widetilde{G}$. The parabolic subgroups of P are of type* (P).

2.1.12. Proposition. *Let S be reduced. Let (P, G, T, M, R) be split of type* (P) *over S. If $R' \subset R$ is a closed subset, there exists a unique smooth subgroup scheme $P' \subset P$ containing T, and such that*

$$\mathrm{Lie}(P') = \mathrm{Lie}(T) \oplus \bigoplus_{r \in R'} \mathrm{Lie}(P)^r.$$

Moreover, we have that P' is of type (P) *and is closed in P. Every parabolic of P is of this form (but not conversely).*

2.1.13. Definition (Quasi-parabolic). Let P be a group scheme of type (P) over S. A closed smooth subgroup scheme Q of P is called **quasi-parabolic** if S is covered by etale neighborhoods $S' \to S$ for each of which there is a splitting (P, G, T, M, R) over S' such that $Q_{S'}$ is of the form given in Proposition 2.1.12 for some closed subset $R' \subset R$ which contains a set of positive roots.

Note, the set R' in the previous definition does not necessarily contain R_W. It follows immediately from the previous proposition that parabolic implies quasi-parabolic. The property of being a parabolic (respectively quasi-parabolic) of P is clearly invariant under conjugation.

2.1.14. Theorem. *Let S be reduced. Let P be a group scheme over S of type* (P). *We let S' vary in schemes over S.*

(1) *The functor*

$$S' \mapsto \{\textit{quasi-parabolic subgroups of } P_{S'}\}$$

is representable by a smooth quasi-projective S-scheme $\mathcal{QPAR}$.

The functor

$$S' \mapsto \{\textit{parabolic subgroups of } P_{S'}\}$$

is representable by a smooth projective S-scheme $\mathcal{PAR}$ which is an open and closed subscheme of $\mathcal{QPAR}$.

(2) *There is an etale sheaf $\mathcal{TYPE}$ of finite sets over S and a surjective morphism of schemes*

$$\mathrm{type} \colon \mathcal{QPAR} \to \mathcal{TYPE},$$

with the property that two quasi-parabolic Q *and* Q' *are locally conjugated in the etale topology, if and only if* $\mathrm{type}(Q) = \mathrm{type}(Q')$.

If (P, G, T, M, R) *is split then there is an isomorphism of schemes*

$$\mathcal{TYPE} \cong \begin{Bmatrix} W(R_G)\text{-orbits of closed subsets of } R \\ \text{containing a set of positive roots} \end{Bmatrix}_S$$

where $W(R_G)$ *is the Weyl group of* R_G.

(3) *Let* Q *be a quasi-parabolic of* P. *The morphism of schemes*

$$P/Q \to \mathcal{QPAR} \quad g \mapsto \mathrm{int}(g)Q$$

is an open and closed immersion onto the fibre above $\mathrm{type}(Q)$ *which is in particular a connected component of* $\mathcal{QPAR}$.

(4) *The functor*

$$S' \mapsto \mathrm{Hom}(\mathbb{G}_{\mathrm{m},S'}, P_{S'})$$

is representable by a smooth affine scheme $\underline{\mathrm{Hom}}(\mathbb{G}_{\mathrm{m}}, P)$ *over* S.

(5) *There is an etale sheaf* $\mathcal{FTYPE}$ *of sets over* S *and a surjective morphism of schemes*

$$\mathrm{ftype}\colon \underline{\mathrm{Hom}}(\mathbb{G}_{\mathrm{m}}, P) \to \mathcal{FTYPE}$$

with the property that α *and* β *in* $\mathrm{Hom}(\mathbb{G}_{\mathrm{m},S'}, P_{S'})$ *are locally conjugated in the etale topology, if and only if* $\mathrm{ftype}(\alpha) = \mathrm{ftype}(\beta)$.

If (P, G, T, M, R) *is split then there is an isomorphism of schemes*

$$\mathcal{FTYPE} \cong \{W(R_G)\text{-orbits in } M^*\}_S.$$

(6) *There is a surjective morphism of schemes*

$$\mathrm{qpar}\colon \underline{\mathrm{Hom}}(\mathbb{G}_{\mathrm{m}}, P) \to \mathcal{QPAR}$$

uniquely determined by the property that for every cocharacter

$$\alpha\colon \mathbb{G}_{\mathrm{m},S'} \to P_{S'}$$

the following holds:

(a) *For every etale morphism* $S'' \to S'$ *with a splitting* $(P_{S''}, G, T, M, R)$ *such that*

$$\alpha_{S''}\colon \mathbb{G}_{\mathrm{m},S''} \to T \subset P_{S''}$$

we have

$$\mathrm{type}\big(\mathrm{qpar}(\alpha)\big) = W(R_G)\{r \in R \mid r \circ \alpha \geq 0\}$$

where the right-hand side denotes the orbit under the action of the Weyl group $W(R_G)$.

(b) *The cocharacter* α *factors through the quasi-parabolic* $\mathrm{qpar}(\alpha) \subset P_{S'}$.

(7) *There is a commutative diagram*

$$\begin{array}{ccc} \underline{\mathrm{Hom}}(\mathbb{G}_{\mathrm{m}}, P) & \xrightarrow{\mathrm{qpar}} & \mathcal{QPAR} \\ \downarrow{\scriptstyle \mathrm{ftype}} & & \downarrow{\scriptstyle \mathrm{type}} \\ \mathcal{FTYPE} & \longrightarrow & \mathcal{TYPE} \end{array}$$

where the definition of the morphism on the bottom may be deduced from the split case. If (P, G, T, M, R) *is split, that morphism is given by*

$$M^* \ni m \mapsto \{r \in R \mid \langle r, m\rangle \geq 0\}.$$

2.2. Mixed Shimura data

In this section we define p-integral mixed Shimura data. This notion is basically an integral version of the one used in Pink's thesis [58]. Proofs of the theorems in this section can be found in [25, Section 2]. Recall the Deligne torus $\mathbb{S} := \mathrm{Res}_{\mathbb{R}}^{\mathbb{C}}(\mathbb{G}_{\mathrm{m}})$ with weight morphism $w\colon \mathbb{G}_{\mathrm{m},\mathbb{R}} \to \mathbb{S}$. Recall also from 2.1.6 the definition of group scheme of type (P).

2.2.1. Definition (Admissible morphism). Let P be a group scheme of type (P) defined over $\mathrm{Spec}(\mathbb{Z}_{(p)})$, and let $\pi\colon P \to P/W$ the projection, where W is its unipotent radical.

A morphism $h\colon \mathbb{S}_{\mathbb{C}} \to P_{\mathbb{C}}$ is called **admissible** if

- $\pi \circ h$ is defined over $\mathbb{R}$;
- $\pi \circ h \circ w$ is a cocharacter of the center of $P_{\mathbf{X},\mathbb{Q}}/W_{\mathbf{X},\mathbb{Q}}$ defined over $\mathbb{Q}$;
- Under the weight filtration on $\mathrm{Lie}(P_{\mathbf{X},\mathbb{Q}})$ defined by $\mathrm{Ad} \circ h \circ w$, we have:

$$W_{-1}\big(\mathrm{Lie}(P_{\mathbf{X},\mathbb{Q}})\big) = \mathrm{Lie}(W_{\mathbf{X},\mathbb{Q}}).$$

Recall from [58, Prop. 1.5] that a morphism $h\colon \mathbb{S} \to \mathrm{GL}(L_{\mathbb{R}})$ is associated with a mixed Hodge structure if and only if h factors through a subgroup $P_{\mathbf{X},\mathbb{Q}} \subset \mathrm{GL}(L_{\mathbb{Q}})$, and the induced morphism is admissible in the sense of Definition 2.2.1.

2.2.2. Lemma. *Let T be a $\mathbb{Q}$-torus. The following conditions are equivalent:*

(1) *$T(\mathbb{Q})$ is discrete in $T(\mathbb{A}^{(\infty)})$;*
(2) *T is an almost direct product of a $\mathbb{Q}$-split torus with a torus T', such that $T'(\mathbb{R})$ is compact.*

2.2.3. Definition (p-MSD). Let p be a prime. A p-integral mixed Shimura datum (p-MSD) $\mathbf{X}$ consists of

(1) a group scheme $P_{\mathbf{X}}$ of type (P) over $\mathrm{Spec}(\mathbb{Z}_{(p)})$;
(2) a closed unipotent subgroup scheme $U_{\mathbf{X}}$;
(3) a homogeneous space $\mathbb{D}_{\mathbf{X}}$ under $P_{\mathbf{X}}(\mathbb{R})U_{\mathbf{X}}(\mathbb{C})$;
(4) a $P_{\mathbf{X}}(\mathbb{R})U_{\mathbf{X}}(\mathbb{C})$-equivariant finite-to-one morphism

$$h_{\mathbf{X}} : \mathbb{D}_{\mathbf{X}} \to \mathrm{Hom}(\mathbb{S}_{\mathbb{C}}, P_{\mathbf{X},\mathbb{C}}), \quad x \mapsto h_x,$$

such that the image consists of *admissible* morphisms (2.2.1),

subject to the following condition: For (one, hence for all) h_x, $x \in \mathbb{D}_{\mathbf{X}}$,

(1) $\mathrm{Ad}_P \circ h_x$ induces on $\mathrm{Lie}(P)$ a mixed Hodge structure of type

$$(-1,1),(0,0),(1,-1) \quad (-1,0),(0,-1) \quad (-1,-1);$$

(2) the weight filtration on $\mathrm{Lie}(P_{\mathbb{Q}})$ is given by

$$W_i\big(\mathrm{Lie}(P_{\mathbb{Q}})\big) = \begin{cases} \mathrm{Lie}(P_{\mathbb{Q}}) & \text{if } i \geq 0, \\ \mathrm{Lie}(W_{\mathbb{Q}}) & \text{if } i = -1, \\ \mathrm{Lie}(U_{\mathbb{Q}}) & \text{if } i = -2, \\ 0 & \text{if } i < -2, \end{cases}$$

where $W_{\mathbb{Q}}$ is the unipotent radical of $P_{\mathbb{Q}}$ and $U_{\mathbb{Q}} = U_{\mathbf{X},\mathbb{Q}}$ is a central subgroup;

(3) $\mathrm{int}\Big(\pi\big(h_x(i)\big)\Big)$ induces a Cartan involution on $G_{\mathbb{R}}^{\mathrm{ad}}$, where $G = P/W$;

(4) $G^{\mathrm{ad}}_{\mathbb{R}}$ possesses no nontrivial factors of compact type that are defined over $\mathbb{Q}$;
(5) the center Z of P satisfies the equivalent conditions of Lemma 2.2.2.

$\mathbf{X}$ is called **pure**, if $W_{\mathbf{X}} = 1$. A **morphism of p-MSD $\mathbf{X} \to \mathbf{Y}$** is a pair of a homomorphism of group schemes $P_{\mathbf{X}} \to P_{\mathbf{Y}}$ and a homogenous map $\mathbb{D}_{\mathbf{X}} \to \mathbb{D}_{\mathbf{Y}}$ respecting the maps h.

We call a morphism an **embedding**, if the morphism of group schemes is a *closed* embedding, and the map $\mathbb{D}_{\mathbf{X}} \to \mathbb{D}_{\mathbf{Y}}$ is *injective*.

If we have a p-MSD $\mathbf{X}$, then an **admissible** compact open subgroup of $P_{\mathbf{X}}(\mathbb{A}^{(\infty)})$ is a group of the form $K^{(p)}P_{\mathbf{X}}(\mathbb{Z}_p)$, where $K^{(p)}$ is a compact open subgroup of $P_{\mathbf{X}}(\mathbb{A}^{(\infty,p)})$.

See [58, 2.1][1] for the rational case. *Note*: The subgroup scheme $U_{\mathbf{X}}$ is determined by the triple $(P_{\mathbf{X}}, \mathbb{D}_{\mathbf{X}}, h_{\mathbf{X}})$.

2.2.4. Example. The main examples of *pure* Shimura data are given by the *symplectic* Shimura data $\mathbf{H}_g$ whose underlying groups are groups of symplectic similitudes $\mathrm{GSp}(L)$. They will be discussed in detail in 2.3.1. Their associated Shimura varieties are moduli spaces of Abelian schemes. In Chapter 3, however, the *orthogonal* Shimura data $\mathbf{O}(L)$ and $\mathbf{S}(L)$ will play the main role. Their underlying groups are special orthogonal, respectively general spin groups. They will be discussed in detail in Section 2.4. The most basic *mixed* Shimura data $\mathbf{H}_{g_0}[(I \otimes I)^s, I \otimes L_0]$ give rise to Shimura varieties which are moduli spaces for 1-motives. Their underlying groups are unipotent extensions of symplectic similitude groups (or even just $\mathbb{G}_{\mathrm{m}}$). Those will be discussed in 2.3.2– 2.3.3. They occur also as *boundary components* (2.2.17) of Shimura data of symplectic type as well as of Shimura data of orthogonal type.

2.2.5. Definition (p-EMSD). A pair ${}^K\mathbf{X}$ is called **p-integral extended mixed Shimura datum (p-EMSD)**, where $\mathbf{X}$ is a p-integral Shimura datum, and K is an admissible compact open subgroup of $P_{\mathbf{X}}(\mathbb{A}^{(\infty)})$.

A **morphism of p-EMSD ${}^{K'}\mathbf{Y} \to {}^K\mathbf{X}$** is a pair (γ, ρ), where γ is a morphism of p-integral mixed Shimura data and $\rho \in P_{\mathbf{X}}(\mathbb{A}^{(\infty,p)})$ such that $\gamma(K')^\rho \subset K$.

If γ is an automorphism then we call the morphism a **Hecke operator**. If γ is an embedding and $\gamma(K')^\rho = K \cap P_{\mathbf{X}}(\mathbb{A}^{(\infty)})$, and such that the map

$$[P_{\mathbf{Y}}(\mathbb{Q})\backslash\mathbb{D}_{\mathbf{Y}} \times (P_{\mathbf{Y}}(\mathbb{A}^{(\infty)})/K')] \to [P_{\mathbf{X}}(\mathbb{Q})\backslash\mathbb{D}_{\mathbf{X}} \times (P_{\mathbf{X}}(\mathbb{A}^{(\infty)})/K)]$$

is a closed embedding then we call the morphism an **embedding**.

2.2.6. Lemma. *For all $x \in \mathbb{D}_{\mathbf{X}}$, $P_{\mathbf{X}}(\mathbb{Q}) \cap \mathrm{Stab}\big(x, P_{\mathbf{X}}(\mathbb{R})U_{\mathbf{X}}(\mathbb{C})\big)K$ is finite for every compact open subgroup $K \subset P_{\mathbf{X}}(\mathbb{A}^{(\infty)})$ and is trivial for sufficiently small K.*

2.2.7. Definition ($K(N)$). Let M be an integer. Let $P_{\mathbf{X}}$ be a group scheme over $\mathbb{Z}[1/M]$ of type (P), $\mathbb{D}_{\mathbf{X}}$, and $h_{\mathbf{X}}$, such that the conditions of Definition 2.2.3 are satisfied. For each prime $p \nmid M$, $P_{\mathbf{X}} \times_{\mathbb{Z}[1/M]} \mathbb{Z}_{(p)}$ will then define *p-integral* mixed Shimura data which we also denote by $\mathbf{X}$. Let $L_{\mathbb{Z}}$ be a lattice with a faithful representation of $P_{\mathbf{X}}$, i.e., a closed embedding $P_{\mathbf{X}} \hookrightarrow \mathrm{GL}(L_{\mathbb{Z}[1/M]})$.

---

[1]Property (5) in the definition is stated in [loc. cit.] in a weaker form, namely it is required that the action on W is through a torus satisfying the conditions of 2.2.2. In [52] this property is called (SV5) for pure Shimura varieties.

For each integer N, we define the following compact open subgroup of $P_{\mathbf{X}}(\mathbb{A}^{(\infty)})$:

$$K(N) := \{g \in P_{\mathbf{X}}(\mathbb{A}^{(\infty)}) \mid gL_{\hat{\mathbb{Z}}} = L_{\hat{\mathbb{Z}}}, g \equiv \mathrm{id} \bmod N\}.$$

Note: If $p \nmid N$ then $K(N)$ is admissible at p.

2.2.8. Lemma. *For each (admissible) compact open subgroup $K \subset P_{\mathbf{X}}(\mathbb{A}^{(\infty)})$, there is some $\gamma \in P_{\mathbf{X}}(\mathbb{A}^{(\infty)})$ (respectively $\in P_{\mathbf{X}}(\mathbb{A}^{(\infty,p)})$) such that*

$$P_{\mathbf{X}}^{\gamma} \subseteq K(1).$$

2.2.9. Lemma. (1) *For each ${}^{K_1}\mathbf{X}_1$ and embedding $\alpha\colon \mathbf{X}_1 \hookrightarrow \mathbf{X}_2$, there is an admissible $K_2 \subset P_{\mathbf{X}_2}(\mathbb{A}^{(\infty)})$ such that ${}^{K_1}\mathbf{X}_1 \hookrightarrow {}^{K_2}\mathbf{X}_2$ is an embedding in the sense of Definition* 2.2.5.

(2) *If ${}^{K_1}\mathbf{X}_1 \hookrightarrow {}^{K_2}\mathbf{X}_2$ is an embedding, and if K_2 is neat, then for each $K_2' \subset K_2$, the map ${}^{K_2' \cap P_{\mathbf{X}_1}(\mathbb{A}^{(\infty)})}\mathbf{X}_1 \hookrightarrow {}^{K_2'}\mathbf{X}_2$ is an embedding.*

(3) *Let $P_{\mathbf{X}_2}$ operate linearly on a lattice $L_{\mathbb{Z}_{(p)}}$, and choose a lattice $L_{\mathbb{Z}} \subset L_{\mathbb{Z}_{(p)}}$. There is an integer N such that for all $(M,p) = 1$, $N \mid M$, ${}^{K(M)_1}\mathbf{X}_1 \hookrightarrow {}^{K(M)_2}\mathbf{X}_2$ is an embedding.*

We have the following structure theorem for p-integral mixed Shimura data:

2.2.10. Proposition. *Let $\mathbf{X}$ be p-MSD. There are smooth closed subgroup schemes of $P_{\mathbf{X}}$: $W_{\mathbf{X}}$ (the unipotent radical) and $U_{\mathbf{X}} \cong \mathbb{W}\big(W_{-2}(\mathrm{Lie}(P_{\mathbf{X}}))\big)$. There is a smooth group scheme $V_{\mathbf{X}} \cong \mathbb{W}(\mathrm{gr}_{-1}(\mathrm{Lie}(P_{\mathbf{X}}))$ and an exact sequence*

$$0 \longrightarrow U_{\mathbf{X}} \longrightarrow W_{\mathbf{X}} \longrightarrow V_{\mathbf{X}} \longrightarrow 0.$$

There is a closed reductive subgroup scheme $G_{\mathbf{X}}$ of $P_{\mathbf{X}}$, such that $P_{\mathbf{X}} = W_{\mathbf{X}} \rtimes G_{\mathbf{X}}$. Any two such subgroup schemes are conjugated by an element in $W_{\mathbf{X}}(\mathbb{Z}_{(p)})$.

A morphism of p-MSD $\alpha\colon \mathbf{X} \to \mathbf{Y}$ induces a diagram

$$\begin{array}{ccccccccc} 0 & \longrightarrow & U_{\mathbf{X}} & \longrightarrow & W_{\mathbf{X}} & \longrightarrow & V_{\mathbf{X}} & \longrightarrow & 0 \\ & & \downarrow & & \downarrow & & \downarrow & & \\ 0 & \longrightarrow & U_{\mathbf{Y}} & \longrightarrow & W_{\mathbf{Y}} & \longrightarrow & V_{\mathbf{Y}} & \longrightarrow & 0. \end{array}$$

If α is an embedding, the vertical maps are closed embeddings, and the outer ones are given by saturated inclusions (i.e., inducing an inclusion mod p *as well) of the corresponding modules over $\mathbb{Z}_{(p)}$. Furthermore, there are closed reductive subgroup schemes $G_{\mathbf{X}}$ and $G_{\mathbf{Y}}$ of $P_{\mathbf{X}}$ and $P_{\mathbf{Y}}$, respectively, such that $P_{\mathbf{X}} = W_{\mathbf{X}} \rtimes G_{\mathbf{X}}$, $P_{\mathbf{Y}} = W_{\mathbf{Y}} \rtimes G_{\mathbf{Y}}$, and such that α maps $G_{\mathbf{X}}$ to $G_{\mathbf{Y}}$.*

We have the following converse to 2.2.10:

2.2.11. Definition/Proposition (Unipotent extension). *Let $\mathbf{X}$ be p-MSD. Two $\mathbb{Z}_{(p)}$-modules V and U acted on by $P_{\mathbf{X}}$ with unimodular* equivariant *symplectic form $\Psi\colon V \times V \to U$ (i.e., inducing an isomorphism $V \cong V^*$), define a group scheme W_0 sitting in an exact sequence:*

$$0 \longrightarrow U_0 \longrightarrow W_0 \longrightarrow V_0 \longrightarrow 0,$$

where $U_0 := \mathbb{W}(U)$, and $V_0 := \mathbb{W}(V)$. Using the action of $P_{\mathbf{X}}$ we may form the semidirect product $P_{\mathbf{X}}' := W_0 \rtimes P_{\mathbf{X}}$. Assume that every subquotient of $\mathrm{Lie}(W_{0,\mathbb{R}})$ *is of type*

$$\{(-1,0),(0,-1)\} \quad or \quad \{(-1,-1)\}.$$

Define $\mathbb{D}'_{\mathbf{X}}$ *as*

$$\{(x,k) \in \mathbb{D}_{\mathbf{X}} \times \mathrm{Hom}(\mathbb{S}_{\mathbb{C}}, P'_{\mathbf{X},\mathbb{C}}) \mid h_x = \phi \circ k;$$
$$\pi' \circ k \colon \mathbb{S}_{\mathbb{C}} \to (P'_{\mathbf{X}}/U'_{\mathbf{X}})_{\mathbb{C}} \text{ is defined over } \mathbb{R}\}.$$

$\mathbf{X}'$ *is then p-MSD, called a* **unipotent extension** *of* $\mathbf{X}$, *denoted by* $\mathbf{X}[U,V]$. $\mathbb{D}'_{\mathbf{X}} \to \mathbb{D}_{\mathbf{X}}$ *is a torsor under* $W_0(\mathbb{R})(W_0 \cap U'_{\mathbf{X}})(\mathbb{C})$. *We have* $\mathbf{X}[U,V]/W_0 \cong \mathbf{X}$.

Proposition 2.2.10 may be strengthened as follows. Every p-MSD satisfies

$$\mathbf{X} \cong (\mathbf{X}/W_{\mathbf{X}})[\mathrm{Lie}(U_{\mathbf{X}}), \mathrm{Lie}(V_{\mathbf{X}})],$$

the symplectic form and action of $G_{\mathbf{X}}$ being determined by $\mathbf{X}$.

2.2.12. We will now describe the theory of boundary components for mixed Shimura data. This is again an integral version of the notion developed by Pink [58, Chapter 4].

2.2.13. Definition (Admissible parabolic). Let $\mathbf{X}$ be (rational) MSD. For $G_{\mathbf{X},\mathbb{Q}} := P_{\mathbf{X},\mathbb{Q}}/W_{\mathbf{X},\mathbb{Q}}$ every $\mathbb{Q}$-parabolic subgroup of $P_{\mathbf{X},\mathbb{Q}}$ is the inverse image of a $\mathbb{Q}$-parabolic subgroup of $G^{\mathrm{ad}}_{\mathbf{X},\mathbb{Q}}$. Let $G^{\mathrm{ad}}_{\mathbf{X},\mathbb{Q}} = G_{1,\mathbb{Q}} \times \cdots \times G_{r,\mathbb{Q}}$ be the decomposition into $\mathbb{Q}$-simple factors. Choose a $\mathbb{Q}$-parabolic subgroup $Q_{i,\mathbb{Q}} \subseteq G_{i,\mathbb{Q}}$ for every i, and let $Q_{\mathbb{Q}}$ be the inverse image of $Q_{1,\mathbb{Q}} \times \cdots \times Q_{r,\mathbb{Q}}$ in $P_{\mathbf{X},\mathbb{Q}}$. We call $Q_{\mathbb{Q}}$ an **admissible** $\mathbb{Q}$-parabolic subgroup of $P_{\mathbf{X},\mathbb{Q}}$, if every $Q_{i,\mathbb{Q}}$ is either equal to $G_{i,\mathbb{Q}}$ or is a maximal proper $\mathbb{Q}$-parabolic subgroup of $G_{i,\mathbb{Q}}$.

2.2.14. Lemma. *If* $\mathbf{X}$ *is p-MSD and* $Q_{\mathbb{Q}}$ *is a parabolic subgroup of* $P_{\mathbf{X},\mathbb{Q}}$, *then there is a parabolic subgroup scheme* Q *of* $P_{\mathbf{X}}$ (2.1.10), *such that* $Q_{\mathbb{Q}} = Q \times_{\mathrm{Spec}(\mathbb{Z}_{(p)})} \mathrm{Spec}(\mathbb{Q})$.

2.2.15. Let S be a maximal $\mathbb{R}$-split torus of $(G_{\mathbb{R}})^{\mathrm{der}}$. Let R be the root system of S (acting on $\mathrm{Lie}(G^{\mathrm{ad}})$). The irreducible components of R are of type (C) or (BC) [18, Corollaire 3.1.7], cf. also [1, §2, Proposition 4]. The *long roots* $L = \{\alpha_1, \ldots, \alpha_g\}$ [18, loc. cit.], cf. also [1, p. 185], form a basis of mutually orthogonal roots of S. An admissible $\mathbb{R}$-parabolic (defined analogously to 2.2.13) $Q_{\mathbb{R}} \supset S$ is determined (in the sense of 2.1.14) by a homomorphism $\lambda \colon \mathbb{G}_{\mathrm{m},\mathbb{R}} \hookrightarrow S$, where

$$\langle \lambda, \alpha \rangle = \begin{cases} 2 & \alpha \in L', \\ 0 & \alpha \notin L'. \end{cases}$$

where L' is any subset of L. If the corresponding cocharacter can be defined over $\mathbb{Q}$, then Q is defined over $\mathbb{Q}$.

2.2.16. Recall [58, 4.3] the group

$$H_0 := \{(z,\alpha) \in \mathbb{S} \times \mathrm{GL}_{2,\mathbb{R}} \mid z\bar{z} = \det(\alpha)\}.$$

Let $h_0 \colon \mathbb{S}_{\mathbb{R}} \to H_{0,\mathbb{R}}$ be the morphism given by $z \mapsto \big(z, h(z)\big)$ for some $h \in \mathbb{D}_{\mathbf{H}_1}$, where $\mathbb{D}_{\mathbf{H}_1}$ is the set of representations of type $(-1,0),(0,-1)$ (see 2.3.1 for the notation). Let $h_\infty \colon \mathbb{S}_{\mathbb{C}} \to H_{0,\mathbb{C}}$ be the morphism $z \mapsto \big(z, h_m(z)\big)$, where $h_m \colon \mathbb{S}_{\mathbb{C}} \to \left(\begin{smallmatrix} * & * \\ & * \end{smallmatrix}\right)$ describes the mixed Hodge structure having the same Hodge filtration as the one described by h, but weight filtration corresponding to (i.e., fixed by) $\left(\begin{smallmatrix} * & * \\ & * \end{smallmatrix}\right)$.

2.2.17. Proposition. *Let* $Q_{\mathbb{Q}}$ *be a* $\mathbb{Q}$*-parabolic subgroup of* $P_{\mathbf{X},\mathbb{Q}}$. *Let* π' *be the projection* $P_{\mathbf{X}} \to P_{\mathbf{X}}/U_{\mathbf{X}}$. *The following are equivalent:*

(1) $Q_{\mathbb{Q}}$ *is admissible.*

(2) *For every* $x \in \mathbb{D}_{\mathbf{X}}$ *there is a unique homomorphism*

$$\omega_x \colon H_{0,\mathbb{C}} \to P_{\mathbf{X},\mathbb{C}}$$

such that

(a) $\pi' \circ \omega_x : H_{0,\mathbb{C}} \to (P_{\mathbf{X}}/U_{\mathbf{X}})_{\mathbb{C}}$ *is already defined over* $\mathbb{R}$,

(b) $h_x = \omega_x \circ h_0$,

(c) $\omega_x \circ h_\infty \circ w \colon \mathbb{G}_{\mathrm{m},\mathbb{C}} \to Q_{\mathbb{C}}$ *is conjugated to* $\mu \cdot \lambda$, *where* λ *is the morphism constructed in* 2.2.15, $\mu = h_x \circ w$, *and* $\mathrm{Lie}(Q_{\mathbb{C}})$ *is the direct sum of all nonnegative weight spaces in* $\mathrm{Lie}(P_{\mathbf{X},\mathbb{C}})$ *under* $\mathrm{Ad}_P \circ \omega_x \circ h_\infty \circ w$.

(3) *There exists an* $x \in \mathbb{D}_{\mathbf{X}}$ *and a homomorphism* ω_x *such that the three conditions in* (2) *are satisfied.*

If $P_{\mathbf{X}}$ is reductive, $Q_{\mathbb{Q}}$ corresponds to a boundary component in the sense of [1, III, p. 220, no. 2]. The morphism $\omega_x \circ h_\infty$ is independent of the choice of h_0.

2.2.18. Definition/Theorem (Boundary component). *Assume that* $\mathbf{X}$ *is p-MSD. Choose an admissible* $Q_{\mathbb{Q}}$ *as above. We will define a p-integral mixed Shimura datum* $\mathbf{B}$ *as follows:*

Let $P_{\mathbf{B},\mathbb{Q}}$ *be the smallest normal* $\mathbb{Q}$*-subgroup of* $Q_{\mathbb{Q}}$*, such that* $\omega_x \circ h_\infty$ *factorizes through it.*

Consider the map

$$\begin{aligned} \mathbb{D}_{\mathbf{X}} &\longrightarrow \pi_0(\mathbb{D}_{\mathbf{X}}) \times \mathrm{Hom}(\mathbb{S}_{\mathbb{C}}, P_{\mathbf{B},\mathbb{C}}) \\ x &\longmapsto ([x], \omega_x \circ h_\infty). \end{aligned} \tag{2.1}$$

Choose a $P_{\mathbf{B}}(\mathbb{R})U_{\mathbf{B}}(\mathbb{C})$*-orbit* $\mathbb{D}_{\mathbf{B}}$ *containing an* $([x], \omega_x \circ h_\infty)$ *in the above image. The image is contained in the union of finitely many such orbits. Each* $\mathbb{D}_{\mathbf{B}}$ *is a finite covering of the corresponding* $P_{\mathbf{B}}(\mathbb{R})U_{\mathbf{B}}(\mathbb{C})$*-orbit* $h(\mathbb{D}_{\mathbf{B}})$ *in* $\mathrm{Hom}(\mathbb{S}_{\mathbb{C}}, P_{\mathbf{B},\mathbb{C}})$. *Let* $\mathbb{D}_{\mathbf{B}\Rightarrow\mathbf{X}}$ *be its inverse image in* $\mathbb{D}_{\mathbf{X}}$.

The closure $P_{\mathbf{B}}$ *of* $P_{\mathbf{B},\mathbb{Q}}$ *in* $P_{\mathbf{X}}$ *is of type* (P) [25, Theorem 2.4.5] *hence* $\mathbf{B}$ *is p-integral mixed Shimura data and called a* **boundary component** *of* $\mathbf{X}$. *It is called* **proper**, *if* $Q_{\mathbb{Q}}$ *is a proper parabolic, otherwise* **improper**.

If $\mathbf{B}'$ *is a p-MSD a* **boundary morphism** $\mathbf{B}' \Longrightarrow \mathbf{X}$ *is an isomorphism of* $\mathbf{B}'$ *with one of the boundary components* $\mathbf{B}$.

For each $Q_{\mathbb{Q}}$ there are finitely many choices of $\mathbb{D}_{\mathbf{B}}$'s and accordingly, finitely many boundary components.

2.2.19. Definition/Proposition (Projection on the imaginary part). *There is a functorial map, called* **projection on the imaginary part**,

$$\mathrm{im} \colon \mathbb{D}_{\mathbf{X}} \to U_{\mathbf{X}}(\mathbb{R})(-1) \quad x \mapsto u_x,$$

where u_x *is the unique element, such that* $\mathrm{int}(u_x^{-1}) \circ h_x$ *is defined over* $\mathbb{R}$.

In general for each abelian group M and $n \in \mathbb{Z}$ we write $M(n)$ for $M \otimes \mathbb{Z}(1)^{\otimes n}$, where $\mathbb{Z}(1)$ is $\mathbb{Z}(2\pi i) \subset \mathbb{C}$.

2.2.20. Proposition. (1) *Let* $\mathbb{D}_{\mathbf{B}\Rightarrow\mathbf{X}}$ *be as in Definition* 2.2.18. *Let* $\mathbb{D}^0_{\mathbf{X}}$ *be a connected component of* $\mathbb{D}_{\mathbf{B}\Rightarrow\mathbf{X}}$ *and* $\mathbb{D}^0_{\mathbf{B}}$ *be corresponding component of* $\mathbb{D}_{\mathbf{B}}$. *Then*

(a) *The map* $\mathbb{D}_{\mathbf{B}\Rightarrow\mathbf{X}} \to \mathbb{D}_{\mathbf{B}}$ *is an open embedding.*

(b) *The image of* $\mathbb{D}^0_{\mathbf{X}}$ *in* $\mathbb{D}^0_{\mathbf{B}}$ *is the inverse image of an open complex cone* $C := C(\mathbb{D}^0_{\mathbf{X}}, P_{\mathbf{B}}) \subset U_{\mathbf{B}}(\mathbb{R})(-1)$ *under the map* $\mathrm{im}|_{\mathbb{D}^0_{\mathbf{B}}}$.

(c) *The cone C is an orbit in $U_{\mathbf{B}}(\mathbb{R})(-1)$ under translation by $U_{\mathbf{X}}(\mathbb{R})(-1)$ and conjugation by $Q(\mathbb{R})^0$. It is also invariant under translation by $(U_{\mathbf{B}} \cap W_{\mathbf{X}})(\mathbb{R})(-1)$.*

(d) *Modulo $(U_{\mathbf{B}} \cap W_{\mathbf{X}})(\mathbb{R})(-1)$ the cone C is a nondegenerate homogeneous self-adjoint cone (in the sense of* [1, II, p. 57, §1.1]).

(2) *Consider a morphism of Shimura data $\iota\colon \mathbf{X} \to \mathbf{X}'$. For each boundary component $\mathbf{B}$ of $\mathbf{X}$ there is a unique boundary component $\mathbf{B}'$ of $\mathbf{X}'$ and a morphism $\tilde{\iota}\colon \mathbf{B} \to \mathbf{B}'$ such that*

$$\begin{array}{ccc} \mathbb{D}_{\mathbf{B} \Rightarrow \mathbf{X}} & \xrightarrow{\iota} & \mathbb{D}_{\mathbf{B}' \Rightarrow \mathbf{X}'} \\ \big\downarrow & & \big\downarrow \\ \mathbb{D}_{\mathbf{B}} & \xrightarrow{\tilde{\iota}} & \mathbb{D}'_{\mathbf{B}} \end{array}$$

commutes. If ι is an embedding, $\tilde{\iota}$ is likewise.

(3) *Each boundary component $\mathbf{B}'$ of $\mathbf{B}$ is naturally a boundary component of $\mathbf{X}$. This defines a partial order on the set of boundary components of $\mathbf{X}$.*

2.2.21. Definition (Conical complex). Let $\mathbf{B}_1$ be a rational boundary component of $\mathbf{X}$ and $\mathbb{D}^0_{\mathbf{X}}$ be a connected component of $\mathbb{D}_{\mathbf{B}_1 \Rightarrow \mathbf{X}}$. Let $C^*(\mathbb{D}^0_{\mathbf{X}}, P_{\mathbf{B}_1}) \subset U_{\mathbf{B}_1}(\mathbb{R})(-1)$ denote the union of the cones $C(\mathbb{D}^0_{\mathbf{X}}, P_{\mathbf{B}_2})$ for all rational boundary components $\mathbf{B}_2$ such that $\mathbf{B}_1 \Longrightarrow \mathbf{B}_2 \Longrightarrow \mathbf{X}$. It is a convex cone. Form the following quotient

$$C_{\mathbf{X}} := \left(\coprod_{(\mathbb{D}^0_{\mathbf{X}}, P_{\mathbf{B}_1})} C^*(\mathbb{D}^0_{\mathbf{X}}, P_{\mathbf{B}_1}) \right) \Big/ \sim$$

by the equivalence relation generated by the graphs of all embeddings

$$C^*(\mathbb{D}^0_{\mathbf{X}}, P_{\mathbf{B}_2}) \hookrightarrow C^*(\mathbb{D}^0_{\mathbf{X}}, P_{\mathbf{B}_1})$$

for $\mathbf{B}_1 \Longrightarrow \mathbf{B}_2 \Longrightarrow \mathbf{X}$. It is called the **conical complex** associated with $\mathbf{X}$ (cf. [58, 4.24]).

2.2.22. Consider the set

$$C_{\mathbf{X}} \times P_{\mathbf{X}}(\mathbb{A}^{(\infty)}).$$

$P_{\mathbf{X}}(\mathbb{Q})$ acts on this set from the left by conjugation of boundary components [58, 4.23] and on the second factor by left multiplication. $P_{\mathbf{X}}(\mathbb{A}^{(\infty)})$ acts via multiplication on the right on the second factor. Furthermore $P_{\mathbf{X}}(\mathbb{A}^{(\infty)})$ acts on the second factor through left multiplication, too.

Let a set Δ of subsets of $C_{\mathbf{X}} \times P_{\mathbf{X}}(\mathbb{A}^{(\infty)})$ be given, such that every $\sigma \in \Delta$ is contained in some $C(\mathbb{D}^0_{\mathbf{X}}, P_{\mathbf{B}}) \times \rho$.

We denote

$$\Delta(\mathbb{D}^0_{\mathbf{X}}, P_{\mathbf{B}}, \rho) := \{\sigma \in \Delta \mid \sigma \subset \overline{C(\mathbb{D}^0_{\mathbf{X}}, P_{\mathbf{B}}) \times \rho}\}.$$

Let K be an (admissible) compact open subgroup of $P_{\mathbf{X}}(\mathbb{A}^{(\infty)})$. For any $\mathbb{D}^0_{\mathbf{X}}$, $\mathbf{B}$ and ρ, we let $\Gamma_U \subset U_{\mathbf{B}}(\mathbb{Q})$ be the image of

$$\left(\{z \in Z(P_{\mathbf{X}}(\mathbb{Q})) \mid z|_{\mathbb{D}_{\mathbf{X}}} = \mathrm{id}\} U_{\mathbf{B}}(\mathbb{Q})\right) \cap {}^{\rho}K$$

under the projection $Z(P_{\mathbf{X}}(\mathbb{Q})) \times U_{\mathbf{B}}(\mathbb{Q}) \to U_{\mathbf{B}}(\mathbb{Q})$.

2.2.23. Definition (Rational polyhedral cone decomposition). Δ is called a ***K*-admissible (partial) rational polyhedral cone decomposition** for $\mathbf{X}$, if

(1) $\Delta(\mathbb{D}^0_{\mathbf{X}}, P_{\mathbf{B}}, \rho)$ is a (partial) rational polyhedral cone decomposition of the closure of $C(\mathbb{D}^0_{\mathbf{X}}, P_{\mathbf{B}}) \times \rho$, for each $\mathbb{D}^0_{\mathbf{X}}$, $\mathbf{B}$, and ρ. *We understand a rational polyhedral cone as open in its closure.*
(2) Δ is invariant under right multiplication of K and under left multiplication of $P_{\mathbf{X}}(\mathbb{Q})$.
(3) For each $\mathbf{B}$ the set $\bigcup_{\rho \in P_{\mathbf{X}}(\mathbb{A}^{(\infty)})} \Delta(\mathbb{D}^0_{\mathbf{X}}, P_{\mathbf{B}}, \rho)$ is invariant under left multiplication of $P_{\mathbf{B}}(\mathbb{A}^{(\infty)})$.

It is called **finite**, if the quotient $P_{\mathbf{X}}(\mathbb{Q}) \backslash \Delta / K$ is finite.

It is called **complete**, if $\Delta(\mathbb{D}^0_{\mathbf{X}}, P_{\mathbf{B}}, \rho)$ is a *complete* rational polyhedral cone decomposition, for each $\mathbb{D}^0_{\mathbf{X}}$, $\mathbf{B}$, and ρ.

It is called **projective**, if on each $\Delta(\mathbb{D}^0_{\mathbf{X}}, P_{\mathbf{B}}, \rho)$ there exists a polarization function ([1, IV, §2.1], cf. [19, IV, 2.4]).

It is called **smooth** with respect to K, if $\Delta(\mathbb{D}^0_{\mathbf{X}}, P_{\mathbf{B}}, \rho)$ is smooth with respect to the corresponding lattice Γ_U, for each $\mathbb{D}^0_{\mathbf{X}}$, $\mathbf{B}$ and ρ.

Condition (3) is called the arithmeticity condition. Without it, the compactification exists over $\mathbb{C}$ but may not descend to the reflex field or to a reflex ring.

2.2.24. Definition (*p*-ECMSD). A triple ${}^K_\Delta\mathbf{X}$ is called ***p*-integral extended compactified mixed Shimura datum (*p*-ECMSD)**, if ${}^K\mathbf{X}$ is as in Definition 2.2.5, and Δ is in addition a K-admissible (partial) rational polyhedral cone decomposition.

Morphisms of p-ECMSD have to satisfy the property that for each $\sigma_1 \in \Delta_1$ there is some $\sigma_2 \in \Delta_2$ with $\gamma(\sigma_1)^\rho \subset \sigma_2$.

Let ${}^K_\Delta\mathbf{X}$ be p-ECMSD and $[\alpha, \rho]\colon {}^{K'}\mathbf{Y} \to {}^K\mathbf{X}$ be a morphism of p-EMSD such that α is a *closed embedding.*

Set $[\alpha, \rho]^*\Delta$ to be the set of all cones $\{(u, \rho') \mid (\alpha(u), \alpha(\rho')\rho) \in \sigma\}$ for all $\sigma \in \Delta$. This is a K'-admissible rational partial cone decomposition for $\mathbf{Y}$. It is finite, respectively complete, respectively projective if Δ is finite, respectively complete, respectively projective.

The association $\Delta \mapsto [\alpha, \rho]^*\Delta$ is functorial. If $[\alpha, \rho]$ is an embedding (this includes a condition on K, K', see 2.2.5), we call ${}_{[\alpha,\rho]^*\Delta}^{K'}\mathbf{Y} \to {}^K_\Delta\mathbf{X}$ an **embedding**.

Be aware that, in general, smoothness is not inherited by $[\alpha, \rho]^*\Delta$, but we have:

2.2.25. Proposition ([25, Theorem 2.4.12])**.** *Let* $\mathbf{X}$ *be p-MSD. Let* K *be an admissible compact open.*

(1) *K-admissible complete rational polyhedral cone decompositions for* $\mathbf{X}$ *exist.*

(2) *If Δ is a K-admissible rational polyhedral cone decomposition for* $\mathbf{X}$, *there is a smooth and projective refinement Δ'.* Any *refinement of Δ' will remain projective.*

(3) *If Δ_1, Δ_2 are 2 rational polyhedral cone decompositions for* $\mathbf{X}$, *there is a common refinement Δ (supported on the intersection of their supports).*

(4) *If $\alpha\colon \mathbf{X} \to \mathbf{X}'$ is an embedding and $\rho \in P'_{\mathbf{X}}(\mathbb{A}^{(\infty,p)})$ is given, there is K' such that we have an embedding $[\alpha, \rho]\colon {}^K\mathbf{X} \to {}^{K'}\mathbf{X}'$, and a K'-admissible rational polyhedral smooth projective cone decomposition Δ' for $\mathbf{X}'$, with the property that*

$\Delta := [\alpha,\rho]^*\Delta'$ is smooth and projective. For every *smooth refinement $\tilde{\Delta}$ of Δ, there is a smooth refinement $\tilde{\Delta}'$ of Δ' with $\tilde{\Delta} := [\alpha,\rho]^*\tilde{\Delta}'$.*

2.2.26. Definition (Boundary morphisms). Let ${}^K_\Delta\mathbf{X}$ be a p-ECMSD and let $\iota\colon \mathbf{B}_1 \Longrightarrow \mathbf{X}$ be a boundary morphism. For any $\rho \in P_{\mathbf{X}}(\mathbb{A}^{(\infty)})$, we define $K' := P_{\mathbf{B}_1}(\mathbb{A}^{(\infty)}) \cap {}^\rho K$, and define Δ' as $([\iota,\rho]^*\Delta)|_{\mathbf{B}_1}$, where restriction is characterized by

$$\Delta|_{\mathbf{B}_1}(\mathbb{D}^0_{\mathbf{B}_1}, P_{\mathbf{B}_2}, \rho_1) = \Delta(\mathbb{D}^0_{\mathbf{X}}, P_{\mathbf{B}_2}, \rho_1)$$

for all $\rho_1 \in P_{\mathbf{B}_1}(\mathbb{A}^{(\infty)})$, every boundary morphism $\mathbf{B}_2 \Longrightarrow \mathbf{B}_1$ and every pair of connected components $\mathbb{D}^0_{\mathbf{X}}$ and $\mathbb{D}^0_{\mathbf{B}_1}$ such that $\mathbb{D}^0_{\mathbf{X}} \hookrightarrow \mathbb{D}^0_{\mathbf{B}_1} \hookrightarrow \mathbb{D}_{\mathbf{B}_2}$. Δ' in general inherits neither completeness nor finiteness. It is K'-admissible.

We write

$$(\iota,\rho) : {}^{K'}_{\Delta'}\mathbf{B}_1 \Longrightarrow {}^K_\Delta\mathbf{X}$$

and call this a **boundary morphism** of p-ECMSD.

We call two boundary morphisms

$$(\iota',\rho')\colon {}^{K'}_{\Delta'}\mathbf{B}' \Longrightarrow {}^K_\Delta\mathbf{X}$$

and

$$(\iota'',\rho'') : {}^{K''}_{\Delta''}\mathbf{B}'' \Longrightarrow {}^K_\Delta\mathbf{X}$$

equivalent, if the images of $P_{\mathbf{B}'}$ and $P_{\mathbf{B}''}$ are conjugated via some $\alpha \in P_{\mathbf{X}}(\mathbb{Q})$ and

$$\alpha\rho' \in \mathrm{Stab}_{Q(\mathbb{Q})}(\mathbb{D}_{\mathbf{B}})P_{\mathbf{B}}(\mathbb{A}^{(\infty)})\rho'' K.$$

(Here Q is the parabolic defining $\mathbf{B}$.) In an equivalence class we may assume $\rho \in P_{\mathbf{X}}(\mathbb{A}^{(\infty,p)})$.

2.2.27. Definition (Δ^0, concentrated in the unipotent fibre). Let ${}^K_\Delta\mathbf{X}$ be p-ECMSD. Define Δ^0 as the set of all $\sigma \in \Delta$, such that $\sigma \subset C(\mathbb{D}^0_{\mathbf{X}}, P_{\mathbf{B}}) \times P_{\mathbf{X}}(\mathbb{A}^{(\infty)})$ for some *improper* rational boundary component $\mathbf{B} \Longrightarrow \mathbf{X}$. We say that Δ is **concentrated in the unipotent fibre**, if $\Delta = \Delta^0$.

2.3. Symplectic, Hodge, and Abelian type Shimura data

2.3.1. Let S be a scheme. Let L be a locally free sheaf on S with unimodular alternating form $\langle v,w\rangle = \Psi(v,w)$, i.e., satisfying $\langle v,v\rangle = 0$ and such that the induced homomorphism $L \to L^*$ is an isomorphism.

Let us assume first that L is *nonzero*. We then have the reductive group schemes $\mathrm{Sp}(L)$ and $\mathrm{GSp}(L)$ over S, the **symplectic group**, respectively the **group of symplectic similitudes** of L.

There is an exact sequence

$$0 \longrightarrow \mathrm{Sp}(L) \longrightarrow \mathrm{GSp}(L) \xrightarrow{\ \lambda\ } \mathbb{G}_m \longrightarrow 0,$$

where λ is the similitude factor. Let ρ denote the standard representation of $\mathrm{GSp}(L)$ on L.

Let now $S = \mathrm{Spec}(\mathbb{Z}_{(p)})$. We define a p-integral pure Shimura datum $\mathbf{H}_g$ associated with L (it depends, up to isomorphism, only on the rank $2g$ of L): We let $P_{\mathbf{H}_g} := \mathrm{GSp}(L)$ and let $\mathbb{D}_{\mathbf{H}_g}$ be *the* conjugacy class of morphisms $h := \mathbb{S} \to \mathrm{GSp}(L_{\mathbb{R}})$, such that they give pure Hodge structures of type $(-1,0),(0,-1)$ on L and which are **polarized**, i.e., such that the form $\langle\cdot, h(i)\cdot\rangle$ is symmetric and (positive or negative) definite.

If L is the *zero* sheaf on S we define $\mathrm{Sp}(L) := 1$ and $\mathrm{GSp}(L) := \mathbb{G}_{\mathrm{m},S}$, and we let $\mathbb{D}_{\mathbf{H}_0}$ be the 2 point set of isomorphisms $\mathbb{Z} \to \mathbb{Z}(1)$ with the nontrivial action of $\mathbb{G}_{\mathrm{m}}(\mathbb{R})$. This defines a ($p$-integral) Shimura datum $\mathbf{H}_0$. We understand the morphism $\lambda : \mathrm{GSp}(0) \to \mathbb{G}_{\mathrm{m}}$ to be the identity.

2.3.2. Let L_0 be a locally free sheaf on S with unimodular alternating form (possibly 0).

Let I be another locally free sheaf on S (also possibly 0). Let I^* be its dual. We define the semidirect product

$$\mathrm{PSp}(L_0, I) := \mathbb{W}(L_0 \otimes I) \rtimes \mathrm{GSp}(L_0)$$

with respect to the following action of $\mathrm{GSp}(L_0)$: the product of the standard representation on L_0 with the trivial one on I. $\mathrm{PSp}(L_0, I)$ is of type (P). It acts on $L := L_0 \oplus I^*$ as follows: The action of GSp is given by the standard representation on L_0 and the trivial action on I^*, an element $X = v' \otimes u' \in L_0 \otimes I$, considered as element of the Lie algebra, acts by

$$X(v, u^*) = ((u^* u)v', 0)$$

and $\mathbb{W}(L_0 \otimes I)$ via the exponential $\exp(X)(v, u^*) = (v, u^*) + X(v, u^*)$ which in this case makes sense over any S. This is compatible with the structure of semidirect product. Let $S = \mathrm{Spec}(\mathbb{Z}_{(p)})$ from now on. The action fixes a weight filtration

$$W_i(L_0 \oplus I^*) := \begin{cases} 0 & i \leq -2 \\ L_0 & i = -1 \\ L_0 \oplus I^* & i \geq 0. \end{cases}$$

We have the unipotent extension $\mathbf{H}_{g_0}[0, I \otimes L_0]$ of the p-integral pure Shimura datum $\mathbf{H}_{g_0}$. Its underlying P is $\mathrm{PSp}(L_0, I)$. Its underlying $\mathbb{D}$ (if $g_0 \neq 0$) may be identified with *the* conjugacy class of morphisms $h\colon \mathbb{S} \to \mathrm{PSp}(L_{0,\mathbb{R}}, I_{\mathbb{R}})$ such that they give mixed Hodge structures of type $(-1, 0), (0, -1), (0, 0)$ with respect to the weight filtration $W_\bullet$, such that on W_{-1} they are polarized. (We have $\mathrm{PSp}(0, I) = \mathrm{GSp}(0)$ and $\mathrm{PSp}(L_0, 0) = \mathrm{GSp}(L_0)$).

2.3.3. Let L_0 and S be as in 2.3.2. Consider the following extension of Abelian unipotent groups (**Heisenberg group**)

$$0 \longrightarrow \mathbb{W}((I \otimes I)^s) \longrightarrow \mathrm{WSp} \longrightarrow \mathbb{W}(L_0 \otimes I) \longrightarrow 0$$

defined (if 2 is invertible in S) by the following group law

$$(u_1 u_2, v u_3)(u_1' u_2', v' u_3') = (u_1 u_2 + u_1' u_2' + \tfrac{1}{2}\langle \boldsymbol{v}, \boldsymbol{v'} \rangle (\boldsymbol{u_3 u_3'} + \boldsymbol{u_3' u_3}), v u_3 + v' u_3').$$

We define the semidirect product

$$\mathrm{USp}(L_0, I) := \mathrm{WSp} \rtimes \mathrm{GSp}(L_0).$$

with respect to the following action of $\mathrm{GSp}(L_0)$ on WSp: Action via λ on $(I \otimes I)^s$ by scalars, and on $L_0 \otimes I$ by the product of the standard representation on L_0 with the trivial one on I.

$\mathrm{USp}(L_0, I)$ is also of type (P).

Denote $\mathfrak{I} := I \oplus I^*$. Choose on $L_0 \oplus \mathfrak{I}$ the symplectic form

$$\langle v_1, u_1, u_1^*; v_2, u_2, u_2^* \rangle := \langle v_1, v_2 \rangle + u_2^* u_1 - u_1^* u_2.$$

We define an action of $\mathrm{USp}(L_0, I)$ on $L := L_0 \oplus \mathfrak{I}$ as follows: The action of $\mathrm{GSp}(L_0)$ is given by the standard representation on L_0, the trivial representation on I^* and λ acting by scalars on I. The element $X = v' \otimes u' \in L_0 \otimes I$, considered as element of the Lie algebra, acts by

$$X(v, u, u^*) = \big((u^* u')v', \langle v', v\rangle u', 0\big),$$

while $X = u_1 \otimes u_2 \in I \otimes I$ acts by

$$X(v, u, u^*) = (0, (u^* u_1)u_2, 0),$$

and $\mathbb{W}(\cdots)$ acts via the exponential

$$\exp(X)(v, u^*) = (v, u^*) + X(v, u^*) + \frac{1}{2}X^2(v, u^*).$$

This is compatible with the group structure given above.

If 2 is not invertible in S, we assume that there is an isomorphism $L_0 \cong L_{00} \oplus L_{00}^*$, such that the alternating form is given by the standard one. In this case we let the groups $\mathbb{W}(L_{00} \otimes I)$, $\mathbb{W}(L_{00}^* \otimes I)$ and $\mathbb{W}\big((I \otimes I)^s\big)$ operate as above via exponential (it terminates after the second term). One checks that

$$\mathbb{W}\big((I \otimes I)^s\big) \times \mathbb{W}(L_{00} \otimes I) \times \mathbb{W}(L_{00}^* \otimes I) \to \mathrm{GSp}(L)$$

is a closed embedding onto a subgroup scheme. Explicitly the group law is given by

$$(X_1, X_2, X_3)(X_1', X_2', X_3') = (X_1 + X_1' + \langle \mathbf{X_3}, \mathbf{X_2'}\rangle - \langle \mathbf{X_2'}, \mathbf{X_3}\rangle, X_2 + X_2', X_3 + X_3').$$

This is in any case compatible with the structure of semidirect product. Let $S = \mathrm{Spec}(\mathbb{Z}_{(p)})$ from now on. The action fixes a weight filtration

$$W_i(L) := \begin{cases} 0 & i \leq -3, \\ I & i = -2, \\ L_0 \oplus I & i = -1, \\ L & i \geq 0. \end{cases}$$

We have the unipotent extension $\mathbf{H}_{g_0}[(I \otimes I)^s, I \otimes L_0]$ of the p-integral pure Shimura datum $\mathbf{H}_{g_0}$. Its underlying P is $\mathrm{USp}(L_0, I)$. Its underlying $\mathbb{D}$ (if $g_0 \neq 0$) may be identified with *the* conjugacy class of morphisms $h\colon \mathbb{S} \to \mathrm{USp}(L_{0,\mathbb{R}}, I_\mathbb{R})$ such that they give (polarized) mixed Hodge structures on $L_\mathbb{R}$ of type $(-1,-1), (-1,0), (0,-1), (0,0)$ with respect to the weight filtration $W_\bullet$.

The action of $\mathrm{USp}(L_0, I)$ on $L = L_0 \oplus \mathfrak{I}$ defined above induces a closed embedding

$$\mathrm{USp}(L_0, I) \hookrightarrow \mathrm{GSp}(L).$$

The image is precisely the subgroup scheme fixing the weight filtration and $I^* = \mathrm{gr}_0(L_0 \oplus \mathfrak{I})$ pointwise.

2.3.4. Definition (Shimura datum of symplectic type). For any saturated submodule $U' \subseteq (I \otimes I)^s$ we call the p-integral mixed Shimura datum

$$\mathbf{H}_{g_0}[(I \otimes I)^s, I \otimes L_0]/\mathbb{W}(U')$$

a Shimura datum of **symplectic type**.

Note that any of the p-integral mixed Shimura data in this section is of this form.

2.3.5. Proposition ([25, Theorem 2.5.4]). *There is a bijection*

$$\{\textit{isotropic subspaces } I_{\mathbb{Q}} \textit{ of } L_{\mathbb{Q}}\} \cong \{\textit{boundary components } \mathbf{B} \textit{ of } \mathbf{H}(L)\}.$$

Let $I_{\mathbb{Q}}$ be an isotropic subspace of dimension $g - g_0$, $I = I_{\mathbb{Z}_{(p)}}$ the associated saturated sublattice and $\mathbf{B} \Longrightarrow \mathbf{H}(L)$ the corresponding boundary component.

For any choice of splitting $L \cong L_0 \oplus \mathfrak{I}$ (where $\mathfrak{I} = I \oplus I^*$, as usual, with natural symplectic form), there is an isomorphism

$$\mathbf{H}(L_0)[(I \otimes I)^s, I \otimes L_0] \cong \mathbf{B},$$

whose underlying morphism is given by the representation of $\mathrm{PSp}(L_0, I)$ on $L = L_0 \oplus I \oplus I^*$ given above (it depends on the choice of splitting).

2.3.6. Corollary. *Every boundary component of a (p-integral) mixed Shimura datum of symplectic type is again of symplectic type.*

2.3.7. We have quotient maps of mixed Shimura data

$$\mathbf{H}_{g_0}[(I \otimes I)^s, I \otimes L_0] \to \mathbf{H}_{g_0}[0, I \otimes L_0] \to \mathbf{H}_{g_0}.$$

The first gives an isomorphism

$$\mathbf{H}_{g_0}[(I \otimes I)^s, I \otimes L_0]/U \cong \mathbf{H}_{g_0}[0, I \otimes L_0]$$

(where $U = \mathbb{W}\big((I \otimes I)^s\big)$), and the second gives an isomorphism

$$\mathbf{H}_{g_0}[0, I \otimes L_0]/W \cong \mathbf{H}_{g_0}$$

(where $W = V = \mathbb{W}(I \otimes L_0)$).

2.3.8. Let P be a group scheme over S. Let P act linearly on a locally free sheaf of $\mathcal{O}_S$-modules L.

2.3.9. Definition ((P, L)-structure). Let E be a locally free sheaf of $\mathcal{O}_S$-modules. A **(P, L)-structure** on E is a section of the etale quotient sheaf

$$\alpha \in H^0(S, \underline{\mathrm{Iso}}(E, L)/P).$$

The preimage under the projection $\underline{\mathrm{Iso}}(E, L) \to \underline{\mathrm{Iso}}(E, L)/P$ is a right P-torsor $\mathrm{tor}(\alpha)$ which we call the **associated right P-torsor**.

2.3.10. Remark. If E is a locally free sheaf over S' for some scheme $S' \to S$, by abuse of notation, we call a $(P \times_S S', L \otimes_{\mathcal{O}_S} \mathcal{O}_{S'})$-structure on E simply a (P, L)-structure.

In the analytic context, we have the following variants of a P-structure:

2.3.11. Definition (($P(R), L$)-structure). Let $R \subset \mathbb{Q}$ be a ring, $S = \mathrm{Spec}(R)$, let P a group scheme over S, and let L_R a free module of finite rank with a linear action of P. Let B be an analytic manifold (or an analytic orbifold).

Let E be a local system of R-modules on B. A $(P(R), L_R)$-structure α on E is a section of the quotient sheaf

$$\alpha \in H^0(B, \underline{\mathrm{Iso}}(E, L_R)/P(R)_B).$$

The preimage of α in $\underline{\mathrm{Iso}}(E, L_R)$ is a local system locally isomorphic to the constant sheaf $P(R)_B$. We call it also the associated right $P(R)$-torsor $\mathrm{tor}(\alpha)$.

Let E be a holomorphic vector bundle on B. A P-structure α on E is a section of the quotient sheaf

$$\alpha \in H^0(B, \underline{\mathrm{Iso}}(E, L_{\mathcal{O}_B})/P_{\mathcal{O}_B}).$$

Here $P_{\mathcal{O}_B}$ is the sheaf of groups on B of analytic maps to $(P_{\mathbb{C}})^{\mathrm{an}}$. We have also an analytic associated right $P_{\mathcal{O}_B}$-torsor $\mathrm{tor}(\alpha)$.

Clearly, a $(P(R), L_R)$-structure α on a local system E induces a (P, L_R)-structure $\alpha \otimes_R \mathcal{O}_B$ on $E \otimes_R \mathcal{O}_B$.

2.3.12. Lemma. *Let S be a scheme with $H^1_{\mathrm{et}}(S, \mathbb{G}_{\mathrm{m}}) = 1$ and M a locally free sheaf on S.*

(1) *A $(\mathrm{GSp}(L), L)$-structure on M is a unimodular symplectic form on M (i.e., inducing an isomorphism $M \cong M^*$) up to multiplication by $H^0(S, \mathcal{O}_S^*)$.*

(2) *A $(\mathrm{PSp}(L_0, I), L)$-structure on M is a saturated filtration*

$$W_0 = M \supset W_{-1} \supset W_{-2} = 0$$

with unimodular symplectic form on W_{-1} up to multiplication by $H^0(S, \mathcal{O}_S^*)$ and an isomorphism $\rho\colon W_0/W_{-1} \cong I_S^*$.

(3) *A $(\mathrm{USp}(L_0, I), L)$-structure on M is a unimodular symplectic form on M up to multiplication by $H^0(S, \mathcal{O}_S^*)$, a saturated filtration*

$$W_0 = M \supset W_{-1} \supset W_{-2} \supset W_{-3} = 0$$

such that W_{-2} is isotropic and $W_{-1} = (W_{-2})^{\perp}$, and an isomorphism $\rho\colon W_0/W_{-1} \cong I_S^*$.

If R is a ring, a statement similar to (1) is true also for a $(\mathrm{GSp}(L)(R), L_R)$-structure on a local system of R-modules. The same holds for the other statements.

2.3.13. Lemma. *Let S be a scheme and let M a locally free sheaf on S. Let $G = \mathbb{W}(U_{\mathbb{Z}_{(p)}}) \rtimes \mathbb{G}_{\mathrm{m}}$ act on $\mathbb{Z}_{(p)} \oplus U^*_{\mathbb{Z}_{(p)}}$ as follows: $\mathbb{G}_{\mathrm{m}}$ acts by scalar multiplication on $\mathbb{Z}_{(p)}$ and $\mathbb{G}_{\mathrm{a}}(U_{\mathbb{Z}_{(p)}})$ acts via $u(x, u^*) = (x - u^*u, u^*)$. Let S be a scheme over $\mathbb{Z}_{(p)}$ and M be a locally free sheaf on S.*

- *A $(G, \mathbb{Z}_{(p)} \oplus U^*_{\mathbb{Z}_{(p)}})$-structure on M is a saturated filtration*

  $$W_0 = M \supset W_{-1} = W_{-2} \supset W_{-3} = 0$$

  with isomorphism

  $$\rho\colon W_0/W_{-1} \cong U_S^*$$

  and such that W_{-1} is locally free of rank 1.

2.3.14. Clearly, given two different faithful representations L, L' of a group scheme G, there is a 1:1 "change of representation correspondence" between P-structures associated with those representations. For a translation of a certain such change of representation correspondence of $\mathrm{USp}(0, I)$ to 1-motives (cf. 2.7.9), we need a fairly explicit construction of this correspondence in terms of linear algebra.

Let S be a $\mathbb{Z}_{(p)}$-scheme. Let $G = \mathrm{USp}(0, I)$ and let L its standard representation. Let M be a locally free sheaf on S with (G, L)-structure. We may associate with it a $(G, \mathbb{Z}_{(p)} \oplus U^*_{\mathbb{Z}_{(p)}})$-structure M'' by the following (local) procedure: Define M' to be the quotient of $M \otimes I^*$ by the kernel of the contraction

$$W_{-1} \otimes I^* \cong (I \otimes I^*)_S \to \mathcal{O}_S$$

(the isomorphism is defined only up to scalars). *Claim*: The set $\Lambda^2 I^*$ has a unique lift to M' under the map

$$M \otimes I^* \to (M/W_{-1}M) \otimes I^* = I^* \otimes I^*$$

defined as follows: Let $i_1^* \wedge i_2^*$ be in $\Lambda^2 I^*$. We choose lifts $\tilde{i}_1^*, \tilde{i}_2^* \in M$ with the property that $\langle \tilde{i}_1^*, \tilde{i}_2^* \rangle = 0$. The projection of the element $\tilde{i}_1^* \otimes i_2^* - \tilde{i}_2^* \otimes i_1^*$ in M' is well-defined and is the sought-for lift of $i_1^* \wedge i_2^*$. (For the well-definition consider a difference $(\tilde{i}_1^* - \tilde{\tilde{i}}_1^*) \otimes i_2^* - (\tilde{i}_2^* - \tilde{\tilde{i}}_2^*) \otimes i_1^*$. Both summands lie already in $W_{-1}M \otimes I^*$ and, since $W_{-1}M$ is isotropic, its contraction may be computed by the expression: $\langle \tilde{i}_1^* - \tilde{\tilde{i}}_1^*, \tilde{\tilde{i}}_2^* \rangle - \langle \tilde{i}_2^* - \tilde{\tilde{i}}_2^*, \tilde{\tilde{i}}_1^* \rangle$. This is zero because of the choice of lifts.) We define $M'' = M'/\Lambda^2(I^*)$ by means of this lift. We get an inherited filtration:

$$0 \subseteq W_{-1}M'' \subseteq M''$$

such that $W_{-1}M'' \cong \mathcal{O}_S$ noncanonically and $M''/W_{-1}M'' \cong \mathrm{Sym}^2(I^*)$ canonically. That is, glueing these local constructions, we get a $(G, \mathbb{Z}_{(p)} \oplus U^*_{\mathbb{Z}_{(p)}})$-structure, where $U_{\mathbb{Z}_{(p)}} = (I \otimes I)^s$. The construction $M \rightsquigarrow M''$ may be reversed.

2.3.15. Definition (Shimura data of Hodge and Abelian type). Let $\mathbf{X}$ be a p-integral mixed Shimura datum (p-MSD). $\mathbf{X}$ is called of **Hodge type**, if there is an embedding into a p-integral mixed Shimura datum of symplectic type (see 2.3.4 for the notation)

$$\mathbf{X} \hookrightarrow \mathbf{H}_{g_0}[(I \otimes I)^s, L_0 \otimes I]/\mathbb{W}(U').$$

A mixed Shimura datum $\mathbf{X}$ is called of **Abelian type**, if there is a mixed Shimura datum of Hodge type $\mathbf{X}'$, and a central isogeny $P_{\mathbf{X}}'^{\mathrm{der}} \to P_{\mathbf{X}}^{\mathrm{der}}$ which induces an isomorphism of Shimura data $\mathbf{X}'^{\mathrm{ad}} \cong \mathbf{X}^{\mathrm{ad}}$.

2.3.16. Proposition. *Let $\mathbf{X}$ be a p-integral mixed Shimura datum and $\mathbf{B}$ a boundary component of $\mathbf{X}$. If $\mathbf{X}$ is of Hodge (respectively Abelian) type then $\mathbf{B}$ is either.*

PROOF. Similar to [25, Theorem 2.6.3]. □

2.4. Orthogonal Shimura data

2.4.1. Let L be a $\mathbb{Z}_{(p)}$-lattice of dimension l with unimodular quadratic form Q_L. We have the reductive group schemes $\mathrm{SO}(L)$ and $\mathrm{GSpin}(L)$. We let $\mathrm{SO}(L)$ act on L by its natural action, and we let $\mathrm{GSpin}(L)$ act on L by $(g, v) \mapsto gvg'$, where $'$ is the main involution of the Clifford algebra $\mathrm{C}(L)$ of L. Note that these actions are *not* compatible with the projection $\pi\colon \mathrm{GSpin}(L) \to \mathrm{SO}(L)$. The group $\mathrm{GSpin}(L)$ is equipped with a morphism $\lambda\colon \mathrm{GSpin}(L) \to \mathbb{G}_\mathrm{m}$ whose kernel is the spin group.

For the rest of this section we assume $p \neq 2$. Then there exists an orthogonal basis $\{v_i\}$ of L, such that all $Q_L(v_i)$ are units. We have the basis

$$v_{i_1} \cdots v_{i_j}, \quad i_1 < \cdots < i_j, \quad j \geq 0 \text{ even}$$

of $\mathrm{C}^+(L)$ and the basis

$$v_{i_1} \cdots v_{i_j}, \quad i_1 < \cdots < i_j, \quad j > 0 \text{ odd}$$

of $\mathrm{C}^-(L)$. The trace of any basis element of $\mathrm{C}^+(L)$ (acting by left multiplication on $\mathrm{C}^+(L)$ or $\mathrm{C}^-(L)$) except 1 is 0. The trace of 1 is 2^{l-1} in both cases.

2.4.2. Lemma. *For an element $\delta \in \mathrm{C}^+(L)^*$ with $\delta' = -\delta$, the form*

$$\langle x, y \rangle_\delta \mapsto \mathrm{tr}(x\delta y')$$

on $\mathrm{C}^+(L)$ is symplectic, unimodular, $\mathrm{Spin}(Q)$-invariant (respectively $\mathrm{GSpin}(Q)$-invariant up to scalar given by λ) where these groups act by left multiplication.

PROOF. The form is invariant because $\mathrm{tr}(AB) = \mathrm{tr}(BA)$. Since δ is invertible the nondegeneracy is equivalent to that of the form $\mathrm{tr}(xy')$ which is given by an invertible diagonal matrix with respect to the basis chosen above. The form is symplectic because tr is invariant under $'$ and $\delta' = -\delta$. □

2.4.3. Suppose again that $L_{\mathbb{Z}_{(p)}}$ is a lattice with unimodular quadratic form of signature $(l-2,2)$ with $l \geq 3$. A polarized Hodge structure of type $(-2,0), (-1,-1), (0,-2)$ on $L_{\mathbb{C}}$ with $\dim(L^{-2,0}) = 1$ is determined by an isotropic subspace $L^{-2,0}$ satisfying $\langle z, \bar{z} \rangle < 0$ for any nonzero $z \in L^{-2,0}$. Then the other spaces in the Hodge decomposition are determined by $L^{0,-2} = \overline{L^{-2,0}}$ and $L^{-1,-1} = (L^{-2,0} + L^{0,-2})^{\perp}$. We define $\mathbb{D}_{\mathbf{S}(L)}$ to be the set of these Hodge structures which, as usual, can be seen as a conjugacy class in $\mathrm{Hom}\big(\mathbb{S}, \mathrm{SO}(L_{\mathbb{R}})\big)$. It is identified with

$$\{\langle z \rangle \in \mathbb{P}(L_{\mathbb{C}}) \mid \langle z, z \rangle = 0, \langle z, \bar{z} \rangle < 0\}.$$

It is equipped with a complex structure coming from $\mathbb{P}(L_{\mathbb{C}})$. We have $\mathbb{D}_{\mathbf{O}(L)} = \mathbb{D}_{\mathbf{S}(L)}$ but with a shift by 2 in the weight.

$\mathbb{D}_{\mathbf{S}(L)}$ has 2 connected components. There is a 2:1 map

$$\begin{aligned} \mathbb{D}_{\mathbf{S}(L)} &\longrightarrow \mathrm{Grass}^-(L_{\mathbb{R}}) \\ \langle z \rangle &\longmapsto D_{\langle z \rangle} := (\langle z, \bar{z} \rangle)^{\mathrm{Gal}(\mathbb{C}|\mathbb{R})} \end{aligned}$$

onto the Grassmannian Grass^- of negative definite subspaces of $L_{\mathbb{R}}$. This map has a section determined by a choice of $\mathbb{C}$-orientation[2] on any of the subspaces D. The images of these sections are the 2 connected components of $\mathbb{D}_{\mathbf{S}(L)}$.

It will be convenient to fix an orientation on some (hence on all) of the negative definite subspaces in $\mathrm{Grass}^-(L_{\mathbb{R}})$. This identifies $\pi_0(\mathbb{D}_{\mathbf{S}(L)})$ with $\mathbb{D}_{\mathbf{H}_0} = \mathrm{Hom}\big(\mathbb{Z}, \mathbb{Z}(1)\big)$ and induces a morphism of p-MSD

$$\mathbf{S}(L) \to \mathbf{H}_0.$$

The space $D_{\langle z \rangle}$ has an orthonormal (with respect to Q_L) basis given by

$$x_1 = \frac{z + \bar{z}}{\sqrt{-\langle z, \bar{z} \rangle}}, \quad x_2 = \frac{z - \bar{z}}{\sqrt{\langle z, \bar{z} \rangle}}. \tag{2.2}$$

In x_1 we understand the positive square root and x_2 depends on a choice of $i \in \mathbb{C}$.

2.4.4. Lemma. (1) $\mathbf{S}(L)$ *is a p-integral pure Shimura datum.*

If L has signature $(0,2)$ and we would define $\mathbb{D}_{\mathbf{S}}$ as above (i.e., consisting of 2 points) then GSpin *would act trivially on it. Therefore, in this case we get 2 different p-integral Shimura data denoted by* $\mathbf{S}(L)_{\pm}$.

(2) *If L' is a saturated sublattice of L of signature $(l-m-2,2)$, $0 \leq m \leq l-2$, it induces an embedding of p-integral Shimura data*

$$\mathbf{S}(L') \hookrightarrow \mathbf{S}(L).$$

(3) *For any $\delta \in \mathrm{C}^+(L)^*$ with $\delta' = -\delta$ there is an embedding*

$$\mathbf{S}(L) \hookrightarrow \mathbf{H}(\mathrm{C}^+(L), \langle\cdot,\cdot\rangle_{\delta})$$

and accordingly $\mathbf{S}(L)$ *is of Hodge type.* $\mathbf{O}(L) := \mathbf{S}(L)/\mathbb{G}_{\mathrm{m}}$ *is of Abelian type.*

[2]i.e., the choice of a $\mathbb{R}$-linear isometry $D \cong \mathbb{C}$

PROOF. (1) To each $\langle z\rangle \in \mathbb{D}_{\mathbf{S}(L)}$ the associated morphism $h = h_{\mathbf{S}(L)}(\langle z\rangle)$ factors through $\mathrm{GSpin}(L_{\mathbb{R}})$. It is explicitly given by

$$\begin{aligned} \mathbb{S} &\longrightarrow \mathrm{C}^+(L_{\mathbb{R}}) \\ w = a + bi &\longmapsto a + bi \cdot \frac{z\bar{z} - \bar{z}z}{\langle z, \bar{z}\rangle} = a + bx_1x_2 \end{aligned}$$

for any choice of $i \in \mathbb{C}$ (coherent with the choice for x_2).

The formula even defines a field isomorphism $\mathbb{C} \cong \mathrm{C}^+(D_{\langle z\rangle}, \mathbb{R})$. It acts on $L^{-2,0}$ by w^2, on $L^{0,-2}$ by $\overline{w}^2$ and on $L^{-1,-1}$ by $w\overline{w}$.

(2) is true because the inclusion $\mathrm{C}^+(L') \subset \mathrm{C}^+(L)$ induces a closed embedding $\mathrm{GSpin}(L') \hookrightarrow \mathrm{GSpin}(L)$, and also a closed embedding $\mathbb{D}_{\mathbf{S}(L')} \to \mathbb{D}_{\mathbf{S}(L)}$.

For (3) let $\langle z\rangle$ be a point in $\mathbb{D}_{\mathbf{S}(L)}$. $P_z^{-1,0} := z\bar{z}/\langle z, \bar{z}\rangle$ respectively its complex conjugate, satisfy

$$P_z^{-1,0} + P_z^{0,-1} = \mathrm{id}, \ (P_z^{i,j})^2 = P_z^{i,j},$$

and on $P_z^{i,j}\,\mathrm{C}^+(L_{\mathbb{C}})$ the morphism h acts as $w^{-i}\overline{w}^{-j}$. Furthermore, for z chosen such that $D_z = D_0$ the form

$$\langle \cdot, x_1x_2 \cdot\rangle_\delta = \langle \cdot, h(i)\cdot\rangle_\delta = \mathrm{tr}(x\delta y'\delta') = -\,\mathrm{tr}(x\delta y'\delta)$$

is symmetric and definite. (Here the choice of $i \in \mathbb{C}$ is determined by the orientation $x_1 \wedge x_2$ of $D_0 = D_z$, see (2.2)). The substitution $x \mapsto gx$, $y \mapsto gy$ for any g in the spin group changes x_1x_2 into $g(x_1)g(x_2)$ but does not affect the properties of symmetry and definiteness. The sign of definiteness, however, is reflected by the chosen orientation of D_0. Hence the map $\langle z\rangle \mapsto P_z^{-1,0}\,\mathrm{C}^+(L_{\mathbb{C}})$ may be seen as a map $\mathbb{D}_{\mathbf{S}(L)} \to \mathbb{D}_{\mathbf{H}(\mathrm{C}^+(L), \langle\cdot,\cdot\rangle_\delta)}$. This map, together with the closed embedding

$$\mathrm{GSpin}(L) \hookrightarrow \mathrm{GSp}(\mathrm{C}^+(L), \langle\cdot,\cdot\rangle_\delta),$$

induces an embedding $\mathbf{S}(L) \hookrightarrow \mathbf{H}(\mathrm{C}^+(L), \langle\cdot,\cdot\rangle_\delta)$.

To see that $\mathbf{S}(L)$ is of Hodge type we have to see that a δ exists: Fix a positive definite $\mathbb{Z}_{(p)}$-sublattice D_0 which is a direct summand, and fix an orthonormal basis x_1, x_2 of D_0 (hence an induced orientation of $D_{0,\mathbb{R}}$). Consider the element $\delta := x_1x_2$. It satisfies the requirements of (3). $\mathbf{O}(L)$ is, as a quotient of $\mathbf{S}(L)$, clearly of Abelian type. □

We have the following proposition (similar to 2.3.5):

2.4.5. Proposition. *There is a bijection*

$$\{\textit{isotropic subspaces } I_{\mathbb{Q}} \textit{ of } L_{\mathbb{Q}}\} \cong \{\textit{boundary components } \mathbf{B} \textit{ of } \mathbf{S}(L) \textit{ or } \mathbf{O}(L)\}.$$

Let $I_{\mathbb{Q}}$ be an isotropic subspace, $I = I_{\mathbb{Z}_{(p)}}$ the associated saturated sublattice and $\mathbf{B}$ the corresponding boundary component of $\mathbf{S}(L)$ (respectively of $\mathbf{O}(L)$).

For any choice of splitting $L = I \oplus I^* \perp L_0$, there is an isomorphism (in both cases, $\mathbf{S}(L)$ or $\mathbf{O}(L)$):

$$\mathbf{B} \cong \begin{cases} \mathbf{H}_0[I \otimes (I^\perp/I), 0] & \textit{if } \dim(I) = 1, \\ \mathbf{H}_1(I)[\Lambda^2 I, I \otimes (I^\perp/I)] & \textit{if } \dim(I) = 2, \end{cases}$$

where, if $\dim(I) = 1$, $\mathbb{G}_m$ acts on $I \otimes (I^\perp/I)$ by scalars and, if $\dim(I) = 2$, $\mathrm{GL}(I)$ acts trivially on $I^\perp/I$ (see the definition of unipotent extension 2.2.11). The isomorphisms depend in addition on a common choice of orientation on the maximal negative-definite subspaces of $L_{\mathbb{R}}$.

PROOF. According to Proposition 2.2.17 we have to classify admissible parabolic subgroups of $\mathrm{SO}(L)$ or $\mathrm{GSpin}(L)$ (this amounts to the same). We could also restrict ourselves to $\mathrm{GSpin}(L_{\mathbb{Q}})$ because of the projectivity of $\mathcal{PAR}$ (2.1.14).

So let Q be any proper *admissible* parabolic of $\mathrm{GSpin}(L)$ (see 2.2.13). It is well-known that Q corresponds to a saturated filtration

$$0 \subset I \subseteq I^{\perp} \subset L,$$

with I isotropic such that Q is the stabilizer. We will now determine the boundary component in the sense of 2.2.18 associated with Q. First note that we can identify $\mathrm{Lie}(\mathrm{SO}) = \mathrm{Lie}(\mathrm{Spin})$ with $\Lambda^2 L$ via the action:

$$(v \wedge w)x := \langle w, x\rangle v - \langle v, x\rangle w.$$

The filtration on L induces the following saturated filtration on $\Lambda^2 L$:

$$0 \subseteq \Lambda^2 I \subseteq I \wedge I^{\perp}$$

consisting of those elements in $\Lambda^2 L$ shifting the filtration by at least -2, respectively -1. We have $\mathrm{Lie}(W_Q) = I \wedge I^{\perp}$, where W_Q is the unipotent radical of Q.

2.4.6. Let L be a unimodular quadratic form over a discrete valuation ring R with 2 invertible. Decompose $L_R = H_{1,R} \perp \cdots \perp H_{k,R} \perp L'_R$, where L' is anisotropic over R. We get an R-split torus of GSpin as follows: Let z_i, z'_i be an isotropic basis of $H_{i,R}$ with $\langle z_i, z'_i\rangle = 1$. We get a 1-parameter subgroup T_i defined by

$$t_i(\alpha) := \alpha z_i z'_i + z'_i z_i = 1 + (\alpha - 1) z_i z'_i$$

in $\mathrm{GSpin}(L)$. π induces an isomorphism of T_i with its image in $\mathrm{SO}(L)$. The product of all T_i is direct and yields a split torus T of $\mathrm{GSpin}(L)$. The product of T with the center of GSpin is a maximal R-split torus. π induces an *isomorphism* of T with $\pi(T)$ which is a *maximal* R-split torus of $\mathrm{SO}(L)$. An element $t_i(\alpha)$ acts, via the standard representation of $\mathrm{GSpin}(L)$ by

$$z_i \mapsto \alpha^2 z_i, \qquad z'_i \mapsto z'_i, \qquad v \mapsto \alpha v \quad \text{for } v \in \langle z, z'\rangle^{\perp}.$$

The element $\pi\big(t_i(\alpha)\big)$ acts, via the standard representation of $\mathrm{SO}(L)$ by

$$z_i \mapsto \alpha z_i, \qquad z'_i \mapsto \alpha^{-1} z'_i, \qquad v \mapsto v \quad \text{for } v \in \langle z, z'\rangle^{\perp}.$$

2.4.7. Let $L_{\mathbb{R}}$ be a quadratic space over $\mathbb{R}$ with signature $(l-2, 2)$. Decompose $L_{\mathbb{R}} = H_{1,\mathbb{R}} \perp H_{2,\mathbb{R}} \perp L'_{\mathbb{R}}$, where $H_{i,\mathbb{R}}$ are hyperbolic planes. A maximal $\mathbb{R}$-split torus of $\mathrm{SO}(L_{\mathbb{R}})$ is $\pi(T_1 \cdot T_2) \cong \mathbb{G}_m^2$. The long roots (2.2.15) are in this case: $\alpha_1 = (1, -1)$ and $\alpha_2 = (1, 1)$ in $\mathbb{Z}^2 = \mathrm{Hom}(\pi(T_1 T_2), \mathbb{G}_{\mathrm{m}})$. The weight morphism $\mu := \mu_x = h_x \circ w$ for any $x \in \mathbb{D}_{\mathbf{S}(L)}$ is given by the natural inclusion $\mathbb{G}_{\mathrm{m}} \hookrightarrow \mathrm{GSpin}$. The homomorphism λ satisfying the criterion in 2.2.15, in particular mapping to $\mathrm{Spin}(L)$, and its product with μ are given in Table 2.1:

Here $(a; b, c)$ denotes the cocharacter of GSpin: $\alpha \mapsto \alpha^a t_1(\alpha^b) t_2(\alpha^c)$.

TABLE 2.1

subset of long roots	λ	$\lambda\cdot\mu$
$\varnothing$	$(0;0,0)$	$(1;0,0)$
$\{\alpha_1\}$	$(0;1,-1)$	$(1;1,-1)$
$\{\alpha_2\}$	$(-1;1,1)$	$(0;1,1)$
$\{\alpha_1,\alpha_2\}$	$(-1;2,0)$	$(0;2,0)$

2.4.8. Case: I is 1-dimensional (boundary point). Choose a splitting $L = I \oplus I^* \perp L_0$. This is possible because the discriminant is a unit at p. We may assume $H_{1,\mathbb{R}} = I_{\mathbb{R}} \oplus I^*_{\mathbb{R}}$. Only the last cocharacter of 2.4.7 is defined over $\mathbb{Q}$ *and* yields a proper parabolic subgroup.

The morphism $\lambda\cdot\mu$ which is a weight morphism for the boundary component has the following eigenspaces:

$$L^{-4} = I \quad L^{-2} = L_0 \quad L^0 = I^*.$$

We get a Levi decomposition

$$Q = \mathbb{W}\big(I\otimes(I^\perp/I)\big) \rtimes G,$$

where $\mathbb{W}\big(I\otimes(I^\perp/I)\big)$ is the unipotent radical and acts via the exponential (note that $I\wedge I^\perp \cong I\otimes(I^\perp/I)$ in this case).

We interpreted $\mathbb{D}_{\mathbf{S}(L)}$ as the set of polarized Hodge structures of weight 2 on $L_{\mathbb{R}}$ with $L_{-2,0}$ isotropic, and therefore $h(\mathbb{D}_{\mathbf{B}})$ can be interpreted as the orbit in the set of mixed Hodge structures with respect to the weight filtration

$$W_i(L) = \begin{cases} 0 & i \le -5 \\ I & i = -4,-3 \\ I^\perp & i = -2,-1 \\ L & i \ge 0 \end{cases}$$

which contains the Hodge filtrations in $\mathbb{D}_{\mathbf{S}(L)}$.

Since the Hodge filtration type stays the same, the induced Hodge structures on $\mathrm{gr}_W(L)$ have to be trivial by weight reasons. Also, we see that a mixed Hodge structure is given by

$$0 \subseteq F^0 = I^*_{\mathbb{C}} \subseteq F^{-1} = (I^*)^\perp \subseteq L_{\mathbb{C}}. \tag{2.3}$$

The associated morphism $h_0\colon \mathbb{S}_{\mathbb{C}} \to \mathrm{GSpin}(L_{\mathbb{C}})$ is given by $z \mapsto t_1(z\bar{z})$. Its weight morphism is precisely the morphism $\lambda\cdot\mu$ constructed above. The product $\mathbb{W}\big(I\otimes(I^\perp/I)\big)T_1$ is already a closed normal subgroup scheme of Q, hence equal to $P_{\mathbf{B}}$. It is the same as the subgroup scheme underlying $\mathbf{H}_0[I\otimes(I^\perp/I),0]$.

Consider

$$\pi_0(\mathbb{D}_{\mathbf{S}}) \times \mathrm{Hom}(\mathbb{S}_{\mathbb{C}}, P_{\mathbf{B},\mathbb{C}}),$$

where $T_1(\mathbb{R})$ acts by the nontrivial operation on the set $\pi_0(\mathbb{D}_{\mathbf{S}})$.

Choose vectors z, z' spanning L and L^* respectively. Via the isomorphism

$$L_0 \to I\otimes I^\perp/I, \quad k \mapsto z\otimes k,$$

a vector $k\in L_{0,\mathbb{C}}$ maps the mixed Hodge structure (2.3) above to the one determined by

$$F^0 = \langle z' + \langle z,z'\rangle k - (\langle k,z'\rangle + \tfrac{1}{2}\langle z,z'\rangle\langle k,k\rangle) z\rangle.$$

If $\Im(k) > 0$, this is the filtration associated with a Hodge structure in $\mathbb{D}_{\mathbf{S}(L)}$ hence by Definition 2.2.18 the $T_1(\mathbb{R})L_{0,\mathbb{C}}$-orbit $\mathbb{D}_{\mathbf{B}}$ in $\pi_0(\mathbb{D}_{\mathbf{S}}) \times \mathrm{Hom}(\mathbb{S}_{\mathbb{C}}, P_{\mathbf{B},\mathbb{C}})$ is equal to $\pi_0(\mathbb{D}_{\mathbf{S}}) \times L_{0,\mathbb{C}} \cdot h_0$.

This is (see Definition 2.2.11) isomorphic to

$$\mathbb{D}_{\mathbf{H}_0[I\otimes(I^\perp/I),0]},$$

where $\pi_0(\mathbb{D}_{\mathbf{S}})$ can be identified (after choice of a common orientation of the negative definite subspaces) with the set of isomorphisms $\mathbb{Z} \cong \mathbb{Z}(1)$ (choice of $i \in \mathbb{C}$). Combining this with the isomorphism $P_{\mathbf{B}} \cong \mathbb{W}(I \otimes (I^\perp/I))T_1$, we get an isomorphism

$$\mathbf{B} \cong \mathbf{H}_0[I \otimes (I^\perp/I), 0],$$

as required.

2.4.9. Case: *I* is 2-dimensional (boundary curve). We may now assume that the whole torus T_1T_2 is defined over $\mathbb{Q}$. We get an associated maximal parabolic for each of the choice of associated morphism $\mu \in \{(0;1,-1),(-1;1,1)\}$. The morphism $\lambda \cdot \mu$ which is a weight morphism for the boundary component has the following eigenspaces:

$$\begin{array}{llll} L^{-3} = \langle z, w' \rangle & L^{-2} = L_0 & L^{-1} = \langle z', w \rangle & \text{if } \lambda\cdot\mu = (1;1,-1) \\ L^{-3} = \langle z, w \rangle & L^{-2} = L_0 & L^{-1} = \langle z', w' \rangle & \text{if } \lambda\cdot\mu = (0;1,1) \end{array}$$

If $m = 4$ these are *not* conjugated in accordance with the fact that $\mathrm{Spin}(L)$ is not simple.

We get in each case for a certain isotropic plane $I \subset L$:

$$W_i(L) = \begin{cases} 0 & i \leq -4 \\ I & i = -3 \\ I^\perp & i = -2 \\ L & i \geq -1. \end{cases}$$

For any isotropic $\langle z \rangle \subset I_{\mathbb{C}}$ and corresponding $\langle z' \rangle = \langle z \rangle^{\perp, I_{\mathbb{C}}^*}$, we get a filtration

$$0 \subseteq F^0 = \langle z' \rangle \subseteq F^{-1} = \langle z' \rangle^\perp \subseteq L_{\mathbb{C}}. \tag{2.4}$$

If $F^0 = \langle z \rangle \subset I_{\mathbb{C}}$ is a Hodge structure on $I_{\mathbb{C}}$, that is if $\langle \bar{z} \rangle \neq \langle z \rangle$, then the filtration (2.4) defines a mixed Hodge structure on $L_{\mathbb{C}}$, together with the weight filtration above.

We have an exact sequence

$$0 \longrightarrow \Lambda^2(I) \longrightarrow I \wedge I^\perp \longrightarrow I \otimes (I^\perp/I) \longrightarrow 0$$

where $\Lambda^2(I)$ are the elements shifting the filtration by -2 and a corresponding central extension

$$0 \longrightarrow U_{\mathbf{B}} := \mathbb{W}(\Lambda^2(I)) \longrightarrow W_{\mathbf{B}} \longrightarrow V_{\mathbf{B}} := \mathbb{W}(I \otimes (I^\perp/I)) \longrightarrow 0.$$

If we choose, in a addition to our splitting, a basis z, w (as before) determining $I \cong (\mathbb{Z}_{(p)})^2$ and dual basis determining $I^* \cong (\mathbb{Z}_{(p)})^2$ we get an induced isomorphism $\det\colon \Lambda^2(I) \cong \mathbb{Z}_{(p)}$ and an isomorphism $\big(I \otimes (I^\perp/I)\big) \cong L_0^2$. An element of

$W_{\mathbf{B}}(\mathbb{R})U_{\mathbf{B}}(\mathbb{C})$ may now be written as (u, k_1, k_2) with $u \in \mathbb{C}$ and $k_i \in L_{0,\mathbb{R}}$. It sends the Hodge filtration (2.4) determined by $z = \left(\begin{smallmatrix} \beta \\ -\alpha \end{smallmatrix}\right)_I$ explicitly to the following one:

$$F^0 = \left\langle \begin{pmatrix} \alpha \\ \beta \end{pmatrix}_{I^*} + u \begin{pmatrix} \beta \\ -\alpha \end{pmatrix}_I + \alpha k_1 + \beta k_2 - \begin{pmatrix} \alpha Q(k_1) \\ \beta Q(k_2) + \alpha \langle k_1, k_2 \rangle \end{pmatrix}_I \right\rangle$$

(the so obtained filtrations again contain all Hodge filtrations in $h(\mathbb{D}_{\mathbf{S}(L)})$.)

We also get a morphism $\rho''\colon \mathrm{SL}(I) \to \mathrm{SO}(L)$, letting a matrix act on I by the standard representation, on I^* by its contragredient and on L_0 trivially. ρ'' lifts to a morphism $\rho'\colon \mathrm{SL}(I) \to \mathrm{Spin}(L) \subset \mathrm{GSpin}(L)$ because $\mathrm{SL}(I)$ is algebraically simply-connected. Note that $t_1(-1)t_2(-1)$ acts in the same way as -1 under the above lift. Hence ρ' extends to a morphism $\rho\colon \mathrm{GL}(I) \to \mathrm{GSpin}(L)$ mapping the diagonal torus to T_1T_2. It hence acts via the standard representation of GSpin as follows on L: by the standard representation times determinant on I; by the contragredient times determinant on I^*; and by the determinant on L_0. The mixed Hodge structure determined by (2.4) is hence given by the morphism $\mathbb{S}_{\mathbb{C}} \to \mathrm{GSpin}(L_{\mathbb{C}})$ obtained composing the morphism $h_0\colon \mathbb{S} \to \mathrm{GL}(I_{\mathbb{R}})$, associated with the Hodge structure $F^0 = \langle z \rangle \subseteq I_{\mathbb{C}}$, with ρ. Its weight morphism is precisely the morphism $\lambda \cdot \mu$ constructed above. The product $W_{\mathbf{B}}\,\mathrm{GL}(L)$ is already a closed normal subgroup scheme of Q hence equal to $P_{\mathbf{B}}$ (see 2.2.18).

Consider

$$\pi_0(\mathbb{D}_{\mathbf{S}}) \times \mathrm{Hom}(\mathbb{S}_{\mathbb{C}}, P_{\mathbf{B},\mathbb{C}}).$$

The stabilizer in $\mathrm{GL}(L_{\mathbb{R}})$ of h_0 is connected (in the real topology) and hence acts trivially on the set $\pi_0(\mathbb{D}_{\mathbf{S}})$. Therefore, by Definition 2.2.18, the $P_{\mathbf{B}}(\mathbb{R})U_{\mathbf{B}}(\mathbb{C})$-orbit $\mathbb{D}_{\mathbf{B}}$ in $\pi_0(\mathbb{D}_{\mathbf{S}}) \times \mathrm{Hom}(\mathbb{S}_{\mathbb{C}}, P_{\mathbf{B},\mathbb{C}})$ is isomorphic to its projection to $\mathrm{Hom}(\mathbb{S}_{\mathbb{C}}, P_{\mathbf{B},\mathbb{C}})$.

Therefore $\mathbb{D}_{\mathbf{B}}$ is also isomorphic to:

$$\mathbb{D}_{\mathbf{H}_1(I)[\Lambda^2 I, I \otimes (I^\perp/I)]}$$

(see Definition 2.2.11).

Combining this with the isomorphism of group schemes $P_{\mathbf{B}} \cong W_{\mathbf{B}}\,\mathrm{GL}(L)$, we get an isomorphism:

$$\mathbf{B} \cong \mathbf{H}_1(I)[\Lambda^2 I, I \otimes (I^\perp/I)]$$

as required. □

2.5. Main theorems on integral models

2.5.1. To prove our main theorems about the arithmetic volume of Shimura varieties of orthogonal type, we use the following statements about their canonical integral models and their toroidal compactifications hold true. The theorems hold true in general for arbitrary Abelian type Shimura varieties but exclusively for the case of good reduction. In this case the theorems have been proven under a certain hypothesis (2.5.6) in the thesis of the author [25]. The hypothesis was recently proven by Madapusi [49].

2.5.2. Definition/Theorem (Reflex ring). *Let $\mathbf{X}$ be a p-integral mixed Shimura datum. To any $x \in \mathbb{D}_{\mathbf{X}}$ there is an associated $h_x : \mathbb{S}_{\mathbb{C}} \to P_{\mathbf{X},\mathbb{C}}$ hence a morphism $u_x\colon \mathbb{G}_{m,\mathbb{C}} \to P_{\mathbf{X},\mathbb{C}}$ via composition with the inclusion of the first factor into $\mathbb{S}_{\mathbb{C}} \cong \mathbb{G}_{m,\mathbb{C}} \times \mathbb{G}_{m,\mathbb{C}}$. The conjugacy class M of these morphisms is defined over a number field $E_{\mathbf{X}}$, called the* **reflex field**. *It is necessarily unramified over p. We call the localization of its ring of integers $\mathcal{O} := \mathcal{O}_E \otimes_{\mathbb{Z}} \mathbb{Z}_{(p)}$ at p the* **reflex ring**.

It is a semilocal ring. The closure $\overline{M}$ *in* $\underline{\mathrm{Hom}}(\mathbb{G}_{\mathrm{m},\mathcal{O}}, P_{\mathbf{X},\mathcal{O}})$ *maps surjectively onto* $\mathrm{Spec}(\mathcal{O})$.

We consider the 2-category of Deligne–Mumford stacks over reflex rings as the 2-category of morphisms $(\pi\colon S \to \mathrm{Spec}(\mathcal{O}))$, *where* $\mathcal{O} \subset \mathbb{C}$ *is the localization at* $\mathbb{Z} \setminus (p)$ *of the ring of integers in a number field and* $\pi\colon S \to \mathrm{Spec}(\mathcal{O})$ *is a morphism of Deligne–Mumford stacks. Morphisms are commutative diagrams where the map* $\mathrm{Spec}(\mathcal{O}) \to \mathrm{Spec}(\mathcal{O}')$ *has to come from the inclusion.*

2.5.3. We begin by describing the functorial theory of (uncompactified) integral models of mixed Shimura varieties in the case of good reduction. The Main Theorem is essentially due to Kisin [27, 28], or Vasiu [62] in the pure case.

2.5.4. Definition (Test scheme). Let $\mathcal{O}$ be a discrete valuation ring with fraction field F. Let a **test scheme** S over $\mathcal{O}$ be as in [54, Def. 3.5], i.e., S has a cover by open affines $\mathrm{Spec}(A)$, such that there exist rings $\mathcal{O} \subseteq \mathcal{O}' \subseteq A_0 \subseteq A$ such that

- $\mathcal{O} \subseteq \mathcal{O}'$ is a faithfully flat and unramified extension of d.v.r. with $\mathcal{O}'/(\pi)$ separable over $\mathcal{O}/(\pi)$.
- A_0 is a smooth $\mathcal{O}'$-algebra.
- $A_0 \subseteq A_1 \subseteq \cdots \subseteq A$ is a countable union of etale extensions.

2.5.5. Main Theorem. *Let* $p \neq 2$ *be a prime. There is a* unique *(up to unique isomorphism) pseudofunctor* M *from the category of* p*-EMSD* ${}^K\mathbf{X}$ *of Abelian type to the* 2*-category of smooth Deligne–Mumford stacks over reflex rings*

$$ {}^K\mathbf{X} \mapsto \left(\mathrm{M}({}^K\mathbf{X}) \to \mathrm{Spec}(\mathcal{O}_{\mathbf{X}})\right) $$

with a pseudonatural isomorphism

$$ [P_{\mathbf{X}}(\mathbb{Q})\backslash \mathbb{D}_{\mathbf{X}} \times (P_{\mathbf{X}}(\mathbb{A}^{(\infty)})/K)] \cong \left(\mathrm{M}({}^K\mathbf{X}) \times_{\mathrm{Spec}(\mathcal{O}_{\mathbf{X}})} \mathrm{Spec}(\mathbb{C})\right)^{\mathrm{an}} $$

satisfying the following properties:

(1) *If* $(\gamma, \rho)\colon {}^{K'}\mathbf{Y} \to {}^K\mathbf{X}$ *is an embedding then*

$$ \mathrm{M}(\gamma, \rho) \times \pi_{\mathbf{Y}}\colon \mathrm{M}({}^{K'}\mathbf{Y}) \to \mathrm{M}({}^K\mathbf{X}) \times_{\mathrm{Spec}(\mathcal{O}_{\mathbf{X}})} \mathrm{Spec}(\mathcal{O}_{\mathbf{Y}}) $$

is a normalization followed by a closed embedding. If (γ, ρ) *is a Hecke operator, then* $\mathrm{M}(\gamma, \rho) \times \pi_{\mathbf{Y}}$ *is etale and finite.*

(2) *(Rational canonicity) For a pure Shimura datum* $\mathbf{Y}$, *where* $P_{\mathbf{Y}}$ *is a torus, the composition of the reciprocity isomorphism (normalized as in* [58, 11.3]*)*

$$ \mathrm{Gal}(\overline{E}|E)^{ab} \cong \pi_0\left(T_E(\mathbb{Q})\backslash T_E(\mathbb{A})\right), $$

(where $T_E = \mathrm{Res}^E_{\mathbb{Q}}(\mathbb{G}_{\mathrm{m}})$*) with* π_0 *of the reflex norm* [58, 11.4]

$$ \pi_0\left(T_E(\mathbb{Q})\backslash T_E(\mathbb{A})\right) \to \pi_0\left(P_{\mathbf{Y}}(\mathbb{Q})\backslash P_{\mathbf{Y}}(\mathbb{A})\right) $$

defines by means of the natural action of $\pi_0\left(P_{\mathbf{Y}}(\mathbb{Q})\backslash P_{\mathbf{Y}}(\mathbb{A})\right)$ *on* $\mathrm{M}({}^K\mathbf{Y})(\mathbb{C})$ *the rational model* $\mathrm{M}({}^K\mathbf{Y})_E$.

(3) *(Integral canonicity) The projective limit*

$$ \mathrm{M}^p(\mathbf{X}) := \varprojlim_K \mathrm{M}({}^K\mathbf{X}) $$

where the limit is taken over all admissible *compact open subgroups satisfies the following extension property:*

For any prime $\wp \mid p$ of $\mathcal{O}_{\mathbf{X}}$, any test scheme (2.5.4) *T over $S = \operatorname{Spec}(\mathcal{O}_{\mathbf{X},\wp})$, and any given morphism*

$$\alpha_E \colon T \times_S Q \to \mathrm{M}^p(\mathbf{X}) \times_S Q$$

where $Q = \operatorname{Spec}(E_{\mathbf{X}})$ is the generic fibre of S there exists a unique morphism

$$\alpha \colon T \to \mathrm{M}^p(\mathbf{X})$$

such that $\alpha_E = \alpha \times_S Q$.

(4) *$M({}^K\mathbf{X})$ is a quasi-projective scheme if K is neat.*

For $p = 2$, it may be necessary to change the notion of 'integral canonicity' as formulated above, cf. [54, 3.4].

The following Main Theorem 2.5.9 on canonical models of *toroidal compactifications* was only proven in [25] under the following hypothesis which has recently been proven by Madapusi [49].

2.5.6. Theorem ([25, Conjecture 3.3.2]). *Let $\mathbf{X}$ be a pure p-integral Shimura datum. Let an embedding ${}^K_\Delta\mathbf{X} \hookrightarrow {}^{K'}_{\Delta'}\mathbf{X}'$ in a datum of symplectic type $\mathbf{X}' = \mathbf{H}_{g_0}[(I \otimes I)^s, L_0 \otimes I]/\mathbb{W}(U')$ be given, such that* Δ and Δ' *are smooth projective and K, K' are neat.*

Consider the toroidal compactification $\mathrm{M}({}^{K'}_{\Delta'}\mathbf{X}')$ over $\operatorname{Spec}(\mathbb{Z}_{(p)})$ *constructed by Faltings–Chai [19], and let D be its boundary divisor. For every prime $\wp \mid p$ of $\mathcal{O}_{\mathbf{X}}$ the following holds:*

If $S =$ $\operatorname{Spec}(\mathcal{O}_{\mathbf{X},\wp})$ and $\mathcal{M}'$ is the Zariski closure of (the rational canonical model of) $P_{\mathbf{X}}(\mathbb{Q})\backslash\mathbb{D}_{\mathbf{X}} \times (P_{\mathbf{X}}(\mathbb{A}^{(\infty)})/K)$ in $\mathrm{M}({}^{K'}_{\Delta'}\mathbf{X}')_S$ then no connected *component of $N(\mathcal{M}') \cap D_S$ lies entirely in the fibre above $\wp$.*

PROOF. Follows from the proof of [49, Theorem 4.2.3.1]. □

2.5.7. The toroidal compactifications will have a stratification which consists of mixed Shimura varieties itself. To describe this we have to set up some notation (cf. [58, 7.3 ff.]). Let $(\iota, \rho) \colon {}^{K'}_{\Delta'}\mathbf{B} \Longrightarrow {}^K_\Delta\mathbf{X}$ be a boundary component. For each cone $\sigma \in \Delta'$ with $\sigma \subset C(\mathbb{D}^0_{\mathbf{X}}, P_{\mathbf{B}}) \times \rho'$ for some ρ' in the equivalence class of ρ we define a p-ECMSD ${}^{K_{[\sigma]}}_{\Delta_{[\sigma]}}\mathbf{B}_{[\sigma]}$ as follows:

$$\begin{aligned}
P_{\mathbf{B}_{[\sigma]}} &:= P_{\mathbf{B}}/\langle\sigma\rangle \\
\mathbb{D}_{\mathbf{B}_{[\sigma]}} &:= \text{the } P_{\mathbf{B}_{[\sigma]}}(\mathbb{R})U_{\mathbf{B}_{[\sigma]}}(\mathbb{C})\text{-orbit generated by } \big(\mathbb{D}^0_{\mathbf{X}}/\langle\sigma\rangle(\mathbb{C})\big) \times \{\sigma P_{\mathbf{X}}(\mathbb{A}^{(\infty)})\} \\
&\qquad \text{in } \big(\mathbb{D}^0_{\mathbf{X}}/\langle\sigma\rangle(\mathbb{C})\big) \times \{[\sigma]/P_{\mathbf{X}}(\mathbb{A}^{(\infty)})\} \text{ for } \sigma \in [\sigma] \cap \Delta(\mathbb{D}^0_{\mathbf{X}}, P_{\mathbf{B}}, 1) \\
K_{[\sigma]} &:= \text{the image of } K' \text{ under the projection } P_{\mathbf{B}} \to P_{\mathbf{B}_{[\sigma]}} \\
\Delta_{[\sigma]} &:= \Delta_{[\sigma]}(\mathbb{D}^0_{\mathbf{B}_{[\sigma]}}, P_{\mathbf{B}_{[\sigma]}}, \overline{\rho}) \\
&:= \{\tau \bmod \langle\sigma\rangle \mid \tau \in \Delta(\mathbb{D}^0_{\mathbf{X}}, P_{\mathbf{B}}, \rho) \text{ such that } \sigma \text{ is a face of } \tau\}
\end{aligned}$$

$\Delta_{[\sigma]}$ is $K_{[\sigma]}$-admissible. Furthermore, we define

$$\Gamma := \Big(\operatorname{Stab}_{Q(\mathbb{Q})}(\mathbb{D}_{\mathbf{B}}) \cap \big(P_{\mathbf{B}}(\mathbb{A}^{(\infty)}) \cdot {}^{\rho}K\big)\Big)\Big/ P_{\mathbf{B}}(\mathbb{Q})$$

where Q is the parabolic which gives rise to $\mathbf{B}$.

$\Sigma(\iota, \rho)$ is defined to be the set of Γ-orbits $\Gamma[\sigma] \in P_{\mathbf{B}}(\mathbb{Q})\backslash(\Delta')^0/P_{\mathbf{B}}(\mathbb{A}^{(\infty)})$ such that $\sigma \subset C(\mathbb{D}^0_{\mathbf{X}}, P_{\mathbf{B}}) \times \rho'$ for some $\mathbb{D}^0_{\mathbf{X}}$.

2.5.8. Recall from 2.2.10: Any morphism

$$\mathbf{X} \to \mathbf{X}/U$$

is a torsor under the *group object*

$$\mathbf{X}/W_{\mathbf{X}}[U,0] \to \mathbf{X}/W_{\mathbf{X}}$$

(cf. also [58, 6.6–6.8]).

In the sequel we assume all K's to be neat. The corresponding map of Shimura varieties:

$$\mathrm{M}({}^{K}(\mathbf{X}/W_{\mathbf{X}})[U,0]) \to \mathrm{M}({}^{K/U}\mathbf{X}/W_{\mathbf{X}})$$

is a split torus with character group (in general non canonically) isomorphic to $U_{\mathbb{Q}} \cap K$. If Δ is a K-admissible polyhedral cone decomposition for $\mathbf{X}$, concentrated in the unipotent fibre (2.2.27), we can define an associated torus embedding. Recall: The rational polyhedral cone decomposition is determined by its restriction to any $(U_{\mathbb{R}})(-1) \times \rho = C(\mathbb{D}^0, P_{\mathbf{X}}) \times \rho$ by the arithmeticity condition 2.2.23(3). Therefore this determines a torus embedding (see [19, IV, §2] for the integral case):

$$\mathrm{M}({}^{K}\mathbf{X}/W_{\mathbf{X}}[U,0]) \hookrightarrow \mathrm{M}({}^{K}_{\Delta}\mathbf{X}/W_{\mathbf{X}}[U,0])$$

and a corresponding toroidal embedding of the torsor:

$$\mathrm{M}({}^{K'}\mathbf{X}) \hookrightarrow \mathrm{M}({}^{K'}_{\Delta}\mathbf{X})$$

where K' and K are related by $K/U = K'/W$ and $K \cap U(\mathbb{A}^{(\infty)}) = K' \cap U(\mathbb{A}^{(\infty)})$.

2.5.9. Main Theorem. *Let $p \neq 2$ be a prime. There is a* unique *(up to unique isomorphism) extension of the pseudofunctor* M *to the category of p-ECMSD of Abelian type—with sufficiently fine smooth (w.r.t. the corresponding K) and projective Δ—to the category of Deligne–Mumford stacks over reflex rings satisfying the following properties:*

(1) *The morphism* $\mathrm{M}(\mathrm{id},1)\colon \mathrm{M}({}^{K}\mathbf{X}) \to \mathrm{M}({}^{K}_{\Delta}\mathbf{X})$ *is an open embedding. If* $(\gamma,\rho)\colon {}^{K'}_{\Delta'}\mathbf{Y} \to {}^{K}_{\Delta}\mathbf{X}$ *is an embedding then*

$$\mathrm{M}(\gamma,\rho) \times \pi_{\mathbf{Y}}\colon \mathrm{M}({}^{K'}_{\Delta'}\mathbf{Y}) \to \mathrm{M}({}^{K}_{\Delta}\mathbf{X}) \times_{\mathrm{Spec}(\mathcal{O}_{\mathbf{X}})} \mathrm{Spec}(\mathcal{O}_{\mathbf{Y}})$$

is a normalization map followed by a closed embedding.

(2) *If* $\Delta = \Delta^0$ *(i.e.,* Δ *is concentrated in the unipotent fibre, see* 2.2.27*) then* $\mathrm{M}({}^{K}_{\Delta}\mathbf{X})$ *is the toroidal embedding described in* 2.5.8.

(3) $\mathrm{M}({}^{K}_{\Delta}\mathbf{X})$ *possesses a stratification into mixed Shimura varieties (stacks)*

$$\coprod_{\substack{[(\iota,\rho)\colon {}^{K'}_{\Delta'}\mathbf{B} \Longrightarrow {}^{K}_{\Delta}\mathbf{X}] \\ [\sigma]\in\Sigma(\iota,\rho)}} [\mathrm{Stab}_{\Gamma}([\sigma])\backslash \mathrm{M}({}^{K_{[\sigma]}}\mathbf{B}_{[\sigma]})].$$

Here $[(\iota,\rho)\colon {}^{K'}_{\Delta'}\mathbf{B} \Longrightarrow {}^{K}_{\Delta}\mathbf{X}]$ *means equivalence classes of boundary components of* ${}^{K}_{\Delta}\mathbf{X}$ (2.2.26).

The stratification is compatible with the functoriality, i.e., the morphisms $\mathrm{M}(\cdot)$ *induce morphisms of the strata of the form* $\mathrm{M}(\cdot)$ *again.*

(4) *For each boundary map* $(\iota,\rho)\colon {}^{K'}_{\Delta'}\mathbf{B} \Longrightarrow {}^{K}_{\Delta}\mathbf{X}$*, and* $\sigma \in \Delta'$ *such that* $\sigma \subset C(\mathbb{D}^0_{\mathbf{X}}, P_{\mathbf{B}}) \times \rho'$ *for some* $\mathbb{D}^0_{\mathbf{X}}$ *and* ρ'*, there is a boundary isomorphism*

$$\mathrm{M}(\iota,\rho)\colon \left[\mathrm{Stab}_{\Gamma}([\sigma])\backslash \widehat{\mathrm{M}({}^{K'}_{\Delta'}\mathbf{B})}\right] \to \widehat{\mathrm{M}({}^{K}_{\Delta}\mathbf{X})}$$

of the formal completions along the boundary stratum

$$\left[\mathrm{Stab}_\Gamma([\sigma])\backslash \mathrm{M}({}^{K_{[\sigma]}}_{\Delta_{[\sigma]}}\mathbf{B}_{[\sigma]})\right].$$

The complexification of $\mathrm{M}(\iota,\rho)$ *converges in a neighborhood of the boundary stratum and is, in the interior, via the identification with the complex analytic mixed Shimura varieties given by the quotient of the map*

$$\mathbb{D}_{\mathbf{B}\Longrightarrow\mathbf{X}} \times (P_{\mathbf{B}}(\mathbb{A}^{(\infty)})/K') \to \mathbb{D}_{\mathbf{B}\Longrightarrow\mathbf{X}} \times (P_{\mathbf{X}}(\mathbb{A}^{(\infty)})/K)$$

induced by the closed embedding $P_{\mathbf{B}} \hookrightarrow P_{\mathbf{X}}$ *where we considered* $\mathbb{D}_{\mathbf{B}\Longrightarrow\mathbf{X}} \subseteq \mathbb{D}_{\mathbf{X}}$ *as a subset of* $\mathbb{D}_{\mathbf{B}}$ *via the analytic boundary map* (2.2.17).

We have in addition:

(5) $\mathrm{M}({}^K_\Delta\mathbf{X})$ *is smooth.*

(6) $\mathrm{M}({}^K_\Delta\mathbf{X})$ *is proper if* Δ *is complete.*

(7) $\mathrm{M}({}^K_\Delta\mathbf{X})$ *is a projective scheme if* Δ *is complete and* K *is neat.*

(8) *The complement* D *of the open stratum* $\mathrm{M}({}^K\mathbf{X})$ *in* $\mathrm{M}({}^K_\Delta\mathbf{X})$*—the union of all lower dimensional strata in* (3)*—is a smooth divisor of normal crossings on* $\mathrm{M}({}^K_\Delta\mathbf{X})$.

2.5.10. Remark. The stratification in (3) is indexed by pairs consisting of an equivalence class of boundary components $[(\iota,\rho) : {}^{K'}_{\Delta'}\mathbf{B} \Longrightarrow {}^K_\Delta\mathbf{X}]$ and a class $[\sigma] \in \Sigma(\iota,\rho)$. The set of these pairs is in bijection with the set of double cosets $P_{\mathbf{X}}(\mathbb{Q})\backslash\Delta/K$. The bijection is as follows: Each $\sigma \in \Delta$ is supported on a $C(\mathbb{D}^0_{\mathbf{X}}, P_{\mathbf{B}}) \times \rho$. $\mathbf{B}$ and ρ determine a boundary component $(\iota,\rho): {}^{K'}_{\Delta'}\mathbf{B} \Longrightarrow {}^K_\Delta\mathbf{X}$ and the class of the image of σ under restriction to Δ' lies in $\Sigma(\iota,\rho)$. The stratum $\mathrm{M}({}^{K_{[\sigma]}}\mathbf{B}_{[\sigma]})$ is contained in the closure of the stratum $\mathrm{M}({}^{K_{[\tau]}}\mathbf{B}_{[\tau]})$, if and only if (up to a change of representatives in $P_{\mathbf{X}}(\mathbb{Q})\backslash\Delta/K$) τ is a face of σ.

2.5.11. There exist integral canonical models of the 'compact' duals as well, which are defined as conjugacy classes of quasi-parabolic subgroup schemes. Let p be a prime and $\mathbf{X}$ be a p-integral mixed Shimura datum. Let $\mathcal{O}_{\mathbf{X}}$ be its reflex ring at p. The closure $\overline{M}$ of the conjugacy class of the morphisms u_x, $x \in \mathbb{D}_{\mathbf{X}}$ is defined over $\mathcal{O}$ by 2.5.2. It corresponds to a section $t' \in \mathcal{FTYPE}(\mathrm{Spec}(\mathcal{O}_{\mathbf{X}}))$.

The image of $\overline{M}$ via qpar is a fibre of the morphism 'type' above a section $t\colon \mathrm{Spec}(\mathcal{O}_{\mathbf{X}}) \to \mathcal{TYPE}$. We define $\mathrm{M}^\vee(\mathbf{X})$ to be this fibre. It may, as a scheme over $\mathrm{Spec}(\mathbb{Z}_{(p)})$, be seen as an open and closed subscheme of $\mathcal{QPAR}$ itself. We understand here the action of $P_{\mathbf{X}}$ on $\mathcal{QPAR}$ by conjugation on the right (contrary to 2.1.14). The action fixes the morphism 'type', hence we have an induced operation on $\mathrm{M}^\vee(\mathbf{X})$.

2.5.12. Main Theorem. *The construction* 2.5.11 *yields a functor*

$$\mathbf{X} \mapsto \left(\mathrm{M}^\vee(\mathbf{X}) \to \mathrm{Spec}(\mathcal{O}_{\mathbf{X}})\right)$$

from p-MSD to schemes over reflex rings (defined analogously). $\mathrm{M}^\vee(\mathbf{X})$ *is a smooth connected quasi-projective right* $P_{\mathbf{X},\mathcal{O}_{\mathbf{X}}}$*-homogeneous scheme over* $\mathrm{Spec}(\mathcal{O}_{\mathbf{X}})$ *called* **the dual** *of* $\mathbf{X}$ *(depending only on* $P_{\mathbf{X}}$ *and* $h(\mathbb{D}_{\mathbf{X}})$*) with the following properties:*

(1) *If* $\mathbf{X}$ *is pure,* $\mathrm{M}^\vee(\mathbf{X})$ *is projective.*

(2) *Let* L *be a free* $\mathbb{Z}_{(p)}$*-module of finite dimension, acted on faithfully by* $P_{\mathbf{X}}$. $\mathrm{M}^\vee(\mathbf{X})$ *(considered as scheme over* $\mathrm{Spec}(\mathcal{O}_{\mathbf{X}})$*) represents the functor*

$$S' \mapsto \{\text{filtrations } \{F^\bullet\} \text{ on } L_{S'} \text{ compatible with } (P_{\mathbf{X},S'}, L_{S'}) \text{ of type } t'\},$$

in a way compatible with the $P_{\mathbf{X},\mathcal{O}_{\mathbf{X}}}$*-action (defined on* L *as* $v \cdot g := g^{-1}v$*).*

(3) *For each morphism of p-MSD* $\gamma\colon \mathbf{Y} \to \mathbf{X}$,

$$\mathrm{M}^\vee(\gamma) \times \pi_{\mathbf{Y}}\colon \mathrm{M}^\vee(\mathbf{Y}) \to \mathrm{M}^\vee(\mathbf{X}) \times_{\mathrm{Spec}(\mathcal{O}_{\mathbf{X}})} \mathrm{Spec}(\mathcal{O}_{\mathbf{Y}})$$

is homogenous and is a closed embedding if γ *is an embedding.*

(4) *For each p-integral boundary morphism* $\iota\colon \mathbf{B} \Longrightarrow \mathbf{X}$, *there is a* $P_{\mathbf{B}}$*-equivariant open embedding*

$$\mathrm{M}^\vee(\iota)\colon \mathrm{M}^\vee(\mathbf{B}) \hookrightarrow \mathrm{M}^\vee(\mathbf{X}).$$

(5) *There is an* $P_{\mathbf{X}}(\mathbb{R})W_{\mathbf{X}}(\mathbb{C})$*-equivariant open embedding (Borel embedding)*

$$h_{\mathbf{X}}(\mathbb{D}_{\mathbf{X}}) \hookrightarrow \left(\mathrm{M}^\vee(\mathbf{X}) \times_{\mathrm{Spec}(\mathcal{O}_{\mathbf{X}})} \mathrm{Spec}(\mathbb{C})\right)^{\mathrm{an}}.$$

It is an isomorphism if $G_{\mathbf{X}}$ *(the reductive part of* $P_{\mathbf{X}}$*) is a torus. These embeddings are compatible with the embeddings in (4), respectively (2.2.20(1)).*[3] *Furthermore they are compatible with morphisms of Shimura data.*

2.5.13. Remark. For a boundary component $\mathbf{B} \Longrightarrow \mathbf{X}$, there is also an embedding

$$\mathrm{M}^\vee(\mathbf{B}/W_{\mathbf{B}}) \hookrightarrow \mathrm{M}^\vee(\mathbf{X})$$

into (in general not onto) the complement of the image of $\mathrm{M}^\vee(\mathbf{B})$. The image of the composition with the Borel embedding $h(\mathbb{D}_{\mathbf{B}}/W_{\mathbf{B}}) \hookrightarrow \mathrm{M}^\vee(\mathbf{B}/W_{\mathbf{B}})(\mathbb{C})$ is the boundary component of $h(\mathbb{D}_{\mathbf{X}})$ associated with Q (the parabolic associated with $\mathbf{B}$) in the sense of [1], if $\mathbf{X}$ is pure.

The following Main Theorem 2.5.14 comprises the arithmetic theory of automorphic vector bundles on toroidal compactifications (see [25, Theorem 3.5.2] for a more classical formulation). We will later derive a classical q-expansion principle (2.8.7). Main Theorem 2.5.14 includes many kinds of classical results on CM-values of integral (at p of good reduction) modular forms on mixed Shimura varieties.

First, by taking quotients, we have a pseudofunctor $\mathbf{X} \to \left[\mathrm{M}^\vee(\mathbf{X})/P_{\mathbf{X},\mathcal{O}_{\mathbf{X}}}\right]$ from p-MSD to the 2-category of Artin stacks over reflex rings (defined analogously). It can be extended to p-ECMSD by forgetting K and Δ.

2.5.14. Main Theorem. *Let* $p \neq 2$ *be a prime. There is a* unique *(up to unique isomorphism) pseudonatural transformation*

$$\Xi({}^K_\Delta\mathbf{X})\colon \mathrm{M}({}^K_\Delta\mathbf{X}) \to \left[\mathrm{M}^\vee(\mathbf{X})/P_{\mathbf{X},\mathcal{O}_{\mathbf{X}}}\right],$$

equipped with a 2-isomorphism (or modification) of its restriction to p-EMSD

$$\left(\Xi({}^K\mathbf{X}) \times_{\mathrm{Spec}(\mathcal{O}_{\mathbf{X}})} \mathrm{Spec}(\mathbb{C})\right)^{\mathrm{an}} \to \Xi_{\mathbb{C}}({}^K\mathbf{X})$$

[3] Note that they go in different directions, however!

where $\Xi_{\mathbb{C}}({}^K\mathbf{X})$ *is the complex analytic morphism described by the diagram*

$$\begin{array}{c} [P_{\mathbf{X}}(\mathbb{Q})\backslash\mathbb{D}_{\mathbf{X}} \times P_{\mathbf{X}}(\mathbb{A}^{(\infty)})/K] \\ \uparrow \\ [P_{\mathbf{X}}(\mathbb{Q})\backslash\mathbb{D}_{\mathbf{X}} \times P_{\mathbf{X}}(\mathbb{C}) \times P_{\mathbf{X}}(\mathbb{A}^{(\infty)})/K] \\ \downarrow \\ \left(\mathrm{M}^{\vee}(\mathbf{X}) \times_{\mathrm{Spec}(\mathcal{O}_X)} \mathrm{Spec}(\mathbb{C})\right)^{\mathrm{an}} \end{array}$$

satisfying:

(1) *Let* $[\iota,\rho]\colon {}^{K'}_{\Delta'}\mathbf{B} \Longrightarrow {}^{K}_{\Delta}\mathbf{X}$ *be a boundary map and let* $\sigma \in \Delta'$ *such that* $\sigma \in C(\mathbb{D}^0_{\mathbf{X}}, P_{\mathbf{B}})$, *as in* 2.5.7.

There is a 2-isomorphism $[\iota,\rho]_{\Xi}$ *fitting into the diagram*

$$\begin{array}{ccc} \left[\mathrm{Stab}_{\Gamma}([\sigma])\backslash \widehat{\mathrm{M}({}^{K'}_{\Delta'}\mathbf{B})}\right] & \xrightarrow{\Xi({}^{K'}_{\Delta'}\mathbf{B})} & [\mathrm{M}^{\vee}(\mathbf{B})/P_{\mathbf{B}}] \\ \Big\downarrow{\scriptstyle \mathrm{M}(\iota,\rho)} & \nearrow^{[\iota,\rho]_{\Xi}} & \Big\downarrow{\scriptstyle \mathrm{M}^{\vee}(\iota)} \\ \widehat{\mathrm{M}({}^{K}_{\Delta}\mathbf{X})} & \xrightarrow[\Xi({}^{K}_{\Delta}\mathbf{X})]{} & [\mathrm{M}^{\vee}(\mathbf{X})/P_{\mathbf{X}}] \end{array}$$

where the formal completion is taken along

$$\left[\mathrm{Stab}_{\Gamma}([\sigma])\Big\backslash \mathrm{M}\Big({}^{K'_{[\sigma]}}_{\Delta'_{[\sigma]}}\mathbf{B}_{[\sigma]}\Big)\right]$$

such that $[\iota,\rho]_{\Xi,\mathbb{C}}$ *converges in the interior and is the obvious 2-morphism over* $\mathbb{C}$.

(2) (*Canonicity*) *For* $\mathbf{X} = \mathbf{H}_{g_0}[(I \otimes I)^s, L_0 \otimes I]$ (*and* $\Delta = 0$), $\Xi({}^K\mathbf{X})$ *is described by the diagram*:

$$\left[{}^K\mathbf{X}\text{-}\boldsymbol{L}\text{-}\mathbf{mot}\right] \leftarrow \left[{}^K\mathbf{X}\text{-}\boldsymbol{L}\text{-}\mathbf{mot}\text{-}\mathbf{triv}\right] \rightarrow \mathrm{M}^{\vee}(\mathbf{X})$$

(*see* 2.7.3) *where the left arrow is given by the forgetful functor* (*which is a right* $P_{\mathbf{X}}$*-torsor*) *and the right arrow given by transport of Hodge filtration. The identification with* $\Xi_{\mathbb{C}}({}^K\mathbf{X})$ *is the one given by the period construction* (2.7.5).

(3) *For* Δ *concentrated in the unipotent fibre* (2.2.27), Ξ *is given as follows*: *Recall from* 2.5.8 *that* $\mathrm{M}({}^{K'}\mathbf{X}) \to \mathrm{M}({}^{K'}\mathbf{X}/U)$ *is a torsor under the torus* $T = \mathrm{M}({}^K\mathbf{X}/W_{\mathbf{X}}[U,0])$ *over* $\mathrm{M}({}^K\mathbf{X}/W_{\mathbf{X}})$ *where* K *and* K' *are as in* 2.5.8. $\Xi({}^{K'}\mathbf{X})$ *may etale locally* (*say on* $S \to \mathrm{M}({}^{K'/U}\mathbf{X}/U_{\mathbf{X}})$) *be trivialized* T*-invariantly.* $\Xi({}^{K'}_{\Delta}\mathbf{X})$ *is then the unique extension of* $\Xi({}^{K'}\mathbf{X})$ *to the torus embedding* $\mathrm{M}({}^{K'}_{\Delta}\mathbf{X})$ *by means of extending any such* T*-invariant trivialization.*

2.5.15. Definition (Standard principal bundle). Let ${}^K_{\Delta}\mathbf{X}$ be p-ECMSD. By definition of quotient stack the morphism

$$\Xi({}^K_{\Delta}\mathbf{X})\colon \mathrm{M}({}^K_{\Delta}\mathbf{X}) \to [\mathrm{M}^{\vee}(\mathbf{X})/P_{\mathbf{X},\mathcal{O}_{\mathbf{X}}}]$$

is given by a right $P_{\mathbf{X},\mathcal{O}_{\mathbf{X}}}$-torsor, denoted $\mathrm{P}({}^K_{\Delta}\mathbf{X})$, over $\mathrm{M}({}^K_{\Delta}\mathbf{X})$ together with an equivariant morphism, denoted Π, to $\mathrm{M}^{\vee}(\mathbf{X})$. The torsor $\mathrm{P}({}^K_{\Delta}\mathbf{X})$ is called the **standard principal bundle**.

2.6. The arithmetic theory of automorphic forms

2.6.1. Let ${}^K_\Delta\mathbf{X}$ be pure p-ECMSD (2.2.24) with smooth (w.r.t. K), projective, and complete Δ, let $E \subset \mathbb{C}$ be the reflex field and let $\mathcal{O}_{\mathbf{X}} = \mathcal{O}_{E,(p)}$ the reflex ring. We have an isomorphism

$$\mathrm{M}({}^K_\Delta\mathbf{X}) \times_{\mathrm{Spec}(\mathbb{Z}_{(p)})} \mathbb{C} \cong \coprod_{\sigma \in \mathrm{Hom}(E,\mathbb{C})} \mathrm{M}({}^K_\Delta\mathbf{X}) \times_{\mathcal{O}_{\mathbf{X}},\sigma} \mathbb{C}.$$

Conjecture A of [53, p. 58] which is proven in [53, Corollary 18.5] states in particular that for each $\sigma \in \mathrm{Hom}(E, \mathbb{C})$ there is a (rational) Shimura datum ${}^\sigma\mathbf{X}$, and we have an isomorphism

$$(\mathrm{M}({}^K\mathbf{X}) \times_{\mathcal{O}_{\mathbf{X}},\sigma} \mathbb{C})^{\mathrm{an}} \to [P_{{}^\sigma\mathbf{X}} \backslash \mathbb{D}_{{}^\sigma\mathbf{X}} \times P_{{}^\sigma\mathbf{X}}(\mathbb{A}^{(\infty)})/K] \tag{2.5}$$

which renders the system $\mathrm{M}({}^K\mathbf{X}) \times_{\mathcal{O}_{\mathbf{X}},\sigma} {}^\sigma\mathcal{O}_{\mathbf{X}}$ with a twisted adelic action into a rational canonical model for the Shimura datum ${}^\sigma\mathbf{X}$.

The compact dual $\mathrm{M}^\vee(\mathbf{X})$, considered as scheme over $\mathrm{Spec}(\mathbb{Z}_{(p)})$, can be identified also with a closed subscheme of $\mathcal{QPAR}$ (2.1.14) and carries an action of $P_{\mathbf{X}}$.

Let $\mathcal{E}$ a locally free sheaf on the stack (over $\mathrm{Spec}(\mathbb{Z}_{(p)})$!)

$$[\mathrm{M}^\vee(\mathbf{X})/P_{\mathbf{X}}]$$

(i.e., a $P_{\mathbf{X}}$-equivariant sheaf on $\mathrm{M}^\vee(\mathbf{X})$) and let $h_{\mathcal{E}} = \{h_{\mathcal{E},\sigma}\}$ for $\sigma \in \mathrm{Hom}(E, \mathbb{C})$ be a collection of $P_{{}^\sigma\mathbf{X}}(\mathbb{R})U_{{}^\sigma\mathbf{X}}(\mathbb{C})$-invariant Hermitian metrics on $E_{\mathbb{C}}|_{h_{{}^\sigma\mathbf{X}}(\mathbb{D}_{{}^\sigma\mathbf{X}})}$ where $\bigcup_\sigma h_{{}^\sigma\mathbf{X}}(\mathbb{D}_{{}^\sigma\mathbf{X}})$ is embedded into $(\mathrm{M}^\vee(\mathbf{X}) \times_{\mathbb{Z}_{(p)}} \mathbb{C})(\mathbb{C})$ via the Borel embeddings (cf. 2.5.12) taking into account the isomorphism (2.5).

Recall from 2.5.14 the 1-morphism

$$\Xi\colon \mathrm{M}({}^K_\Delta\mathbf{X}) \to [\mathrm{M}^\vee(\mathbf{X})/P_{\mathbf{X}}].$$

2.6.2. Let D be the boundary divisor (2.5.9). It is a divisor of normal crossings. We will define a Hermitian metric $\Xi^* h_{\mathcal{E}}$ on $(\Xi^*\mathcal{E})_{\mathbb{C}}$, singular along $D_{\mathbb{C}}$, as follows[4]: We do this for each ${}^\sigma\mathbf{X}$ separately so let $\mathbf{X}$ be one of them. By definition $\Xi^*\mathcal{E}_{\mathbb{C}}(U)$ is given, for any open $U \subset \mathrm{M}({}^K_\Delta\mathbf{X})(\mathbb{C})$, by the $P(\mathbb{Q})$-invariant sections of the pullback of $\mathcal{E}_{\mathbb{C}}$ to the standard principal bundle $\mathrm{P}({}^K_\Delta\mathbf{X})$ (the right $P_{\mathbf{X}}$-torsor describing the morphism Ξ) via the compatibility isomorphism of Ξ over $\mathbb{C}$ with $\Xi_{\mathbb{C}}$ (2.5.15). Now consider the diagram of analytic manifolds (we assumed K to be neat):

$$\begin{array}{ccccc}
P_{\mathbf{X}}(\mathbb{Q})\backslash\mathbb{D}_{\mathbf{X}}\times(P_{\mathbf{X}}(\mathbb{A}^{(\infty)})/K) & \longleftarrow & \mathbb{D}_{\mathbf{X}}\times(P_{\mathbf{X}}(\mathbb{A}^{(\infty)})/K) & \xrightarrow{\ \pi\ } & h(\mathbb{D}_{\mathbf{X}}) \\
\Big\| & & \big\downarrow & & \big\downarrow \\
P_{\mathbf{X}}(\mathbb{Q})\backslash\mathbb{D}_{\mathbf{X}}\times(P_{\mathbf{X}}(\mathbb{A}^{(\infty)})/K) & \longleftarrow & P_{\mathbf{X}}(\mathbb{Q})\backslash\mathbb{D}_{\mathbf{X}}\times P_{\mathbf{X}}(\mathbb{C})\times(P_{\mathbf{X}}(\mathbb{A}^{(\infty)})/K) & \longrightarrow & \mathrm{M}^\vee(\mathbf{X})(\mathbb{C}).
\end{array}$$

Let $U \cong \mathrm{M}({}^K\mathbf{X})(\mathbb{C})$ be the complement of $D_{\mathbb{C}}$. The diagram shows that $\Xi^*\mathcal{E}_{\mathbb{C}}|_U$ can be canonically identified with the $P_{\mathbf{X}}(\mathbb{Q})$-invariant sections of the pullback $\pi^*(\mathcal{E}_{\mathbb{C}}|_{h_{\mathbf{X}}(\mathbb{D}_{\mathbf{X}})})$ to the standard local system $\mathbb{D}_{\mathbf{X}} \times P_{\mathbf{X}}(\mathbb{A}^{(\infty)})/K$. This pullback carries the pullback of the Hermitian metric $h_{\mathcal{E}}$. For the induced $P_{\mathbf{X}}(\mathbb{Q})$-action on $\pi^*(\mathcal{E}_{\mathbb{C}}|_{h_{\mathbf{X}}(\mathbb{D}_{\mathbf{X}})})$ compatible with the usual action on the standard local system, $h_{\mathcal{E}}$ is invariant as well and descends to a smooth Hermitian metric on U.

[4]This is a slight abuse of notation, since the construction seems to depend on the particular presentation of the target stack as a quotient of $\mathrm{M}^\vee$

2.6.3. Theorem. *Let* ${}^K_\Delta\mathbf{X}$ *be a* pure *p-ECSD, let* $(\mathcal{E}, h_\mathcal{E})$ *be as in* 2.6.1 *with fully-decomposed* $\mathcal{E}$, *i.e., the unipotent radical of a stabilizer group under the action of* $P_\mathbf{X}(\mathbb{C})$ *on* $\mathrm{M}^\vee(\mathbf{X})(\mathbb{C})$ *acts trivially on* $\mathcal{E}$.

Then the vector bundles $\Xi^*\mathcal{E}_\mathbb{C}$ *are the unique extensions to* $\mathrm{M}({}^K_\Delta\mathbf{X})$ *of their restriction to* $\mathrm{M}({}^K\mathbf{X})$ *such that the metric* $\Xi^* h_\mathcal{E}$ *is* good *along the boundary divisor* $D_\mathbb{C}$ *in the sense of* [55, p. 242].

PROOF. The assertion follows from the construction of the extension in [loc. cit.], cf. also [23, §3.8]. □

2.6.4. Definition (Hermitian automorphic vector bundle). Let $(\mathcal{E}, h_\mathcal{E})$ be as in 2.6.1. The locally free sheaf $\Xi^*\mathcal{E}$, together with the singular (along $D_\mathbb{C}$) Hermitian metric $\Xi^* h_\mathcal{E}$ is called the **Hermitian automorphic vector bundle** associated to $(\mathcal{E}, h_\mathcal{E})$ on $\mathrm{M}({}^K_\Delta\mathbf{X})$.

2.6.5. Remark. It follows directly from the construction of metrics in 2.6.2 that for a morphism of p-ECMSD

$$(\alpha, \rho)\colon {}^{K_1}_{\Delta_1}\mathbf{X}_1 \to {}^{K_2}_{\Delta_2}\mathbf{X}_2$$

and for a given $(\mathcal{E}, h_\mathcal{E})$, with $\mathcal{E}$ a locally free sheaf on $\left[\mathrm{M}^\vee(\mathbf{X}_1)/P_{\mathbf{X}_1}\right]$ and for $h_\mathcal{E} = \{h_{\mathcal{E},\sigma}\}$ a collection of invariant Hermitian metrics, we have a canonical isomorphism

$$\mathrm{M}(\alpha, \rho)^*(\Xi^*\mathcal{E}, \Xi^* h_\mathcal{E}) \cong (\Xi^*\,\mathrm{M}^\vee(\alpha)^*\mathcal{E}, \Xi^*\,\mathrm{M}^\vee(\alpha)^* h_\mathcal{E}),$$

in particular these isomorphisms are compatible with composition.

(For the metrics this follows from the existence and functoriality of the 2-isomorphism

$$\left(\Xi({}^K\mathbf{X}) \times_{\mathrm{Spec}(\mathcal{O}_\mathbf{X})} \mathrm{Spec}(\mathbb{C})\right)^{\mathrm{an}} \to \Xi_\mathbb{C}({}^K\mathbf{X})$$

given by 2.5.14.)

Similarly, in the case of a boundary component

$$(\iota, \rho)\colon {}^{K_1}_{\Delta_1}\mathbf{B} \Longrightarrow {}^K_\Delta\mathbf{X},$$

and in the case $\mathcal{O}_\mathbf{X} = \mathbb{Z}_{(p)}$ (for simplicity) if $h_\mathcal{E}$ is the restriction of a $P_\mathbf{B}(\mathbb{R})U_\mathbf{X}(\mathbb{C})$-invariant metric $h_{\mathcal{E},1}$ on $\mathbb{D}_\mathbf{B}$ to $\mathbb{D}_{\mathbf{B}\Longrightarrow\mathbf{X}} \subset \mathbb{D}_\mathbf{B}$, we have on the formal completion a canonical isomorphism

$$\mathrm{M}(\iota, \rho)^*(\Xi^*\mathcal{E}, \Xi^* h_\mathcal{E}) \cong (\Xi^*\mathcal{E}|_{\mathrm{M}^\vee(\mathbf{B})}, \Xi^* h_{\mathcal{E},1})$$

taking into account that the formal isomorphism converges over $\mathbb{C}$. However, in general, the metrics $h_\mathcal{E}$ do not come from some $h_{\mathcal{E},1}$ as above (they are not even defined everywhere on $\mathbb{D}_\mathbf{B}$), but of course $\Xi^* h_\mathcal{E}$ is always defined in a neighborhood of $D_\mathbb{C}$ on $\mathrm{M}({}^{K_1}_{\Delta_1}\mathbf{B})(\mathbb{C})$ and the above statement makes sense.

2.6.6. In the reminder of this section we will describe the necessary tools from Arakelov theory that will be needed to define heights on (respectively arithmetic volumes of) Shimura varieties w.r.t. automorphic line bundles with their singular metrics. We will be as minimalist as possible, stating only the properties of the objects involved that are needed in the sequel and which uniquely characterize the resulting values. We assume familiarity with classical Arakelov theory as presented in [61].

2.6.7. We consider the category (X, D), where X is a smooth complex algebraic variety equipped with a divisor of normal crossings D. Morphisms $\alpha\colon (X, D_X) \to (Y, D_Y)$ are morphisms $\alpha\colon X \to Y$ which satisfy $\alpha^{-1}D_Y \subseteq D_X$.

For a smooth complex algebraic variety U, we denote by $A^{i,j}(U)$ the space of complex valued $\mathcal{C}^\infty$-differential forms of type (i,j) on U. We consider an association which maps a pair (X, D) to a subspace of differential forms

$$A^{i,j}(X, D) \subset A^{i,j}(X - D)$$

for every i, j, stable w.r.t. the differentials d and d^c. In addition, for any vector bundle E on X, we consider a set $\mathcal{M}_E$ of smooth Hermitian metrics on $E|_{X-D}$, such that the following holds

(F0) For $D = \emptyset$, we have $A^{i,j}(X, \emptyset) = A^{i,j}(X)$ and for all E, $\mathcal{M}_E$ consists of the set of all smooth metrics on E.

(F1) For any morphism $f\colon (X, D) \to (Y, D')$, f^* restricts to a map $A^{i,j}(Y, D') \to A^{i,j}(X, D)$ and f^*h for a metric $h \in \mathcal{M}_E$ is a metric in $\mathcal{M}_{f^*E}$.

(F2) Each $\xi \in A^{i,j}(X, D)$ defines a current on X, and we have $\mathrm{d}[\xi] = [\mathrm{d}\xi]$, respectively $\mathrm{d}^c[\xi] = [\mathrm{d}^c\xi]$.

(F3) The Chern character form of any vector bundle E with metric in $\mathcal{M}_E$ is in $\bigoplus_p A^{p,p}(X, D)$.

(F4) There is a unique way to associate with each (X, D) and exact sequence

$$E_\bullet\colon 0 \to E_{-1} \to E_0 \to E_1 \to 0$$

of vector bundles on X and metrics $h_i \in \mathcal{M}_{E_i}$ for $i = -1, 0, 1$ a Bott–Chern character

$$\mathrm{ch}(E_\bullet, h_{-1}, h_0, h_1) \in \bigoplus_p A^{p,p}(X, D)/(\mathrm{im}\,\mathrm{d} + \mathrm{im}\,\mathrm{d}^c)$$

such that the following holds:

(a) $\mathrm{d}\,\mathrm{d}^c\,\mathrm{ch}(E_\bullet, h_{-1}, h_0, h_1) = \sum_i (-1)^{i+1}\,\mathrm{ch}(E_i, h_i)$,

(b) the formation is compatible with pullback,

(c) $\mathrm{ch}(E_\bullet, h_{-1}, h_0, h_1) = 0$ if $E_\bullet$ splits and $h_0 = h_{-1} \oplus h_1$,

(d) for an exact square consisting of vector bundles $E_{i,j}$, $i, j = -1, 0, 1$ equipped with metrics $h_{i,j}$ in $\mathcal{M}_{E_{i,j}}$, we have

$$\sum_i (-1)^i\,\mathrm{ch}(E_{i,\bullet}, h_{i,-1}, h_{i,0}, h_{i,1}) = \sum_j (-1)^j\,\mathrm{ch}(E_{\bullet,j}, h_{-1,j}, h_{0,j}, h_{1,j}).$$

The following property, which however we will not need, is often convenient:

(F5) If Y is a cycle of codimension p which intersects D properly then for any Green's function g of logarithmic type along Y [61, 2.1, Definition 3], and for all $\xi \in A^{p,p}(X, D)$, $g\xi$ is locally integrable on X.

Note that by the properties (F4)(a)–(c) the Bott–Chern characters involving only smooth metrics have to be necessarily the same as those considered in [61, 3.1, Theorem 2]. We will write also $\mathrm{ch}(E, h_{-1}, h_0)$ for $\mathrm{ch}(E_\bullet, h_{-1}, h_0, 0)$ where $E_\bullet$ is the sequence $0 \to E \to E \to 0 \to 0$.

2.6.8. From closedness under d and d^c follows in particular that for $e_1 \in A^{p,p}(X, D)$ and $e_2 \in A^{q,q}(X, D)$ the (star-)product $e_1 * e_2 := (\mathrm{d}\,\mathrm{d}^c e_1)e_2$ is associative and commutative modulo the image of d, d^c. If g is a current on $(X - D)(\mathbb{C})$ which can be expressed as a sum $\tilde{g} + g_0$, where $\tilde{g}$ is a Green's form of logarithmic type [61, Definition 3, p. 43] and g_0 is in $A^{p,p}(X, D)$, and h is a similar form, we define

$g * h \in A^{p+q+1,p+q+1}(X - D)$ to be $\tilde{g} * \tilde{h} + g_0 \,\mathrm{d}\,\mathrm{d}^c \tilde{h} + h_0 \,\mathrm{d}\,\mathrm{d}^c \tilde{g} + h_0 \,\mathrm{d}\,\mathrm{d}^c g_0$, which is clearly (modulo the image of d, d^c) independent of the way of expressing the sum.

2.6.9. If Y_1 and Y_2 are cycles on X of codimension p and q, respectively, *without intersection*, and $\tilde{g}_1$, $\tilde{g}_2$ are Green's forms of logarithmic type for them, we can represent their star-product (modulo the image of d, d^c) by

$$\tilde{g}_1 * \tilde{g}_2 = \sigma_1 \tilde{g}_1 (\mathrm{d}\,\mathrm{d}^c \tilde{g}_2) + \mathrm{d}\,\mathrm{d}^c(\sigma_2 \tilde{g}_1)\tilde{g}_2,$$

where $\sigma_1 + \sigma_2 = 1$ is a partition of unity such that $\sigma_1 = 0$ on Y_1 and $\sigma_2 = 0$ on Y_2. If g_1 and g_2 are the sums of the previous Green's forms with forms $g_{1,0} \in A^{p,p}(X, D)$, respectively $g_{2,0} \in A^{q,q}(X, D)$, we have the same formula for the star product defined in 2.6.8:

$$g_1 * g_2 = \sigma_1 g_1 (\mathrm{d}\,\mathrm{d}^c g_2) + \mathrm{d}\,\mathrm{d}^c(\sigma_2 g_1) g_2.$$

For this note that, e.g.,

$$\sigma_1 g_1 (\mathrm{d}\,\mathrm{d}^c g_{2,0}) + \mathrm{d}\,\mathrm{d}^c(\sigma_2 g_1) g_{2,0} = g_{2,0} \,\mathrm{d}\,\mathrm{d}^c g_1 + \sigma_1 g_1 (\mathrm{d}\,\mathrm{d}^c g_{2,0}) - \mathrm{d}\,\mathrm{d}^c(\sigma_1 g_1) g_{2,0}.$$

However in the expession

$$\sigma_1 g_1 (\mathrm{d}\,\mathrm{d}^c g_{2,0}) - \mathrm{d}\,\mathrm{d}^c(\sigma_1 g_1) g_{2,0}$$

$\sigma_1 g_1$ and $g_{2,0}$ are in $A^{q,q}(X, D)$. Therefore it is contained in the image of d, d^c. Using an embedded resolution of singularities and properties (F1) and (F2), one can extend these explicit formulas to nonempty intersections. We will not need this case, however. See also [13] for generalizations.

2.6.10. Theorem (Burgos, Kramer, Kühn). *A collection $A^{i,j}(\cdot,\cdot)$ of differential forms and a collection of metrics $\mathcal{M}$ satisfying the axioms* (F0)–(F4) *such that $\mathcal{M}$ includes the natural metrics $\Xi^* h_{\mathcal{E}}$ of fully decomposed automorphic vector bundles $\Xi^*\mathcal{E}$ on pure Shimura varieties exist.*

PROOF. The collection of differential forms is the collection of log-log-forms [14, Section 2.2]. Those are in particular pre-log-log forms [15, Section 7.1]. The collection of metrics is the collection of log-singular ones [15, 4.29]. The requested properties are shown respectively at the following places: (F1) [14, Proposition 2.24] for forms, [14, Proposition 4.35] for the metrics, (F2) [15, [Proposition 7.6], (F0) and (F3) are part of their definition of log-singular metric, (F4) [14, Proposition 4.64]. The statement about fully decomposed automorphic vector bundles is proven in [14, Theorem 6.3]. □

Note that (F5) (not needed here) follows from the proof of [15, Lemma 7.35].

2.6.11. Let X be a projective regular and flat scheme over $\mathrm{Spec}(R)$, where R is any localization of $\mathbb{Z}$. Let D be a divisor on X such that $D_{\mathbb{C}}$ is a divisor of normal crossings. From now on we assume that collections $A^{i,j}(\cdot,\cdot)$ and $\mathcal{M}$ satisfying (F1)–(F4) have been chosen and define in analogy to [61, p. 55]:

$$A^{p,p}(X, D) := \{\omega \in A^{p,p}(X(\mathbb{C}), D(\mathbb{C})) \mid \omega \text{ real}, F_\infty^* \omega = (-1)^p \omega\}$$
$$\tilde{A}^{p,p}(X, D) := A^{p,p}(X, D)/(\mathrm{im}\,\partial + \mathrm{im}\,\overline{\partial}).$$

2.6.12. Definition (Arithmetic Chow group). We define the **arithmetic Chow group** $\widehat{\mathrm{CH}}^p(X, D)$ by the following pushout

$$\begin{array}{ccccccccc}
\mathrm{CH}^{p-1,p}(X) & \longrightarrow & \tilde{A}^{p-1,p-1}(X,D) & \xrightarrow{a} & \widehat{\mathrm{CH}}^p(X,D) & \longrightarrow & \mathrm{CH}^p(X) & \longrightarrow & 0 \\
\| & & \uparrow & & \uparrow & & \| & & \\
\mathrm{CH}^{p-1,p}(X) & \longrightarrow & \tilde{A}^{p-1,p-1}_{\mathrm{GS}}(X) & \xrightarrow{a_{\mathrm{GS}}} & \widehat{\mathrm{CH}}^p_{\mathrm{GS}}(X) & \longrightarrow & \mathrm{CH}^p(X) & \longrightarrow & 0
\end{array}$$

where an index GS means the corresponding object is defined as in [61].

We define a product

$$\widehat{\mathrm{CH}}^p(X, D) \times \widehat{\mathrm{CH}}^q(X, D) \to \widehat{\mathrm{CH}}^{p+q}_{\mathbb{Q}}(X, D)$$

by setting $a(b_1)a(b_2) := a(b_1 \,\mathrm{d}\,\mathrm{d}^c b_2)$ and $a(b)\hat{z} := a\big(b\omega(\hat{z})\big)$ (for the map ω of [loc. cit.]). This is well-defined because the formulas hold for a_{GS}. This product is obviously associative and commutative. We also form the direct sum $\widehat{\mathrm{CH}}^\bullet(X, D)_{\mathbb{Q}} := \bigoplus_p \widehat{\mathrm{CH}}^p(X, D)_{\mathbb{Q}}$, which yields a contravariant functor (w.r.t. pullback) from the category of pairs (X, D), where X is projective, regular, and flat over $\mathrm{Spec}(R)$ and D is a divisor on X such that $D_{\mathbb{C}}$ has only normal crossings, to the category of graded $\mathbb{Q}$-algebras.

2.6.13. If X is projective over $\mathrm{Spec}(R)$ of relative dimension d, where R is any localization of $\mathbb{Z}$, we have a pushforward

$$\pi_*\colon \widehat{\mathrm{CH}}^p(X, D) \to \widehat{\mathrm{CH}}^{p-d}\big(\mathrm{Spec}(R)\big),$$

which on the image of a is defined by integration (the integrals exist because of (F2)).

2.6.14. Definition (Arithmetic Chern character). For any vector bundle $\mathcal{E}$ with metric $h \in \mathcal{M}_{\mathcal{E}_{\mathbb{C}}}$ we define an arithmetic Chern character as

$$\widehat{\mathrm{ch}}(\mathcal{E}, h) := \widehat{\mathrm{ch}}_{\mathrm{GS}}(\mathcal{E}, \tilde{h}) + a\big(\mathrm{ch}(\mathcal{E}_{\mathbb{C}}, h, \tilde{h})\big),$$

where $\tilde{h}$ is any smooth metric on $\mathcal{E}_{\mathbb{C}}$. In particular the graded components define Chern classes $\hat{\mathrm{c}}_i(\mathcal{E}, h)$ of every degree.

2.6.15. Lemma. $\widehat{\mathrm{ch}}(\mathcal{E}, h)$ *does not depend on the choice of* $\tilde{h}$.

PROOF. We have the property $\widehat{\mathrm{ch}}_{\mathrm{GS}}(\mathcal{E}_{\mathbb{C}}, \tilde{h}_2) - \widehat{\mathrm{ch}}_{\mathrm{GS}}(\mathcal{E}_{\mathbb{C}}, \tilde{h}_1) = a\big(\mathrm{ch}(\mathcal{E}_{\mathbb{C}}, \tilde{h}_2, \tilde{h}_1)\big)$ [61, Proposition 1, p. 86]. From properties (F4)(c), (d) it follows that $\mathrm{ch}(\mathcal{E}_{\mathbb{C}}, \tilde{h}_2, \tilde{h}_1) = \mathrm{ch}(\mathcal{E}_{\mathbb{C}}, \tilde{h}_2, h) - \mathrm{ch}(\mathcal{E}_{\mathbb{C}}, \tilde{h}_1, h)$. □

The formation of Chern characters (hence of Chern classes) is compatible with pullback because this is true for $\widehat{\mathrm{ch}}_{\mathrm{GS}}(\mathcal{E}, \tilde{h})$.

2.6.16. For any localization R of $\mathbb{Z}$ we have a map

$$\widehat{\deg}\colon \widehat{\mathrm{CH}}^1\big(\mathrm{Spec}(R)\big)_{\mathbb{Q}} \to \mathbb{R}_R,$$

where $\mathbb{R}_R$ is $\mathbb{R}$ modulo rational multiples of $\log(p)$ for all p with $p^{-1} \in R$. We denote $\mathbb{R}_{\mathbb{Z}_{(p)}}$ also by $\mathbb{R}^{(p)}$ and $\mathbb{R}_{\mathbb{Z}[1/N]}$ by $\mathbb{R}_N$.

Obviously $\mathbb{R}_N$ *injects* into the fibre product of all $\mathbb{R}^{(p)}$ $p \nmid N$ over $\mathbb{R}_{\mathbb{Q}}$.

2.6.17. Definition (Arithmetic volume, height). Let X be as in 2.6.12. If $\overline{\mathcal{E}} = (\mathcal{E}, h)$ is a line bundle with metric in $\mathcal{M}$ and Y is a cycle on X such that $Y_{\mathbb{Q}}$ intersects $D_{\mathbb{Q}}$ properly, we define the **arithmetic volume** (or **height**) of Y with respect to $\overline{\mathcal{E}}$ as follows: Choose a Green's form g of logarithmic type for $Y_{\mathbb{C}}$ (it exists by [61, Theorem 3, p. 44]) and a smooth metric $\tilde{h}$ on $\mathcal{E}$, and set

$$\begin{aligned}\widehat{\mathrm{vol}}_{\overline{\mathcal{E}}}(Y) &:= \widehat{\deg}\pi_*\left((Y,g)\cdot \hat{\mathrm{c}}_1(\mathcal{E},\tilde{h})^{n+1-d}\right) - \frac{1}{2}\int_X g\,\mathrm{c}_1(\mathcal{E}_{\mathbb{C}},\tilde{h})^{n+1-d} + \frac{1}{2}\int_Y \phi \\ &= \widehat{\mathrm{vol}}_{(\mathcal{E},\tilde{h}),GS}(Y) + \frac{1}{2}\int_Y \phi,\end{aligned}$$

where $\phi \in \tilde{A}^{n-d,n-d}(X,D)$ is such that $a(\phi) = \hat{\mathrm{c}}_1(\mathcal{E},h)^{n+1-d} - \hat{\mathrm{c}}_1(\mathcal{E},\tilde{h})^{n+1-d}$. The integral over Y exists because of property (F2) (consider a desingularization of Y). Here it is crucial that Y and D intersect properly.

2.6.18. Lemma. *$\widehat{\mathrm{vol}}_{\overline{\mathcal{E}}}(Y)$ does not depend on the choice of g and $\tilde{h}$.*

PROOF. The expression is independent of g by means of the definition of the star product [61, Definition 4, p. 50]. We have to show that it is independent of the choice of smooth metric $\tilde{h}$. Let $\tilde{h}_i$, $i = 1, 2$ be such choices. Let ϕ_i be such that $a(\phi_i) = \hat{\mathrm{c}}_1(\mathcal{E},h)^{n+1-d} - \hat{\mathrm{c}}_1(\mathcal{E},\tilde{h}_i)^{n+1-d}$. Note that $\phi_2 - \phi_1$ is smooth. We have

$$\widehat{\deg}\pi_*\left((Y,g)\cdot \hat{\mathrm{c}}_1(\mathcal{E},\tilde{h}_2)^{n+1-d} - (Y,g)\cdot \hat{\mathrm{c}}_1(\mathcal{E},\tilde{h}_1)^{n+1-d}\right) = \frac{1}{2}\int_{X(\mathbb{C})} (\phi_1 - \phi_2)\,\mathrm{d}\,\mathrm{d}^c g;$$

$$-\frac{1}{2}\int_{X(\mathbb{C})} g\,\mathrm{c}_1(\mathcal{E}_{\mathbb{C}},\tilde{h}_2)^{n+1-d} + \frac{1}{2}\int_{X(\mathbb{C})} g\,\mathrm{c}_1(\mathcal{E}_{\mathbb{C}},\tilde{h}_1)^{n+1-d} = \frac{1}{2}\int_{X(\mathbb{C})} g\,\mathrm{d}\,\mathrm{d}^c(\phi_2 - \phi_1);$$

$$\frac{1}{2}\int_{Y(\mathbb{C})} \phi_2 - \frac{1}{2}\int_{Y(\mathbb{C})} \phi_1 = \frac{1}{2}\int_{X(\mathbb{C})} (\phi_2 - \phi_1)\,\mathrm{d}\,\mathrm{d}^c g - \frac{1}{2}\int_{X(\mathbb{C})} g\,\mathrm{d}\,\mathrm{d}^c(\phi_2 - \phi_1);$$

so the sum over these differences is zero. □

From (F1) and (F2) together with (F5) the following equality follows formally—using an embedded resolution of singularities (see the proof of [15, Theorem 7.33] with $\sigma_{YZ} = 0$):

$$\int_X g\,\mathrm{d}\,\mathrm{d}^c g_2 = \int_Y g_2 + \int_X (\mathrm{d}\,\mathrm{d}^c g) g_2$$

for any form g_2 in our class. Hence, if (F5) holds, we could simply define

$$\widehat{\mathrm{vol}}_{\overline{\mathcal{E}}}(Y) := \widehat{\deg}\left(\pi_*\left((Y,g)\cdot \hat{\mathrm{c}}_1(\mathcal{E},h)^{n+1-d}\right)\right) - \frac{1}{2}\int_{X(\mathbb{C})} g\,\mathrm{c}_1(\mathcal{E}_{\mathbb{C}},h)^{n+1-d}$$

which is literally the same definition as $\mathrm{ht}_{\overline{\mathcal{E}},GS}$.

2.6.19. In any case, if there is a desingularization $(\widetilde{Y}, \widetilde{D})$ of $(Y, Y \cap D)$ which is itself a pair as considered in 2.6.12, we get from the projection formula [61, Theorem 3(iii), p. 64] that

$$\widehat{\mathrm{vol}}_{\overline{\mathcal{E}}}(Y) = \widehat{\deg}\pi_*\left(f^*\hat{\mathrm{c}}_1(\mathcal{E},h)\right)^{n+1-d}$$

for the corresponding morphism $f\colon (\widetilde{Y}, \widetilde{D}) \to (X, D)$.

2.6.20. We now apply the constructions above to (pure) Shimura varieties. Assume first that all occurring K's are neat, see 2.6.25 for the general case. For a pure Shimura variety (considered as defined over $\mathbb{Z}_{(p)}$ as in 2.6.1) $\mathrm{M}({}^K_\Delta\mathbf{X})$ with complete, smooth w.r.t. K and projective Δ, the exceptional divisor D has normal crossings (2.5.9) and we define $\widehat{\mathrm{CH}}^\bullet(\mathrm{M}({}^K_\Delta\mathbf{X}))$ as $\widehat{\mathrm{CH}}^\bullet(\mathrm{M}({}^K_\Delta\mathbf{X}), D)$ as defined in 2.6.12 w.r.t. any collection of forms and metrics satisfying (F0)–(F4), such that the metrics of fully decomposed automorphic vector bundles are in the collection (such collections exist by Theorem 2.6.10). We assume that Δ is sufficiently fine and hence $\mathrm{M}({}^K_\Delta\mathbf{X})$ exists by 2.5.9.

2.6.21. Definition (Geometric and arithmetic volume of Shimura varieties). Let ${}^K_\Delta\mathbf{X}$ be pure p-ECMSD, with complete, smooth w.r.t. K and projective Δ, and let $\overline{\mathcal{E}} = (\mathcal{E}, h_{\mathcal{E}})$ consist of an *invertible* sheaf $\mathcal{E}$ on $[\mathrm{M}^\vee(\mathbf{X})/P_{\mathbf{X}}]$ (where the stack is considered over $\operatorname{Spec}\mathbb{Z}_{(p)}$) and $h_{\mathcal{E}} = \{h_{\mathcal{E},\sigma}\}$ a collection of invariant Hermitian metrics as in 2.6.1.

The **geometric volume**

$$\mathrm{vol}_{\mathcal{E}}\big(\mathrm{M}({}^K_\Delta\mathbf{X})\big)$$

is defined as the volume of $\mathrm{M}({}^K_\Delta\mathbf{X})(\mathbb{C})$ with respect to the volume form $\big(\mathrm{c}_1(\Xi^*\overline{\mathcal{E}}_{\mathbb{C}})\big)^d$, where d is the dimension of $\mathrm{M}({}^K_\Delta\mathbf{X})_{\mathbb{C}}$.

The **arithmetic volume** at p

$$\widehat{\mathrm{vol}}_{\overline{\mathcal{E}},p}\big(\mathrm{M}({}^K_\Delta\mathbf{X})\big)$$

is defined as the image of

$$\pi_*(\hat{\mathrm{c}}_1(\Xi^*\mathcal{E}, \Xi^* h_{\mathcal{E}})^{d+1})$$

under the natural map $\widehat{\mathrm{CH}}^1(\operatorname{Spec}(\mathbb{Z}_{(p)}))_{\mathbb{Q}} \to \mathbb{R}^{(p)}$, where $\pi\colon \mathrm{M}({}^K_\Delta\mathbf{X}) \to \operatorname{Spec}(\mathbb{Z}_{(p)})$ is the structural morphism.

If $P_{\mathbf{X}}$ is a group scheme of type (P) over $\operatorname{Spec}(\mathbb{Z}[1/N])$ and K is admissible *for all* $p \nmid N$, then we define the global **arithmetic volume**

$$\widehat{\mathrm{vol}}_{\overline{\mathcal{E}}}\big(\mathrm{M}({}^K_\Delta\mathbf{X})\big)$$

in $\mathbb{R}_N$ as the value determined by the arithmetic volumes at p (cf. 2.6.16, see also 2.6.26).

If $(\alpha, \rho)\colon {}^{K'}_{\Delta'}\mathbf{Y}_{\mathbb{Q}} \hookrightarrow {}^K_\Delta\mathbf{X}_{\mathbb{Q}}$ is an embedding of *rational* pure Shimura data, we define the **arithmetic volume** at p:

$$\widehat{\mathrm{vol}}_{\overline{\mathcal{E}},p}\big(\mathrm{M}({}^{K'}_{\Delta'}\mathbf{Y})\big) = \widehat{\mathrm{vol}}_{\Xi^*\overline{\mathcal{E}}}\Big(\overline{\mathrm{M}(\alpha,\rho)\big(\mathrm{M}({}^{K'}_{\Delta'}\mathbf{Y}_{\mathbb{Q}})\big)}^{\mathrm{Zar}}\Big),$$

where $\widehat{\mathrm{vol}}_{\Xi^*\overline{\mathcal{E}}}$ is the height (2.6.17). We similarly define the (global) arithmetic volume.

2.6.22. Question. We mention that there is a **proportionality principle** [55] in the geometric case. This means that, if $\mathcal{E}$ is fully-decomposed, *all* polynomials of degree $\dim(\mathrm{M}({}^K_\Delta\mathbf{X})_{\mathbb{C}})$ in the Chern classes of the automorphic vector bundle $(\Xi^*\mathcal{E})_{\mathbb{C}}$, considered as a number, are proportional to the same expression in the Chern classes of the original bundle $\mathcal{E}$, computed on $\mathrm{M}^\vee(\mathbf{X})$.

Is there an analogue in the arithmetic case?—cf. also [29].

2.6.23. Remark. If (α, ρ) was in fact an embedding of *p-integral* data: ${}^{K'}_{\Delta'}\mathbf{Y} \hookrightarrow {}^{K}_{\Delta}\mathbf{X}$, we have, by the projection formula (cf. 2.6.19) and 2.6.5:

$$\widehat{\mathrm{vol}}_{\overline{\mathcal{E}},p}(\mathrm{M}({}^{K'}_{\Delta'}\mathbf{Y})) = \widehat{\mathrm{vol}}_{\mathrm{M}^\vee(\alpha)^*\overline{\mathcal{E}},p}(\mathrm{M}({}^{K'}_{\Delta'}\mathbf{Y})).$$

2.6.24. Lemma. *The geometric and arithmetic volumes do not depend on the rational polyhedral cone decomposition Δ, i.e., they do not depend on the chosen toroidal compactification.*

Proof. For each pair Δ_i, $i = 1, 2$, there is a common refinement Δ, cf. 2.2.25, and we have two projections (2.5.5):

$$\begin{array}{ccc} & \mathrm{M}({}^{K}_{\Delta}\mathbf{X}) & \\ \swarrow & & \searrow \\ \mathrm{M}({}^{K}_{\Delta_1}\mathbf{X}) & & \mathrm{M}({}^{K}_{\Delta_2}\mathbf{X}). \end{array}$$

Furthermore this diagram is compatible with formation of $\Xi^*\mathcal{E}, \Xi^* h_E$ by 2.6.5. Its Chern forms are therefore transported into each other by pullback along the arrows in the diagram. Since the forms are integrable on every $\mathrm{M}({}^{K}_{\Delta}\mathbf{X})$ (see [55], cf. also 2.6.3), the two geometric volumes coincide.

Now, there is pullback map

$$\alpha^* : \widehat{\mathrm{CH}}^\bullet(\mathrm{M}({}^{K}_{\Delta_i}\mathbf{X}))_{\mathbb{Q}} \to \widehat{\mathrm{CH}}^\bullet(\mathrm{M}({}^{K}_{\Delta}\mathbf{X}))_{\mathbb{Q}}$$

as well, which is a ring homomorphism and is compatible with Chern classes (see (F4)(b) and [61, Theorem 3, p. 84]).

Therefore the assertion for the arithmetic volume boils down to the fact that for a class $x = (z, g) \in \widehat{\mathrm{CH}}^{d+1}_{\mathrm{GS}}({}^{K}_{\Delta}\mathbf{X})$ or $x = a(g')$, where z is a zero-cycle in the fibre above p and g is a smooth top-degree Green's form and g' is a top-degree form on $\mathrm{M}({}^{K}\mathbf{X})(\mathbb{C})$ which is integrable on the whole $\mathrm{M}({}^{K}_{\Delta}\mathbf{X})(\mathbb{C})$, we have $\widehat{\deg}(\pi_*\alpha^*(x)) = \widehat{\deg}(\pi_* x)$, where the π are the respective structural morphisms. This is true because the pushforwards of g, g' are computed by an integral. $\square$

2.6.25. Remark. If K is not neat, $\mathrm{M}({}^{K}_{\Delta}\mathbf{X})$ is only a Deligne–Mumford stack. Instead of extending the theory of [14, 15] to stacks like, e.g., in [20], we define the arithmetic volume in this case as follows: We take some neat admissible $K' \subset K$ and define

$$\widehat{\mathrm{vol}}_{\overline{\mathcal{E}},p}(\mathrm{M}({}^{K}_{\Delta}\mathbf{X})) := \frac{1}{[K : K']}\widehat{\mathrm{vol}}_{\overline{\mathcal{E}},p}(\mathrm{M}({}^{K'}_{\Delta}\mathbf{X}))$$

as value in $\mathbb{R}^{(p)}$. The global arithmetic volume in $\mathbb{R}_N$, assuming that K is admissible for all $p \nmid N$, will be again the value determined by the $\widehat{\mathrm{vol}}_{\overline{\mathcal{E}},p}$ for $p \nmid N$ (here the neat K' that will be used to define $\widehat{\mathrm{vol}}_{\overline{\mathcal{E}},p}$ may depend on p!)

It follows from 2.6.5, the definition of pullback, and the same reasoning as in Lemma 2.6.24 that this is independent of the choice of K'. In particular, if K is neat—or using any reasonable extension of Arakelov theory to stacks—this definition agrees with the previous one.

In this sense the geometric and arithmetic volume do not depend essentially on K, as they are both multiplied by the index, if K is changed. For the arithmetic case this is of course only true for admissible K, and the more primes are considered (i.e. the smaller N is) the less admissible groups there are.

2.6.26. Remark. Since the canonical models can be defined as the normalization of a Zariski closure in a variety defined over some $\mathbb{Z}[1/N]$ (compactification of a moduli space of Abelian varieties), it follows that there is even a model of $\mathrm{M}({}^K_\Delta\mathbf{X})$ defined over some $\mathcal{O}_\mathbf{X}[1/N]$, yielding the canonical ones over the various reflex rings $\mathcal{O}_{\mathbf{X},(p)}$ for $p \nmid N$, such that the arithmetic volume of this model coincides with the local values in Definition 2.6.21. This justifies a priori that the *global* arithmetic volume in the sense of Definition 2.6.21 actually lies in $\mathbb{R}_N$.

2.7. Mixed Shimura varieties of symplectic type

2.7.1. If ${}^K\mathbf{X}$ is p-EMSD and L is a representation of $P_\mathbf{X}$, there should exist a category fibred in groupoids (over $\mathrm{Spec}(\mathcal{O}_\mathbf{X})$)

$$\left[{}^K\mathbf{X}\text{-}\boldsymbol{L}\text{-}\mathbf{mot}\right]$$

of motives with "$(P_\mathbf{X}, L)$-structure" (see 1.5). It should yield (up to taking several connected components) a canonical integral model of the Shimura variety described by 2.5.5. This can be made precise for P.E.L.-type Shimura data and the associated standard representation. We will need only the case of symplectic Shimura data and certain representations (sometimes different ones, however) of their groups.

For each of the symplectic mixed Shimura data (2.3.4) we will explicitly describe a moduli problem, in terms of 1-motives, which belongs to the standard representation of the underlying group scheme. These moduli problems are representable by a Deligne–Mumford stack and give an explicit description of the canonical integral models. The corresponding explicit model was already used to normalize the morphism Ξ, encoding the theory of automorphic vector bundles, for all mixed Shimura varieties of Abelian type (cf. Main Theorem 2.5.14).

For the language of 1-motives as used here, we refer the reader to [25, Chapter 5]. We will denote a motive M by both $[\underline{Y} \xrightarrow{\mu} G]$ and the symmetric description $(A, A^\vee, \underline{X}, \underline{Y}, \alpha, \alpha^\vee, \nu)$.

2.7.2. Lemma. *The reflex field of the symplectic mixed Shimura data* (2.3.4) *is* $\mathbb{Q}$.

2.7.3. Definition (Categories of 1-motives). We define the following stacks over $\mathbb{Z}_{(p)}$:

(1) Recall the mixed Shimura datum $\mathbf{X} = \mathbf{H}_{g_0}[(I \otimes I)^s, L_0 \otimes I]$ from 2.3.4. Consider $L = I^* \oplus L_0 \oplus I$ with standard representation (cf. 2.3.3) of $P_\mathbf{X}$ on it, where $L_{0,\mathbb{Z}_{(p)}}$ is a lattice of dimension $2g_0$ with unimodular symplectic form (if $g_0 \neq 0$), $I_{\mathbb{Z}_{(p)}}$ is any lattice, and $K \subset \mathrm{PSp}(L_{\mathbb{A}^{(\infty)}})$ is an admissible compact open subgroup. Assume that $L \neq 0$. We define

$$\left[{}^K\mathbf{H}_{g_0}[(\boldsymbol{I} \otimes \boldsymbol{I})^s, \boldsymbol{L_0} \otimes \boldsymbol{I}]\text{-}\boldsymbol{L}\text{-}\mathbf{mot}\right]$$

as the stack of the following data:

(a) a $\mathbb{Z}_{(p)}$-scheme S,
(b) a 1-motive $M = (A, A^\vee, \underline{X}, \underline{Y}, \alpha, \alpha^\vee, \nu)$ over S, with $\dim(A) = g_0$,
(c) a p-polarization of some degree $d \in \mathbb{Z}_{(p)}^*$

$$\psi_1 \colon A \to A^\vee$$

and an isomorphism

$$\psi_2 \colon \underline{X} \otimes \mathbb{Z}_{(p)} \to \underline{Y} \otimes \mathbb{Z}_{(p)}$$

of the same degree d, such that they give a p-polarization of M,

(d) an isomorphism $\rho\colon (I^*_{\mathbb{Z}_{(p)}})_S \cong \underline{Y} \otimes \mathbb{Z}_{(p)}$,

(e) a $K^{(p)}$-level structure

$$\xi \in \underline{\mathrm{Iso}}_{(\mathrm{USp}(L_0,I),L)}\big(L_{\mathbb{A}^{(\infty,p)}}, H^{\mathrm{et}}(M, \mathbb{A}^{(\infty,p)})\big)/K^{(p)},$$

where the isomorphisms in the etale sheaf on the right have to be compatible with the following $(\mathrm{USp}(L_{0,\mathbb{Z}_{(p)}}, I_{\mathbb{Z}_{(p)}}), L_{\mathbb{Z}_{(p)}})$-structure on both parameters.
The $(\mathrm{USp}(L_{0,\mathbb{A}^{(\infty,p)}}, I_{\mathbb{A}^{(\infty,p)}}), L_{\mathbb{A}^{(\infty,p)}})$-structure is given as follows: The p-polarization induces an isomorphism

$$H^{\mathrm{et}}(M, \mathbb{A}^{(\infty,p)}) \to H^{\mathrm{et}}(M, \mathbb{A}^{(\infty,p)})^*(1).$$

Choosing some isomorphism $\mathbb{Z}^{\mathrm{et}}_{(p)}(1) \cong \mathbb{A}^{(\infty,p)}$, we get an alternating form on $H^{\mathrm{et}}(M, \mathbb{A}^{(\infty,p)})$ up to a scalar. For the weight filtration W_{-2} totally isotropic and $W_{-1} = (W_{-2})^{\perp}$. Furthermore, we have (via ρ) an isomorphism

$$\mathrm{gr}^0\big(H^{\mathrm{et}}(M, \mathbb{A}^{(\infty,p)})\big) = \underline{Y} \otimes_{\mathbb{Z}} \mathbb{A}^{(\infty,p)} = I^*_{\mathbb{A}^{(\infty,p)}}$$

(i.e., a USp-structure, see 2.3.12).
Morphisms are $\mathbb{Z}_{(p)}$-morphisms of schemes $S \to S'$, together with a $\mathbb{Z}_{(p)}$-isomorphism of M with the pullback of M' to S, compatible with polarizations (up to scalar), the respective morphisms ρ, and level structures.

(2) With the same notation as in (1) we define

$$\big[{}^K\mathbf{H}_{g_0}[(I \otimes I)^s, L_0 \otimes I]\text{-}L\text{-mot-triv}\big]$$

as the stack consisting of the same data $(S, M, \dots)$ as in (1), but in addition with a trivialization

$$\gamma\colon L \otimes \mathcal{O}_S \to H^{\mathrm{dR}}(M)$$

of the de Rham realization of M compatible with the $\mathrm{USp}(L_0, I)$-structure on both sides. $P_{\mathbf{H}_{g_0}[(I\otimes I)^s, L_0 \otimes I]} = \mathrm{USp}(L_0, I)$ acts on this stack *on the right* by translation of γ. This renders the stack

$$\big[{}^K\mathbf{H}_{g_0}[(I \otimes I)^s, L_0 \otimes I]\text{-}L\text{-mot-triv}\big]$$

into a right $\mathrm{USp}(L_0, I)$-torsor over $\big[{}^K\mathbf{H}_{g_0}[(I \otimes I)^s, L_0 \otimes I]\text{-}L\text{-mot}\big]$.

(3) Consider the Shimura datum $\mathbf{X} = \mathbf{H}_0[U, 0]$, a simple unipotent extension of $\mathbf{H}_0$, where U is any lattice of dimension k. We have $P_{\mathbf{X}} = \mathbb{W}(U) \rtimes \mathbb{G}_m$. $P_{\mathbf{X}}$ acts on $L' = \mathbb{Z}_{(p)} \oplus U^*$ as in 2.3.13 (If $U = (I \otimes I)^s$ we get one of the basic symplectic Shimura data but L' is *not* the standard representation). Let $K \subset \mathrm{PSp}(L_{\mathbb{A}^{(\infty)}})$ be an admissible compact open subgroup. We define

$$\big[{}^K\mathbf{H}_0[U, 0]\text{-}L'\text{-mot}\big]$$

as the stack of the following data:

(a) a $\mathbb{Z}_{(p)}$-scheme S,

(b) a 1-motive of the form $M = [\underline{Y} \to T]$ (without part of weight -1),

(c) an isomorphism $\rho\colon (U^*_{\mathbb{Z}_{(p)}})_S \cong \underline{Y} \otimes \mathbb{Z}_{(p)}$,

(d) an isomorphism $\rho'\colon (\mathbb{Z}_{(p)})_S \to \underline{X} \otimes \mathbb{Z}_{(p)}$,

(e) a $K^{(p)}$-level structure

$$\xi \in \underline{\mathrm{Iso}}_{(P_{\mathbf{X}}, L_{\mathbb{Z}_{(p)}})}\big(L'_{\mathbb{A}^{(\infty,p)}}, H^{\mathrm{et}}(M, \mathbb{A}^{(\infty,p)})\big)/K^{(p)},$$

where the isomorphisms have to be compatible with the obvious $(P_{\mathbf{X}}, L'_{\mathbb{Z}_{(p)}})$-structure on both parameters.
Isomorphisms are morphisms over $\mathrm{Spec}(\mathbb{Z}_{(p)})$ of schemes $S \to S'$ together with an isomorphism of M with the pullback of M' compatible with the respective morphisms ρ and ρ', and level structures.

(4) With the same notation as in (3) we define

$$\left[{}^{\boldsymbol{K}}\mathbf{H}_0[\boldsymbol{U}, \mathbf{0}]\text{-}\boldsymbol{L}'\text{-}\mathbf{mot\text{-}triv}\right]$$

consisting of the same data $(S, M, \dots)$ as in (3), but in addition with a trivialization

$$\gamma\colon L' \otimes \mathcal{O}_S \to H^{\mathrm{dR}}(M)$$

of the de Rham realization of M compatible with the $(P_{\mathbf{H}_0[U,0]}, L')$-structure on both sides.

$P_{\mathbf{H}_0[U,0]}$ acts on this stack *on the right* by translation of γ.

This renders the stack $\left[{}^{\boldsymbol{K}}\mathbf{H}_0[\boldsymbol{U}, \mathbf{0}]\text{-}\boldsymbol{L}'\text{-}\mathbf{mot\text{-}triv}\right]$ into a right $P_{\mathbf{X}}$-torsor over $\left[{}^{\boldsymbol{K}}\mathbf{H}_0[\boldsymbol{U}, \mathbf{0}]\text{-}\boldsymbol{L}'\text{-}\mathbf{mot}\right]$.

Actually there is a similar description of stacks

$$\left[{}^{\boldsymbol{K}}\mathbf{X}\text{-}\boldsymbol{L}\text{-}\mathbf{mot\text{-}triv}\right] \quad \left[{}^{\boldsymbol{K}}\mathbf{X}\text{-}\boldsymbol{L}\text{-}\mathbf{mot}\right]$$

where $\mathbf{X}$ is any of the symplectic Shimura data. We leave it to the reader to work out the details.

2.7.4. Definition. Let $\mathbf{X}$ be one of the symplectic mixed Shimura data with natural representation of $P_{\mathbf{X}}$ on $L_{\mathbb{Z}_{(p)}}$. Furthermore for each $\gamma \in P_{\mathbf{X}}(\mathbb{A}^{(\infty,p)})$ and admissible compact open subgroups K_1, K_2, such that $K_1^{\gamma} \subseteq K_2$ we have a map

$$\left[{}^{\boldsymbol{K_1}}\mathbf{X}\text{-}\boldsymbol{L}\text{-}\mathbf{mot}\right] \to \left[{}^{\boldsymbol{K_2}}\mathbf{X}\text{-}\boldsymbol{L}\text{-}\mathbf{mot}\right]$$

by multiplication of the level-structure by γ from the right.

2.7.5. For $\mathbf{X} = \mathbf{H}_{g_0}[(I \otimes I)^s, L_0 \otimes I]$ with $g_0 \neq 0$, $\left[{}^{\boldsymbol{K}}\mathbf{X}\text{-}\boldsymbol{L}\text{-}\mathbf{mot}\right]$ yields a model of the analytic mixed Shimura variety:

$$[P_{\mathbf{X}}(\mathbb{Q})\backslash \mathbb{D}_{\mathbf{X}} \times P_{\mathbf{X}}(\mathbb{A}^{(\infty)})/K].$$

The identification is given as follows. Let an object

$$(M, \dots) \in \left[\mathbf{Spec}(\mathbb{C})\text{-}{}^{\boldsymbol{K}}\mathbf{X}\text{-}\boldsymbol{L}\text{-}\mathbf{mot}\right]$$

be given. Choose an isomorphism

$$\beta\colon H^{\mathrm{B}}(M, \mathbb{Q}) \to L_{\mathbb{Q}}$$

(compatible with $(P_{\mathbf{X}}(\mathbb{Q}), L_{\mathbb{Q}})$-structures on both sides). Consider the mixed Hodge structure $F^\bullet, W_\bullet$ on $H^{\mathrm{B}}(M, \mathbb{C})$, where $F^\bullet$ comes via the comparison isomorphism

$$\rho_{\mathrm{dR,B}}\colon H^{\mathrm{dR}}(M, \mathbb{C}) \to H^{\mathrm{B}}(M, \mathbb{C})$$

from the Hodge filtration of H^{dR}. The Hodge structure $\beta(F^\bullet), \beta(W_\bullet)$ is given by a morphism $h\colon \mathbb{S}_{\mathbb{C}} \to \mathrm{GL}(L_{\mathbb{C}})$ which factors through the representation of $P_{\mathbf{X},\mathbb{C}}$ on $L_{\mathbb{C}}$ because β was chosen to be compatible with the $P_{\mathbf{X}}$-structures. The level structure ξ gives an element $\beta\rho_{\mathrm{et,B}}\xi \in P_{\mathbf{X}}(\mathbb{A}^{(\infty)})$. The class of $(h, \beta\rho_{\mathrm{et,B}}\xi)$ is well-defined.

Similarly the stack $\left[{}^{\boldsymbol{K}}\mathbf{X}\text{-}\boldsymbol{L}\text{-}\mathbf{mot\text{-}triv}\right]$ yields a model of the analytic standard principal bundle, i.e., the space

$$P_{\mathbf{X}}(\mathbb{Q})\backslash \mathbb{D}_{\mathbf{X}} \times P_{\mathbf{X}}(\mathbb{C}) \times P_{\mathbf{X}}(\mathbb{A}^{(\infty)})/K.$$

The identification is given as follows. An object

$$(M, \dots, \gamma) \in \left[\mathbf{Spec}(\mathbb{C})\text{-}{}^K\mathbf{X}\text{-}\boldsymbol{L}\text{-}\mathbf{mot}\text{-}\mathbf{triv}\right],$$

where $(M, \dots)$ is as before, and $\gamma \colon L_{\mathbb{C}} \to H_1^{\mathrm{dR}}(M)$ is a trivialization compatible with the $(P_{\mathbf{X}}, L_{\mathbb{Z}_{(p)}})$-structure, is mapped to the (well-defined) point

$$(h, \beta \rho_{\mathrm{dR,B}} \gamma, \beta \rho_{\mathrm{et,B}} \xi).$$

If $g_0 = 0$, but $I \neq 0$, i.e,. in the case of the mixed Shimura datum $\mathbf{X} = \mathbf{H}_0[(I \otimes I)^s, 0]$, we have two descriptions as a moduli problem, according to the representations $L = I \oplus I^*$ or $L' = \mathbb{Z}_{(p)} \oplus (I \otimes I)^s$.

In the first case (representation L) $\mathbb{D}_{\mathbf{X}}$ is given as the product of the set of Hodge structures as above and $\mathbb{D}_{\mathbf{H}_0} = \mathrm{Hom}(\mathbb{Z}, \mathbb{Z}(1))$. We get the point

$$(\alpha, h, \beta \rho_{\mathrm{dR,B}} \gamma, \beta \rho_{\mathrm{et,B}} \xi)$$

where α is determined as follows: The chosen isomorphism $\beta \colon H_1(M, \mathbb{Q}) \cong L_{\mathbb{Q}}$, pins down a symplectic form on $H_1(M, \mathbb{Q})$. The comparison with the natural $\mathbb{Q}(1)$-valued form on $H_1(M, \mathbb{Q})$ (they coincide up to scalar) induces an isomorphism $\mathbb{Q} \to \mathbb{Q}(1)$ of the form $q\alpha$ with $q \in \mathbb{Q}_{>0}$.

In the second case (representation L')—and this even includes $I = 0$—the isomorphism $W_{-1}(\beta)^{-1} \colon \mathbb{Q} \cong H_1(\mathbb{G}_{\mathrm{m}}, \mathbb{Q}) = \mathbb{Q}(1)$ is of the form $q\alpha$, as before.

This determines in both cases an $\alpha \in \mathrm{Hom}(\mathbb{Z}, \mathbb{Z}(1))$.

The following theorem is due to Riemann, Mumford, Artin, Deligne, etc. in the pure case, is due to Pink in the mixed case (rational case), and was extended to the integral mixed case in the thesis of the author [25]:

2.7.6. Theorem. *For each of the p-integral mixed Shimura data of symplectic type* $\mathbf{X}$ *and every admissible compact open subgroup* $K \subset P_{\mathbf{X}}(\mathbb{A}^{(\infty)})$

$$\left[{}^K\mathbf{X}\text{-}\boldsymbol{L}\text{-}\mathbf{mot}\right],$$

considered as a model of the analytic mixed Shimura variety via construction 2.7.5, *is representable by the canonical model* $\mathrm{M}({}^K\mathbf{X})$ (*compatible with the* $P_{\mathbf{X}}(\mathbb{A}^{(\infty,p)})$*-action) which is a smooth Deligne–Mumford stack of finite type over* $\mathrm{Spec}(\mathbb{Z}_{(p)})$. *It is a quasi-projective scheme, if* K *is neat.*

The stack

$$\left[{}^K\mathbf{X}\text{-}\boldsymbol{L}\text{-}\mathbf{mot}\text{-}\mathbf{triv}\right]$$

is representable by a right $P_{\mathbf{X}}$*-torsor over* $\left[{}^K\mathbf{X}\text{-}\boldsymbol{L}\text{-}\mathbf{mot}\right]$.

2.7.7. Note that the torsor $\left[{}^K\mathbf{X}\text{-}\boldsymbol{L}\text{-}\mathbf{mot}\text{-}\mathbf{triv}\right]$ was used to "normalize" the theory of the of the morphism Ξ in 2.5.14. Recall (2.5.15) that Ξ is given by a right $P_{\mathbf{X}}$-torsor $\mathrm{P}({}^K\mathbf{X})$ over $\mathrm{M}({}^K\mathbf{X})$ and a $P_{\mathbf{X}}$-equivariant morphism $\mathrm{P}({}^K\mathbf{X}) \to \mathrm{M}^\vee(\mathbf{X})$. The "normalization" requires that, for $\mathbf{X}$ one of the symplectic Shimura data, $\mathrm{P}({}^K\mathbf{X})$ represents

$$\left[{}^K\mathbf{X}\text{-}\boldsymbol{L}\text{-}\mathbf{mot}\text{-}\mathbf{triv}\right]$$

and that the equivariant morphism be given by transport of the Hodge filtration to $L_{\mathcal{O}_S}$ by means of the trivialization β (and then taking its fixing quasi-parabolic, cf. 2.5.12(2)). See also Section 1.5.

We have the following 'integral' variants of the stacks of Definition 2.7.3:

2.7.8. Definition (Categories of 1-motives; integral variant). Let $N \in \mathbb{N}$.

(1) Let $L_{0,\mathbb{Z}}$ be a lattice with unimodular symplectic form of dimension $2g_0$. Let $I_{\mathbb{Z}}$ be another lattice and $L_{\mathbb{Z}} = I_{\mathbb{Z}}^* \oplus L_{0,\mathbb{Z}} \oplus I_{\mathbb{Z}}$ as above. Assume $L_{\mathbb{Z}} \neq 0$. We define

$$\left[{}^N\mathbf{H}_{g_0}[(\boldsymbol{I}_{\mathbb{Z}} \otimes \boldsymbol{I}_{\mathbb{Z}})^s, \boldsymbol{L}_{0,\mathbb{Z}} \otimes \boldsymbol{I}_{\mathbb{Z}}]\text{-}\boldsymbol{L}_{\mathbb{Z}}\text{-}\mathbf{mot}\right]$$

as the stack of the following data:

(a) a scheme S over $\mathrm{Spec}(\mathbb{Z}[1/N])$,
(b) a 1-motive $M = (A, A^\vee, \underline{X}, \underline{Y}, \alpha, \alpha^\vee, \nu)$ over S, where $\dim(A) = g$,
(c) a principal polarization

$$\psi_1 \colon A \to A^\vee$$

and an isomorphism

$$\psi_2 \colon \underline{X} \to \underline{Y}$$

such that they induce a polarization of M,
(d) an isomorphism $\rho\colon (I_{\mathbb{Z}}^*)_S \cong \underline{Y}$,
(e) a level-N-structure

$$\xi \in \mathrm{Iso}_{(\mathrm{USp}(L_{0,\mathbb{Z}/N\mathbb{Z}}, I_{\mathbb{Z}/N\mathbb{Z}}), L_{\mathbb{Z}/N\mathbb{Z}})}\left(L_{\mathbb{Z}/N\mathbb{Z}}, H^{\mathrm{et}}(M, \mathbb{Z}/N\mathbb{Z})\right).$$

This means that the isomorphisms have to be compatible with the $(\mathrm{USp}(L_{0,\mathbb{Z}/N\mathbb{Z}}, I_{\mathbb{Z}/N\mathbb{Z}}), L_{\mathbb{Z}/N\mathbb{Z}})$-structure (see the adelic case) on both parameters.

Morphisms are as in the adelic case with the only difference, that the $\mathbb{Z}_{(p)}$-morphism of the 1-motives has to be an isomorphism. (*Note*: In particular ψ_Y is determined by the respective ρ's and ψ_X by them and the polarization.)

(2) We define the stack

$$\left[{}^N\mathbf{H}_{g_0}[(\boldsymbol{I}_{\mathbb{Z}} \otimes \boldsymbol{I}_{\mathbb{Z}})^s, \boldsymbol{L}_{0,\mathbb{Z}} \otimes \boldsymbol{I}_{\mathbb{Z}}]\text{-}\boldsymbol{L}_{\mathbb{Z}}\text{-}\mathbf{mot}\text{-}\mathbf{triv}\right]$$

as before with a trivialization of de Rham.

(3) Let $L'_{\mathbb{Z}} = \mathbb{Z} \oplus U_{\mathbb{Z}}^*$ be as in 2.7.3(3). We define the stack

$$\left[{}^N\mathbf{H}_0[\boldsymbol{U}_{\mathbb{Z}}, 0]\text{-}\boldsymbol{L}'_{\mathbb{Z}}\text{-}\mathbf{mot}\right]$$

as the category of the following data:

(a) a scheme S over $\mathrm{Spec}(\mathbb{Z}[1/N])$,
(b) a 1-motive of the form $M = [\underline{Y} \to T]$,
(c) an isomorphism $\rho\colon (U_{\mathbb{Z}}^*)_S \cong \underline{Y}$,
(d) an isomorphism $\rho'\colon (\mathbb{Z})_S \cong \underline{X}$,
(e) a level-N-structure

$$\xi \in \underline{\mathrm{Iso}}_{(P_{\mathbf{X}}, L'_{\mathbb{Z}/N\mathbb{Z}})}\left(L'_{\mathbb{Z}/N\mathbb{Z}}, H^{\mathrm{et}}(M, \mathbb{Z}/N\mathbb{Z})\right),$$

where the isomorphisms in the etale sheaf on the right have to be compatible with the obvious $(P_{\mathbf{X}}, L'_{\mathbb{Z}/N\mathbb{Z}})$-structure on both parameters. Morphisms are as in the adelic case, with the only difference that the $\mathbb{Z}_{(p)}$-morphism of the 1-motives has to be an isomorphism.

(4) We define the stack

$$\left[{}^N\mathbf{H}_0[\boldsymbol{U}_{\mathbb{Z}}, 0]\text{-}\boldsymbol{L}'_{\mathbb{Z}}\text{-}\mathbf{mot}\text{-}\mathbf{triv}\right]$$

as before with a trivialization of de Rham.

(5) We define the stack

$$\left[\,{}^{N,0}\mathbf{H}_0[U_{\mathbb{Z}},0]\text{-}L'_{\mathbb{Z}}\text{-}\mathbf{mot}\,\right],$$

where $L'_{\mathbb{Z}} = \mathbb{Z} \oplus U^*_{\mathbb{Z}}$, as the stack of data as in (3), but with level structure

$$\xi \in \underline{\mathrm{Iso}}\big(\mathbb{Z}/N\mathbb{Z}, H^{\mathrm{et}}(\mathrm{gr}_{-2} M, \mathbb{Z}/N\mathbb{Z})\big),$$

only.

(6) We define the stack

$$\left[\,{}^{N,0}\mathbf{H}_0[U_{\mathbb{Z}},0]\text{-}L'_{\mathbb{Z}}\text{-}\mathbf{mot\text{-}triv}\,\right]$$

as before with a trivialization of de Rham.

2.7.9. Remark. For $U = (I \otimes I)^s \neq 0$ recall the two representations of $\mathrm{USp}(0,I)$: $L = I^* \oplus I$ and $L' = \mathbb{Z}_{(p)} \oplus U^*$. We explicitly describe the "change-of-representation equivalences"

$$\left[\,{}^{K}\mathbf{H}_0[(I \otimes I)^s,0]\text{-}L\text{-}\mathbf{mot}\,\right] \to \left[\,{}^{K}\mathbf{H}_0[U,0]\text{-}L'\text{-}\mathbf{mot}\,\right]$$

and

$$\left[\,{}^{K}\mathbf{H}_0[(I \otimes I)^s,0]\text{-}L\text{-}\mathbf{mot\text{-}triv}\,\right] \to \left[\,{}^{K}\mathbf{H}_0[U,0]\text{-}L'\text{-}\mathbf{mot\text{-}triv}\,\right]$$

as follows: Let an object $(M = (0,0,\underline{X},\underline{Y},0,0,\nu),0,\psi_2,\rho,\xi)$ over S be given. ν can, via polarization and via ρ, be considered as a morphism $\mathrm{Sym}^2(I^*)_S \to \mathbb{G}_{\mathrm{m},S}$. This gives rise to a motive M'' and a ρ required for the second category (ρ' is given by the canonical identification $\mathrm{Hom}(\mathbb{G}_{\mathrm{m}},\mathbb{G}_{\mathrm{m}}) \cong \mathbb{Z}$). The level structures translate as follows: It is easy to check that the realizations $H^{\mathrm{et}}(M'')$, $H^{\mathrm{dR}}(M'')$ and $H^{\mathrm{B}}(M'')$ (when defined), together with their $\big(\mathrm{USp}(0,I), \mathbb{Z}_{(p)} \oplus \mathrm{Sym}^2(I^*)\big)$-structures, are obtained from the corresponding realizations of M with their $(\mathrm{USp}(0,I), I \oplus I^*)$-structures by the procedure described in 2.3.13. Since the procedure described in 2.3.13 applied to $I \oplus I^*$ gives $\mathbb{Z}_{(p)} \oplus \mathrm{Sym}^2(I^*)$ canonically, we may translate the K-level structure or trivialization of H^{dR} accordingly.

2.7.10. Proposition. *Let S be a scheme over $\mathrm{Spec}(\mathbb{Z}_{(p)})$ and let $N \in \mathbb{N}$ be given with $(N,p) = 1$. For each of the p-integral mixed symplectic Shimura data as above, we have equivalences*

$$\left[\,S\text{-}{}^{N}\mathbf{X}\text{-}L_{\mathbb{Z}}\text{-}\mathbf{mot}\,\right] \to \left[\,S\text{-}{}^{K(N)}\mathbf{X}\text{-}L\text{-}\mathbf{mot}\,\right],$$

where

$$K(N) = \{g \in P_{\mathbf{X}}(\mathbb{A}^{(\infty)}) \mid gL_{\hat{\mathbb{Z}}} = L_{\hat{\mathbb{Z}}}, g \equiv \mathrm{id} \quad (N)\}.$$

We have also a similar equivalences for $\mathbf{X} = \mathbf{H}_0[U,0]$ and the representation L', as well as

$$\left[\,S\text{-}{}^{N,0}\mathbf{X}\text{-}L'_{\mathbb{Z}}\text{-}\mathbf{mot}\,\right] \to \left[\,S\text{-}{}^{U_{\hat{\mathbb{Z}}} \rtimes K(\mathbb{G}_{\mathrm{m}},N)}\mathbf{X}\text{-}L'\text{-}\mathbf{mot}\,\right].$$

2.7.11. Remark. Since the $K(N)$ form a cofinal system of compact open groups, the integral categories give a complete description of the canonical models. In the case $\mathbf{X} = \mathbf{H}_0[U,0]$, each $K(N)$ is conjugated to $U_{\hat{\mathbb{Z}}} \rtimes K(\mathbb{G}_{\mathrm{m}},N)$, therefore already the integral categories with $(N,0)$-level structure suffice to describe the canonical models.

PROOF OF PROPOSITION 2.7.10. We will prove the statement for the case $\mathbf{H}_{g_0}[(I_{\mathbb{Z}} \otimes I_{\mathbb{Z}})^s, L_{0,\mathbb{Z}} \otimes I_{\mathbb{Z}}]$, where $I_{\mathbb{Z}} \neq 0$, $L_{0,\mathbb{Z}} \neq 0$—the other cases are degenerate special cases of this construction. We may assume that S is connected.

We first describe the functor. Let $[M, \psi_1, \psi_2, \rho, \xi]$ be an object of $\left[{}^N\mathbf{X}\text{-}\boldsymbol{L}_{\mathbb{Z}}\text{-}\mathbf{mot}\right]$ over S. Choose a geometric point $\bar{s}$. ξ can be considered as an isomorphism

$$\xi\colon L_{\mathbb{Z}/N\mathbb{Z}} \to H^{\mathrm{et}}(M_{\bar{s}}, \mathbb{Z}/N\mathbb{Z})$$

which is invariant under the action of $\pi_1^{\mathrm{et}}(S, \bar{s})$, i.e., $H^{\mathrm{et}}(M, \mathbb{Z}/N\mathbb{Z})$ has to be constant. Choose some isomorphism

$$\delta\colon H^{\mathrm{et}}(M_{\bar{s}}, \widehat{\mathbb{Z}}^{(p)}) \to L_{\widehat{\mathbb{Z}}^{(p)}}$$

compatible with USp-structures. Composing the reduction modulo N with ξ, we get an element of $\mathrm{USp}(L_{0,\mathbb{Z}/N\mathbb{Z}}, I_{\mathbb{Z}/N\mathbb{Z}})$. Since USp is a smooth group scheme over $\mathbb{Z}$, by Hensel's Lemma we get a lift to $\mathrm{USp}(L_{\widehat{\mathbb{Z}}^{(p)}})$. Taking composition again with δ^{-1} we get a morphism

$$\xi'\colon L_{\widehat{\mathbb{Z}}^{(p)}} \to H^{\mathrm{et}}(M_{\bar{s}}, \widehat{\mathbb{Z}}^{(p)})$$

that reduces to ξ modulo N, and which is well-defined modulo $K(N)$ and its class is, by construction, invariant under $\pi_1^{\mathrm{et}}(S, \bar{s})$. This functor is faithful. It is full because a p-morphism of 1-motives which induces an isomorphism $H^{\mathrm{et}}(M_{\bar{s}}, \widehat{\mathbb{Z}}^{(p)}) \cong H^{\mathrm{et}}(M'_{\bar{s}}, \widehat{\mathbb{Z}}^{(p)})$ must be a morphism. This follows from [25, Theorem 5.1.7].

Let on the other hand $[M, \psi_1, \psi_2, \rho, \xi']$ be an object of $\left[{}^{K(N)}\mathbf{X}\text{-}\boldsymbol{L}\text{-}\mathbf{mot}\right]$ over S. Choose a geometric point $\bar{s}$. If ξ' is represented by an isomorphism

$$L_{\widehat{\mathbb{Z}}^{(p)}} \to H^{\mathrm{et}}(M_{\bar{s}}, \widehat{\mathbb{Z}}^{(p)}) \tag{2.6}$$

then the object is in the image of the functor because of the following remarks:

(1) ξ' is actually a lift of a ξ as above.
(2) There is a principal polarization in the class of ψ'. For there is a $d \in \mathbb{A}^{(\infty,p)}$ such that Ψ induces an isomorphism

$$H^{\mathrm{et}}(M_{\bar{s}}, \widehat{\mathbb{Z}}^{(p)}) \to dH^{\mathrm{et}}(M_{\bar{s}}^{\vee}, \widehat{\mathbb{Z}}^{(p)})$$

because ξ' is a morphism of USp-structures, hence a symplectic similitude. So we get a principal polarization by changing ψ' by $+d$ or $-d$, which must lie in ${\mathbb{Z}_{(p)}}^*$. Only one sign leads to a *polarization*.
(3) ρ maps $I_{\mathbb{Z}}^*$ to $\underline{Y}$.

If ξ' is not represented by an isomorphism like in (2.6), we have to show that there exists an isogenous object with this property.

Composing with a p-isogeny in $\mathbb{Z}_{(p)} \setminus \{0\}$, we may assume that there is some c such that $p \nmid c$, and a diagram

$$\begin{array}{ccccccccc}
0 & \longrightarrow & H^{\mathrm{et}}(M_{\bar{s}}, \widehat{\mathbb{Z}}^{(p)}) & \xhookrightarrow{\xi^{-1}} & L_{\widehat{\mathbb{Z}}^{(p)}} & \longrightarrow & K_{\bar{s}} & \longrightarrow & 0 \\
 & & \| & & \downarrow & & \downarrow & & \\
0 & \longrightarrow & H^{\mathrm{et}}(M_{\bar{s}}, \widehat{\mathbb{Z}}^{(p)}) & \xrightarrow{[c]} & H^{\mathrm{et}}(M_{\bar{s}}, \widehat{\mathbb{Z}}^{(p)}) & \longrightarrow & \ker([c]) & \longrightarrow & 0.
\end{array}$$

Furthermore the action of $\pi_1^{\mathrm{et}}(S, \bar{s})$ induces one on $K_{\bar{s}}$, hence there is a finite etale group scheme $K \subset \ker([c])$ with fibre $K_{\bar{s}}$, and we have [25, Theorem 5.1.8] an isogeny $\psi\colon M \to M'$ with $\ker(\psi) = K$ hence a diagram

$$\begin{array}{ccccccccc} 0 & \longrightarrow & H^{\mathrm{et}}(M_{\bar{s}}, \widehat{\mathbb{Z}}^{(p)}) & \xhookrightarrow{\xi^{-1}} & L_{\widehat{\mathbb{Z}}^{(p)}} & \longrightarrow & K_{\bar{s}} & \longrightarrow & 0 \\ & & \| & & \uparrow (\xi')^{-1} & & \| & & \\ 0 & \longrightarrow & H^{\mathrm{et}}(M_{\bar{s}}, \widehat{\mathbb{Z}}^{(p)}) & \xrightarrow{\psi} & H^{\mathrm{et}}(M'_{\bar{s}}, \widehat{\mathbb{Z}}^{(p)}) & \longrightarrow & K_{\bar{s}} & \longrightarrow & 0. \end{array}$$

There is an ρ' because ξ is a morphism of USp-structures and hence ψ_Y has to be an isomorphism of $\mathbb{Z}$-lattices! Therefore we get an isogenous object with the property that ξ' is an isomorphism as in (2.6). □

2.7.12. Let U be any $\mathbb{Z}_{(p)}$-lattice. Consider the mixed Shimura datum $\mathbf{X} = \mathbf{H}_0[U, 0]$. It plays the mayor role in the description of point-like boundary components of many pure Shimura varieties, in particular of those symplectic and also orthogonal type (cf. 2.3.5 and 2.4.5).

Consider the category (cf. Definition 2.7.3(3))

$$\left[{}^{K}\mathbf{X}\text{-}\boldsymbol{U}^* \oplus \mathbb{Z}_{(\boldsymbol{p})}\text{-}\mathbf{mot}\right]$$

where K is any admissible compact open subgroup of $P_{\mathbf{H}_0[U,0]}(\mathbb{A}^{(\infty)})$.

Recall that, by definition,

$$P_{\mathbf{H}_0[U,0]} = \mathbb{W}(U) \rtimes \mathbb{G}_{\mathrm{m}} \tag{2.7}$$

and that K contains the subgroup $U_{\widehat{\mathbb{Z}}} \rtimes K(M)$ for some $M \in \mathbb{Z}$ and some (saturated at p) lattice $U_{\mathbb{Z}} \subset U$. The splitting (2.7) determines and is determined by a $\mathbb{Z}_{(p)}$-valued point v of the dual $\mathrm{M}^\vee(\mathbf{X})$. (The preimage of $\mathbb{G}_{\mathrm{m}}$ under the splitting is the quasi-parabolic corresponding to the point.)

The category

$$\left[{}^{U_{\widehat{\mathbb{Z}}} \rtimes K(M)}\mathbf{H}_0[\boldsymbol{U}, \mathbf{0}]\text{-}\boldsymbol{U}^* \oplus \mathbb{Z}_{(\boldsymbol{p})}\text{-}\mathbf{mot}\right]$$

is isomorphic to the integral category

$$\left[{}^{(\boldsymbol{M},\mathbf{0})}\mathbf{H}_0[\boldsymbol{U}, \mathbf{0}]\text{-}\boldsymbol{U}^*_{\mathbb{Z}} \oplus \mathbb{Z}\text{-}\mathbf{mot}\right],$$

see 2.7.10. An object in the latter is isomorphic to (M, ξ), where M is of the form

$$[\mu\colon U^*_{\mathbb{Z}} \to \mathbb{G}_{\mathrm{m}}],$$

with a level structure $\xi\colon \mathbb{Z}/M\mathbb{Z} \to H^{\mathrm{et}}_1(\mathbb{G}_{\mathrm{m}}, \mathbb{Z}/M\mathbb{Z}) \cong \mathbb{G}_{\mathrm{m}}[M]$ only.

We choose an isomorphism

$$\mathbb{D}_{\mathbf{H}_0} \times U_{\mathbb{C}} \to \mathbb{D}_{\mathbf{H}_0[U,0]}$$

such that an element (α, u) is mapped to $\big(\alpha, u \circ h_0 \circ (-u)\big)$, where $u \circ h_0 \circ (-u)$ is the u-conjugate (from the left) of the morphism $h_0\colon \mathbb{S}_{\mathbb{C}} \to P_{\mathbf{H}_0[U,0]} = \mathbb{W}(U) \rtimes \mathbb{G}_{\mathrm{m}}$ mapping z to $z\bar{z}$ in the second factor. The associated mixed Hodge structure is the u-translate of the one given by $F^0 = U_{\mathbb{C}}$.

The following Proposition—which is quite tautological—describes the associated canonical models. It will be used in the proof of the q-expansion principle (Theorem 2.8.7).

2.7.13. Proposition. *The S-valued points of the dual* $\mathrm{M}^\vee(\mathbf{X})(S)$ *are in natural bijection with the set of splittings* $P_{\mathbf{X},S} \cong \mathbb{W}(U_S) \rtimes \mathbb{G}_{\mathrm{m},S}$.

We have a commutative diagram (where all maps are functorial in S)

$$\begin{array}{ccc}
\left[\boldsymbol{S}\text{-}^{(M,0)}\mathbf{X}\text{-}\mathbb{Z} \oplus \boldsymbol{U}_{\mathbb{Z}}^*\text{-}\mathbf{mot\text{-}triv}\right] & \xrightarrow[i_3]{\sim} & (P \times_{\mathbb{Z}_{(p)}} \mathbb{G}_{\mathrm{m},\mathbb{Z}_{(p)}(\zeta_M)} \otimes U_{\mathbb{Z}})(S) \\
p \downarrow \uparrow \text{can.} & & p \downarrow \uparrow \text{can.} \\
\left[\boldsymbol{S}\text{-}^{(M,0)}\mathbf{X}\text{-}\mathbb{Z} \oplus \boldsymbol{U}_{\mathbb{Z}}^*\text{-}\mathbf{mot}\right] & \xrightarrow[i_2]{\sim} & (\mathbb{G}_{\mathrm{m},\mathbb{Z}_{(p)}(\zeta_M)} \otimes U_{\mathbb{Z}})(S) \\
p \downarrow & & p \downarrow \\
\left[\boldsymbol{S}\text{-}^{M}\mathbf{H}_0\text{-}\mathbb{Z}\text{-}\mathbf{mot}\right] & \xrightarrow[i_1]{\sim} & \mathrm{Spec}(\mathbb{Z}_{(p)}(\zeta_M))(S)
\end{array} \tag{2.8}$$

where ζ_M an (abstract) primitive Mth root of unity. Here the maps i_2 define a morphism of tori, and the maps i_3 define a morphism of right P_S-torsors.

Over $\mathbb{C}$, the identifications as in 2.7.5 *render this into the following commutative diagram*

$$\begin{array}{ccc}
P(\mathbb{Q})\backslash\mathbb{D}_{\mathbf{H}_0} \times U_{\mathbb{C}} \times P(\mathbb{C}) \times P(\mathbb{A}^{(\infty)})/K & \xrightarrow[j_3]{\sim} & (P \times_{\mathbb{Z}_{(p)}} \mathbb{G}_{\mathrm{m},\mathbb{Z}_{(p)}(\zeta_M)} \otimes U_{\mathbb{Z}})(\mathbb{C}) \\
p \downarrow \uparrow \text{can.} & & p \downarrow \uparrow \text{can.} \\
P(\mathbb{Q})\backslash\mathbb{D}_{\mathbf{H}_0} \times U_{\mathbb{C}} \times P(\mathbb{A}^{(\infty)})/K & \xrightarrow[j_2]{\sim} & (\mathbb{G}_{\mathrm{m},\mathbb{Z}_{(p)}(\zeta_M)} \otimes U_{\mathbb{Z}})(\mathbb{C}) \\
p \downarrow & & p \downarrow \\
\mathbb{Q}^*\backslash\mathbb{D}_{\mathbf{H}_0} \times (\mathbb{A}^{(\infty)})^*/K(M) & \xrightarrow[j_1]{\sim} & \mathrm{Spec}(\mathbb{Z}_{(p)}(\zeta_M))(\mathbb{C})
\end{array} \tag{2.9}$$

where $K = U_{\widehat{\mathbb{Z}}} \rtimes K(M)$ w.r.t. the splitting (2.7). *The morphism* can. *on the left-hand side is given as follows:*

$$[\alpha, \mu_{\log}, \rho] \mapsto [\alpha, \mu_{\log}, \mu_{\log}\alpha^{-1}(1), \rho]$$

where ρ must be chosen to be in $U_{\widehat{\mathbb{Z}}} \rtimes K(1)$ or even in $K(1) \cong \widehat{\mathbb{Z}}^*$ (this amounts to the same).

This yields a description of the canonical integral models $\mathrm{M}({}^K\mathbf{X}), \Xi({}^K\mathbf{X})$ *and* $\mathrm{M}({}^{K(M)}\mathbf{H}_0)$, *respectively.*

2.7.14. The lower morphisms p describe in all diagrams obvious structures of a relative group object, which are the same as those induced by the unipotent extension morphism of Shimura data $\mathbf{H}_0[U, 0] \to \mathbf{H}_0$, which is a group object (with the unit section given by the splitting γ, chosen above). The trivialization can. extends to partial compactifications, which are given by torus embeddings of $\mathbb{G}_{\mathrm{m},\mathbb{Z}_{(p)}[\zeta_M]} \otimes U_{\mathbb{Z}}$ over $\mathrm{Spec}(\mathbb{Z}_{(p)}[\zeta_M])$.

More precisely: By the arithmeticity condition and left invariance under conjugation by $P_{\mathbf{X}}(\mathbb{Q})$, a rational (w.r.t. $U_{\mathbb{Q}}$) polyhedral cone decomposition of $U_{\mathbb{R}}$ into cones $\{\sigma\}$ determines a rational polyhedral cone decomposition Δ of the conical complex (here $\mathbf{H}_0 \times U_{\mathbb{R}}(-1)$). On $\alpha \times U_{\mathbb{R}}(-1)$, the decomposition is given by $\{\alpha^{-1}(o)\}$.

Hence, for each such σ the corresponding torus embedding is

$$\mathrm{Spec}(\mathbb{Z}_{(p)}[\zeta_M][U_{\mathbb{Z}}^*]) \hookrightarrow \mathrm{Spec}(\mathbb{Z}_{(p)}[\zeta_M][U_{\mathbb{Z}}^* \cap \sigma^\vee]).$$

Those embeddings are glued in the usual way using the topology of the decomposition.

PROOF OF PROPOSITION 2.7.13. We begin by describing the horizontal morphisms in the diagram (2.8):

i_1: Up to isomorphism any element in the category

$$\big[\, \boldsymbol{S}\text{-}^{\boldsymbol{M}}\mathbf{H}_0\text{-}\mathbb{Z}\text{-}\mathbf{mot} \,\big]$$

is of the form $X = (\mathbb{G}_{\mathrm{m},S}, 0, \mathrm{id}, \xi)$ with

$$\xi\colon (\mathbb{Z}/N\mathbb{Z})_S \to \mathbb{G}_{\mathrm{m}}[M]_S.$$

$i_1(X)$ is the map determined by $1 \in \mathbb{Z}/N\mathbb{Z}$, which determines the level-structure uniquely.

i_2: Up to isomorphism any element in the category

$$\big[\, \boldsymbol{S}\text{-}^{(\boldsymbol{M},\mathbf{0})}\mathbf{H}_0[\boldsymbol{U}_{\mathbb{Z}}, \mathbf{0}]\text{-}\boldsymbol{L}_{\mathbb{Z}}\text{-}\mathbf{mot} \,\big]$$

is of the form $X = (\mu\colon (U_{\mathbb{Z}})_S \to \mathbb{G}_{\mathrm{m},S}, \mathrm{id}, \mathrm{id}, \xi)$. i_2 is defined as the fibre product of $i_1(X)$ (defined as before) with μ.

i_3: Up to isomorphism any element in the category

$$\big[\, \boldsymbol{S}\text{-}^{(\boldsymbol{M},\mathbf{0})}\mathbf{H}_0[\boldsymbol{U}_{\mathbb{Z}}, \mathbf{0}]\text{-}\boldsymbol{L}_{\mathbb{Z}}\text{-}\mathbf{mot}\text{-}\mathbf{triv} \,\big]$$

is of the form $X = (\mu\colon (U_{\mathbb{Z}})_S \to \mathbb{G}_{\mathrm{m},S}, \mathrm{id}, \mathrm{id}, \xi, \gamma)$, where $\gamma\colon L_S \to H_1^{\mathrm{dR}}(M)$ is a trivialization (compatible with the $P_{\mathbf{X},S}$-structure). We have a canonical trivialization

$$\mathrm{can.}\colon F^0\big(H^{\mathrm{dR}}(M)\big) \oplus \mathcal{O}_S\left(\frac{\mathrm{d}z}{z}\right)^* \cong H^{\mathrm{dR}}(M)$$

where $\mathcal{O}_S(\frac{\mathrm{d}z}{z})^*$ is $\mathrm{Lie}(\mathbb{G}_{\mathrm{m}})$ and $U_S^* \cong F^0\big(H^{\mathrm{dR}}(M)\big)$ (canonically via ρ) over any S. Mapping 1 to $(\mathrm{d}z/z)^*$ the domain of can. can be identified with L_S. In other words, we have

$$\mathrm{can.}\colon L_S \to H^{\mathrm{dR}}(M).$$

Equipping a point

$$X = (\mu\colon (U_{\mathbb{Z}})_S \to \mathbb{G}_{\mathrm{m},S}, \mathrm{id}, \mathrm{id}, \xi) \in \big[\, \boldsymbol{S}\text{-}^{(\boldsymbol{M},\mathbf{0})}\mathbf{H}_0[\boldsymbol{U}_{\mathbb{Z}}, \mathbf{0}]\text{-}\boldsymbol{L}_{\mathbb{Z}}\text{-}\mathbf{mot} \,\big]$$

with the trivialization can. defines the vertical map can. on the left hand side in (2.8). The map can. on the right-hand side is the obvious trivialization. The map i_3 is defined as the only $P_{\mathbf{X},S}$-equivariant map that makes the diagram commutative: it maps X to

$$\big(\mathrm{can.}^{-1} \circ \gamma, i_2(X)\big) \in P_{\mathbf{X}}(S) \times (\mathbb{G}_{\mathrm{m},\mathbb{Z}_{(p)}[\zeta_M]} \otimes U_{\mathbb{Z}})(S).$$

We now describe the resulting horizontal morphisms in diagram (2.9):

j_1: We choose an isomorphism

$$\beta\colon H_1^{\mathrm{B}}(\mathbb{G}_{\mathrm{m}}, \mathbb{Z}) \to \mathbb{Z}.$$

Since $H_1^{\mathrm{B}}(\mathbb{G}_{\mathrm{m}}, \mathbb{Z}) = \mathbb{Z}(1)$, we have an $\alpha \in \mathbb{D}_{\mathbf{H}_0}$ with $\alpha^{-1} = \beta$. Let ξ be a level structure. Its image in $\widehat{\mathbb{Z}}^*$ is given modulo $K(M)$ by the composite (see 2.7.5):

$$\mathbb{Z}/N\mathbb{Z} \xrightarrow{\ \xi\ } H_1^{\mathrm{et}}(\mathbb{G}_{\mathrm{m}}, \mathbb{Z}/N\mathbb{Z}) = \mathbb{G}_{\mathrm{m}}[N](\mathbb{C}) \xrightarrow{\rho_{\mathrm{et,B}}} H_1^{\mathrm{B}}(\mathbb{G}_{\mathrm{m}}, \mathbb{Z}/N\mathbb{Z}) \xrightarrow{\ \beta\ } \mathbb{Z}/N\mathbb{Z}\,.$$

Note that $\rho_{\mathrm{et,B}}$ is given by the logarithm. We get an element

$$[\alpha, \rho] \in \mathbb{Q}^* \backslash \mathbb{D}_{\mathbf{H}_0} \times \mathbb{A}^{(\infty)^*}/K(M)$$

where $\rho \in K(1)$ and

$$j_1([\alpha,\rho]) := \exp\left(\alpha\left(\frac{1}{M}\rho\right)\right) \in \mathbb{G}_{\mathrm{m}}(\mathbb{C})[M].$$

j_2: Choose an isomorphism $\beta\colon H^{\mathrm{B}}(M) \to \mathbb{Z} \oplus U^*_{\mathbb{Z}}$ (in the $P_{\mathbf{X}}(\mathbb{Z})$-structure) such that $W_{-1}(\beta) = \alpha^{-1}$ as morphisms $H_1^{\mathrm{B}}(\mathbb{G}_{\mathrm{m}}) = \mathbb{Z}(1) \to \mathbb{Z}$.

Recall that $L_{\mathbb{Z}} = \mathbb{Z} \oplus U^*_{\mathbb{Z}}$. The composite

$$L_{\mathbb{C}} \xrightarrow{\text{can.}} H^{\mathrm{dR}}(M) \xrightarrow{\rho_{dR,B}} H^B(M) \otimes \mathbb{C} \xrightarrow{\beta} L_{\mathbb{C}}\ .$$

maps $(1,0)$ to $(\alpha^{-1}(1),0)$ and $(0,u^*)$ to $\Big(-\alpha^{-1}\big(\log(\mu(u^*))\big),u^*\Big)$. This is the same as the operation of $\mu_{\log}\alpha^{-1}(1) \in P_{\mathbf{X}}(\mathbb{C})$ under the action defined in 2.7.3(4), where $\mu_{\log} \in U_{\mathbb{C}}/U_{\mathbb{Z}}$ is such that $u^*(\mu_{\log}) = \alpha^{-1}\big(\log(\mu(u^*))\big)$ for all $u^* \in U^*_{\mathbb{Z}}$.

We have then $\beta\big(F^0(M)\big) = \mu_{\log}U^*_{\mathbb{C}}$. So the moduli point of

$$([\mu\colon U^*_{\mathbb{Z}} \to \mathbb{G}_{\mathrm{m}}], \mathrm{id}, \mathrm{id}, \xi)$$

in the complex analytic Shimura variety is:

$$[\alpha, \mu_{\log}, \rho] \in P(\mathbb{Q})\backslash\mathbb{D}_{\mathbf{H}_0} \times U_{\mathbb{C}} \times P_{\mathbf{X}}(\mathbb{A}^{(\infty)})/K$$

j_3: The image of the point $[\alpha,\mu_{\log},\rho]$ via can. in the standard principal bundle is given by

$$[\alpha, \mu_{\log}, \mu_{\log}\alpha(1)^{-1}, \rho] \in P(\mathbb{Q})\backslash\mathbb{D}_{\mathbf{H}_0} \times U_{\mathbb{C}} \times P_{\mathbf{X}}(\mathbb{C}) \times P_{\mathbf{X}}(\mathbb{A}^{(\infty)})/K,$$

therefore the trivialization can. on the left in diagram (2.9) is given by

$$\begin{aligned} P(\mathbb{Q})\backslash\mathbb{D}_{\mathbf{H}_0} \times U_{\mathbb{C}} \times P(\mathbb{A}^{(\infty)})/K &\longrightarrow P(\mathbb{Q})\backslash\mathbb{D}_{\mathbf{H}_0} \times U_{\mathbb{C}} \times P(\mathbb{C}) \times P(\mathbb{A}^{(\infty)})/K \\ [\alpha,\mu_{\log},\rho] &\longmapsto [\alpha,\mu_{\log},\mu_{\log}\alpha(1)^{-1},\rho] \end{aligned}$$

for $\rho \in U_{\widehat{\mathbb{Z}}}K(1)$ (!), as claimed. This also defines the morphism j_3 by $P_{\mathbf{X}}(\mathbb{C})$-equivariance.

It is easy to see that this morphism can. is compatible with the action of the torus multiplication in the relative torus $\mathrm{M}({}^K\mathbf{X})$ over $\mathrm{M}({}^{K(N)}\mathbf{H}_0)$. It is hence by the characterization 2.5.14(3) the trivialization which extends to every partial compactification $\mathrm{M}({}^K_\Delta\mathbf{X})$.

The model of the Shimura variety (and the model of the standard principal bundle, respectively) described by the schemes representing

$$\left[{}^K\mathbf{H}_0[U,0]\text{-}\boldsymbol{L}\text{-}\mathbf{mot}\right], \quad \text{and} \quad \left[{}^K\mathbf{H}_0[U,0]\text{-}\boldsymbol{L}\text{-}\mathbf{mot\text{-}triv}\right],$$

respectively, are the canonical ones by 2.7.6 and 2.7.9. □

2.8. The q-expansion principle

2.8.1. From the abstract theory of canonical models of toroidal compactifications of Shimura varieties as described by the Main Theorems 2.5.9 and 2.5.14, a classical "q-expansion principle" can be derived. In this section we explain this for a wide class of Shimura varieties with boundary.

2.8.2. Resume the discussion from 2.7.12. Let $\mathcal{E}$ be a $P_{\mathbf{X}}$-equivariant vector bundle on $\mathrm{M}^{\vee}(\mathbf{X}) = \mathbb{W}(U)v$. Here v is the $\mathbb{Z}_{(p)}$-valued point corresponding to the splitting (2.7). The space $H^0(\mathrm{M}({}^K\mathbf{X})_{\mathbb{C}}, \Xi^*\mathcal{E})$ is identified with the $P_{\mathbb{Q}}$-invariant functions

$$\mathbb{D}_{\mathbf{X}} \times P(\mathbb{A}^{(\infty)})/K \to \mathcal{E}_{\mathbb{C}}$$

Choose a basis $v_1, \ldots, v_n$ of the fibre $\mathcal{E}_v$.

The canonical trivialization (can. of the Proposition 2.7.13) of the standard principal bundle (cf. 2.5.15) trivializes also $\Xi^*\mathcal{E}$. For this consider the following diagram

$$\mathrm{M}({}^K\mathbf{X}) \underset{p}{\overset{\text{can.}}{\rightleftarrows}} \mathrm{P}({}^K\mathbf{X}) \xrightarrow{\ \Pi\ } \mathrm{M}^{\vee}(\mathbf{X}).$$

Recall that $\Xi^*\mathcal{E}$ can be identified with the sheaf of P-invariants in the pullback $\Pi^*\mathcal{E}$. The morphism can. identifies those with $\text{can.}^*\,\Pi^*\mathcal{E} = (\mathcal{E}_v)_{\mathrm{M}({}^K\mathbf{X})}$. This last identification comes from the fact, that $\Pi \circ \text{can.}$ is constant with image v by construction. By Proposition 2.7.13 this trivialization, base changed to $\mathbb{C}$, is given by the functions

$$\begin{aligned} s_i \colon \mathbb{D}_{\mathbf{H}_0} \times U_{\mathbb{C}} \times P_{\mathbf{X}}(\mathbb{A}^{(\infty)})/K &\longrightarrow \mathcal{E}_{\mathbb{C}} \\ [\alpha, \mu_{\log}, \rho] &\longmapsto \mu_{\log}\alpha(1)^{-1}v_i \quad \text{for } \rho \in K_U K(1). \end{aligned}$$

(Extend these functions to other $\rho \in P_{\mathbf{X}}(\mathbb{A}^{(\infty)})$ by requiring $P_{\mathbf{X}}(\mathbb{Q})$-invariance.)

A general section may be written as

$$[\alpha, \mu_{\log}, \rho] \longmapsto \sum_{k \in U_{\mathbb{Z}}^*, i} a_{k,\alpha,\rho,i} \exp\big(\alpha(\mu_{\log}k)\big) s_i,$$

where $\rho \in K(1) = \widehat{\mathbb{Z}}^*$, and the corresponding ith coordinate-function on

$$(\mathbb{G}_{\mathrm{m},\mathbb{C}} \otimes U_{\mathbb{Z}}) \times_{\mathbb{C}} \mathbb{C}[\zeta_M] = \mathrm{Spec}(\bigoplus_{\zeta_{M,\mathbb{C}}} \mathbb{C}[U_{\mathbb{Z}}^*])$$

is given by

$$\sum_{k \in U_{\mathbb{Z}}^*} a_{k,\alpha,\rho,i}[k]$$

in the fibre above $\zeta_{M,\mathbb{C}} = \exp\Big(\alpha\big((1/M)\rho\big)\Big)$ for some M depending on K (cf. 2.7.12).

Let σ now be a maximal dimensional rational polyhedral cone of $U_{\mathbb{R}}$. The completion at the corresponding boundary stratum is given by

$$\mathrm{spf}\left(\bigoplus_{\zeta_{M,\mathbb{C}}} \mathbb{C}[\![U_{\mathbb{Z}}^* \cap \sigma^{\vee}]\!]\right).$$

If a formal function f given by

$$\sum_{k \in U_{\mathbb{Z}}^* \cap \sigma^{\vee}} a_{k,\alpha,\rho}(f)[k],$$

in the fibre above the respective $\zeta_{M,\mathbb{C}} = \exp\Big(\alpha\big((1/M)\rho\big)\Big)$, converges in a neighborhood of the boundary points, it has the form

$$[\alpha, \mu_{\log}, \rho] \mapsto \sum_{k \in U_{\mathbb{Z}}^*} a_{k,\alpha,\rho}(f) \exp\big(\alpha(\mu_{\log}k)\big),$$

there, too.

The following observation is important for the q-expansion principle: The function ${}^{\tau}f$ for an automorphism $\tau \in \mathrm{Aut}(\mathbb{C})$ has the expansion

$$\sum_{k \in U_{\mathbb{Z}}^{*} \cap \sigma^{\vee}} {}^{\tau}a_{k,\alpha,\rho}(f)[k]$$

in the fibre over ${}^{\tau}(\zeta_{M,\mathbb{C}}) = \exp\Big(\alpha\big((1/M)\rho\kappa_{\tau}\big)\Big)$, where $\kappa_{\tau} \in K(1)/K(M)$ is the image of τ under the class field theory isomorphism.

Therefore the Fourier coefficients satisfy:

$$a_{k,\alpha,\rho}({}^{\tau}f) = {}^{\tau}a_{k,\alpha,\rho\kappa_{\tau}^{-1}}(f). \tag{2.10}$$

2.8.3. We continue by discussing a formal q-expansion principle, stating roughly that—under certain hypotheses—a regular function is integral, if after completion at some point, the corresponding formal function is integral.

For a commutative ring R and ideal $I \subseteq R$, we denote by $C_I(R)$ the completion of R w.r.t. the I-adic topology. For an I-adically complete ring R and a multiplicatively closed subset $S \subset R$, we denote by $R\{S^{-1}\}$ the completion w.r.t. $I[S^{-1}]$ of the localization $R[S^{-1}]$.

Let R be an excellent normal integral domain and $I \subset R$ an ideal, such that $\mathrm{spf}\big(C_I(R)\big)$ is connected (or equivalently, such that $R/\sqrt{I}$ has no nontrivial idempotents).

Let s be a prime element of R, neither a zero divisor nor a unit of R/I. Let M be a projective R-module.

2.8.4. Lemma. *The diagram*

$$\begin{array}{ccc} M & \hookrightarrow & M[s^{-1}] \\ \big\downarrow & & \big\downarrow \\ C_I(M) & \hookrightarrow & C_I(M)\{s^{-1}\} = C_{I[s^{-1}]}(M[s^{-1}]) \end{array}$$

is Cartesian.

Proof. Since M is projective this follows from the case $M = R$, which we assume from now on. The top horizontal map is injective because R is integral. The vertical maps are injective by Krull's Theorem. Because of the assumptions, the ring $C_I(R)$ is an integral domain, hence the bottom horizontal arrow is injective. Therefore also the map $C_I(R)[s^{-1}] \to C_I(R)\{s^{-1}\}$ is injective, and we are left to show that for arbitrary n, the left square in the following diagram with exact rows is Cartesian:

$$\begin{array}{ccccccccc} 0 & \longrightarrow & R & \hookrightarrow & (1/s^n)R & \xrightarrow{\cdot s^n} & R/s^nR & \longrightarrow & 0 \\ & & \big\downarrow & & \big\downarrow & & \big\downarrow & & \\ 0 & \longrightarrow & \hat{R} & \longrightarrow & \widehat{(1/s^n)R} & \longrightarrow & \widehat{R/s^nR} & \longrightarrow & 0 \end{array}$$

where in the bottom line we mean the I-adic completions of the respective f.g. R-modules. (The rows are exact because of the exactness of the completion functor on f.g. modules. This is, however, not used in the sequel.) We have to show that the right vertical map is injective. This is the case if $0 \in R/s^nR$ is closed w.r.t. the

I-adic topology. Its closure is (by an extension of Krull's Theorem [5, Chap. III, §3, 2., Prop. 5]) the set of those $x \in R/s^nR$ for which there exists an $m \in I$ such that $(1-m)x = 0$ holds true. Therefore the above map is injective, if no $m \in I$ and $x \in R$ exist such that $s^n|(1-m)x$ and $s^n \nmid x$. Since s is prime, this is the case if no $m \in I$ exists such that $s|(1-m)$. Since by assumption s is not a unit modulo I, this is indeed impossible. $\square$

2.8.5. Let ${}^K_\Delta\mathbf{X}$ be p-ECMSD such that an integral canonical model $\mathrm{M}({}^K_\Delta\mathbf{X})$ exists. We consider Δ "up to refinements" in this section, i.e., we feel free to refine it whenever necessary. Assume that there is a boundary morphism:

$$\iota\colon \mathbf{H}_0[U,0] \Longrightarrow \mathbf{X},$$

which we fix once and for all in this section.

By [58, Proposition 12.1], the existence of this boundary morphism forces the reflex field of $\mathbf{X}$ to be $\mathbb{Q}$. We write also shorthand $\mathbf{B} = \mathbf{H}_0[U,0]$. It induces a dual boundary morphism (open embedding):

$$\mathrm{M}^\vee(\iota)\colon \mathrm{M}^\vee(\mathbf{B}) \hookrightarrow \mathrm{M}^\vee(\mathbf{X})$$

between $\mathbb{Z}_{(p)}$-schemes and therefore a base point $\mathrm{M}^\vee(\iota)(v) \in \mathrm{M}^\vee(\mathbf{X})(\mathbb{Z}_{(p)})$ which we denote by v, too, ι being fixed.

Let $\mathcal{E}$ be a $P_\mathbf{X}$-equivariant line bundle on $\mathrm{M}^\vee(\mathbf{X})$. For any $\rho \in P_\mathbf{X}(\mathbb{A}^{(\infty)})$ we have an associated boundary morphism

$$(\iota,\rho)\colon {}^{K_1}_{\Delta_1}\mathbf{B} \Longrightarrow {}^K_\Delta\mathbf{X},$$

where $K_1 = K^\rho \cap \iota(P_\mathbf{B}(\mathbb{A}^{(\infty)}))$ and Δ_1 is defined as pullback (2.2.24). Refining the original Δ, if necessary, we can assume that Δ_1 is smooth.

The pullback of the automorphic line bundle

$$\mathrm{M}(\iota,\rho)^*\Xi^*\mathcal{E} \cong \Xi^*\,\mathrm{M}^\vee(\iota)^*\mathcal{E}$$

(cf. 2.5.14(1) for this functorility) may be trivialized as in 2.8.2, using a basis vector z' of $(\mathrm{M}^\vee(\iota)^*\mathcal{E})_v = \mathcal{E}_v$. K_1 always contains a group of the form $U_{\widehat{\mathbb{Z}}}K(M)$ for a saturated (at p) lattice $U_\mathbb{Z} \subseteq U$ and $p \nmid M$.

Let $s_{z'}$ denote the corresponding trivializing section (2.8.2). Thus every section $f \in H^0(\mathrm{M}({}^K_\Delta\mathbf{X})_\mathbb{C}, \Xi^*\mathcal{E})$ yields Fourier coefficients[5] (see 2.8.2):

$$a_{\iota,z',k,\alpha,\rho}(f) := a_{k,\alpha,1}\left(\frac{\mathrm{M}(\iota,\rho)^*f}{s_{z'}}\right).$$

2.8.6. Lemma. *If $f \in H^0(\mathrm{M}({}^K_\Delta\mathbf{X})_\mathbb{C}, \Xi^*\mathcal{E}_{merom.})$ is a meromorphic section (for any Δ, including $\{0\}$, the meromorphic sections are the same), we get Fourier coefficients*

$$a_{\iota,z',k,\alpha,\rho,\sigma}(f) := a_{k,\alpha,\rho,\sigma}\left(\frac{\mathrm{M}(\iota,\rho)^*f}{s_{z'}}\right),$$

which are defined in the same way (2.8.2) but only on a set of the form:

$$\{\alpha\} \times (x + U_\mathbb{R} + \sigma) \times \{\rho\} \subset \mathbb{D}_\mathbf{B} \times P_\mathbf{B}(\mathbb{A}^{(\infty)}), \tag{2.11}$$

[5]This definition includes also the Fourier expansions defined in 2.8.2 for nontrivial $\rho \in K(1)$ because we have for $\rho' \in \widehat{\mathbb{Z}}^* = K(1) \subseteq \mathbb{A}^{(\infty)*}$:

$$a_{\iota,z',k,\alpha,\rho\rho'}(f) = a_{k,\alpha,\rho'}\left(\frac{\mathrm{M}(\iota,\rho)^*f}{s_{z'}}\right).$$

where $x \in U_{\mathbb{C}}$*, and* σ *is a maximal dimensional rational polyhedral cone in* $U_{\mathbb{R}}(-1)$.

PROOF. We show that there is a boundary point on *some* toroidal compactification over $\mathbb{C}$ at which locally the divisor of f consists only of components of the boundary divisor itself. The statement then follows because a small neighborhood around such a boundary point is mapped to a set of the form (2.11).

For this it suffices to investigate the divisor in a formal neighborhood of the boundary point. Writing f as a quotient of formal functions, we are reduced to the case where f is a formal function in the formal completion at some boundary point associated with $\iota, \rho, \alpha \in \mathbb{D}_{\mathbf{H}_0}$ and a rational polyhedral cone σ of maximal dimension (now considered in $U_{\mathbb{R}}$, see 2.7.14). Let σ be generated by $\{\lambda_i \in U_{\mathbb{Z}}\}$ and call $\{z_i\}$ the variables corresponding to the vertices of the dual cone.

This formal completion is

$$\operatorname{spf}\Big(\bigoplus_{\zeta_{M,\mathbb{C}}} \mathbb{C}[\![U_{\mathbb{Z}}^* \cap \sigma^\vee]\!]\Big),$$

hence the formal function is an element in

$$\bigoplus_{\zeta_{M,\mathbb{C}}} \mathbb{C}[\![U_{\mathbb{Z}}^* \cap \sigma^\vee]\!] = \bigoplus_{\zeta_{M,\mathbb{C}}} \mathbb{C}[\![z_1, \dots, z_n]\!].$$

We consider its components separately and call f any of them. We may now refine the polyhedral cone decomposition in a way that it induces barycentric subdivisions of faces of σ. The faces τ of σ correspond to subsets $S \subseteq \{1, \dots, n\}$ (the vertices of τ). There are then $|S|$ new maximal dimensional cones σ_j in the corresponding subdivison, indexed by the respective vertex j which does not belong to σ_j. Fix such a j and let $\lambda'_1, \dots, \lambda'_n$ be the generators of the corresponding cone. We have $\lambda'_j = \sum_{i \in S} \lambda_i$ and $\lambda'_i = \lambda_i$, $i \neq j$. We get a corresponding homomorphism:

$$\mathbb{C}[\![z_1, \dots, z_n]\!] \to \mathbb{C}[\![z'_1, \dots, z'_n]\!].$$

given by $z_j \mapsto \prod_{i \in S} z'_i$ and $z_i \mapsto z'_i$, $i \neq j$.

Since both rings have the same form, we may consider this homomorphism as an automorphism $T_{j,S}$ of the ring on the left, with the obvious identification. We have to show that repeatedly applying these automorphisms to f and clearing factors of z_i, $i = 1, \dots, n$ one arrives at an element $\tilde{f}$ containing a constant term. For this we proceed by induction on the minimal degree of the occurring monomials in f. Let S_j be the composition of the $T_{i,\{i,j\}}$ for $i \neq j$. It maps $z_j \mapsto z_j$ and $z_i \mapsto z_i z_j$ for $i \neq j$. Let $g = \prod z_i^{n_i}$ be a monomial of minimal degree in f so $\deg(g) = \sum_i n_i$. Let j be such that $n_j \neq 0$. By applying S_j to f, a monomial $h = \prod z_i^{m_i}$ gets transformed to $z_j^{\deg(h)} \prod_{i \neq j} z_i^{m_i}$. Since $\deg(g)$ is minimal, we may define $\tilde{f}$ by $S_j(f)/z_j^{\deg(g)}$. We have $\deg\big(S_j(g)/z_j^{\deg(g)}\big) = \sum_{i \neq j} n_i < \deg(g)$, hence $\tilde{f}$ has a term of smaller minimal degree. □

2.8.7. Theorem (q-expansion principle). *With the notations of* 2.8.5*, there is some* $M \in \mathbb{N}$*, with* $p \nmid M$*, and a finite subset* $\{\rho_1, \dots, \rho_n\} \subset P_{\mathbf{X}}(\mathbb{A}^{(\infty)})$*, such that the following holds: For every* N *such that* $M \mid N$*, we have for* $R :=$

$\mathbb{Z}_{(p)}[\zeta_N]$ *that*

$$H^0(\mathrm{M}({}^K_\Delta\mathbf{O})_R, \Xi^*\mathcal{E}) \\ = \{f \in H^0(\mathrm{M}({}^K_\Delta\mathbf{O})_\mathbb{C}, \Xi^*\mathcal{E}) \mid \forall i\ \forall k \in U^*_\mathbb{Q}, \forall\alpha\ a_{\iota,z',k,\alpha,\rho_i}(f) \in R\}$$

and similarly[6]

$$H^0(\mathrm{M}({}^K_\Delta\mathbf{O})_R, \Xi^*\mathcal{E}_{\mathrm{horz.}}) = \\ = \left\{f \in H^0(\mathrm{M}({}^K\mathbf{O})_\mathbb{C}, \Xi^*\mathcal{E}_{\mathrm{merom.}}) \;\middle|\; \begin{matrix}\forall i\ \forall\alpha\ \exists\sigma \subset U_\mathbb{R} \text{ open r.p. cone} \\ \forall k \in U^*_\mathbb{Q}\ a_{\iota,z',k,\alpha,\rho_i,\upsilon}(f) \in R\end{matrix}\right\}.$$

Moreover, we have

$$a_{\iota,z',k,\alpha,\rho,\sigma}({}^\tau f) = {}^\tau\left(a_{\iota,z',k,\alpha,\rho k_\tau^{-1},\sigma}(f)\right) \quad \forall\tau \in \mathrm{Gal}(\mathbb{Q}(\zeta_N) \mid \mathbb{Q}),$$

hence f is (in both cases) defined over $\mathbb{Z}_{(p)}$ instead of R, if it satisfies

$$a_{\iota,z',k,\alpha,\rho,\sigma}(f) = {}^\tau\left(a_{\iota,z',k,\alpha,\rho k_\tau^{-1},\sigma}(f)\right)$$

for all $\tau \in \mathrm{Gal}(\mathbb{Q}(\zeta_N) \mid \mathbb{Q})$, for all ρ, α, k and for some σ (depending on ρ and α). Here k_τ is the image of τ under the natural isomorphism

$$\mathrm{Gal}(\mathbb{Q}(\zeta_N) \mid \mathbb{Q}) \xrightarrow{\sim} K(1)/K(N).$$

PROOF. We may show the theorem for a particular Δ and all its refinements. A posteriori, it will then be true for any Δ', using a common refinement of Δ and Δ'. Let f be a complex section of the bundle $\Xi^*(\mathcal{E})$ over $\mathrm{M}({}^K_\Delta\mathbf{X})$. By the abstract q-expansion principle 2.8.4, a section of a locally free sheaf is defined over $\mathrm{M}({}^K_\Delta\mathbf{X})_R$ if on any connected component (over R) there is a point p defined over R, such that after passing to the completion at p, the corresponding formal series is defined over R. Now on any connected component i of $\mathrm{M}({}^K\mathbf{X})_\mathbb{C}$, there is a boundary point corresponding to a boundary component (ι, ρ_i) and some $\sigma \subset U_\mathbb{R}$. Using the boundary isomorphism $\mathrm{M}(\iota, \rho_i)$ (2.5.9), defined over $\mathbb{Z}_{(p)}$, it suffices to show that the pullback of f to $\Xi^*(\mathrm{M}^\vee(\iota)^*\mathcal{E})$ over $\widehat{\mathrm{M}\left({}^{K_i}_{\Delta_i}\mathbf{H}_0[U,0]\right)}$ (completion with respect to the boundary point corresponding to σ) is defined over R. We may choose the integer M such that all connected components of $\mathrm{M}({}^{K_i}_{\Delta_i}\mathbf{H}_0[U,0])$ are defined over R, for all i, and such that we may test f by pulling it further back to $\widehat{\mathrm{M}\left({}^{K'_i}_{\Delta_i}\mathbf{H}_0[U,0]\right)}$, where $K'_i = U_{\widehat{\mathbb{Z}},i}K(M)$ for some lattice $U_{\mathbb{Z},i} \subset U$ as in 2.8.2. We refine Δ such that Δ_i is smooth w.r.t. K'_i for all i.

By the consideration in 2.8.5 above, the formal function $f/s_{z'}$ on

$$\widehat{\mathrm{M}\left({}^{U_{\widehat{\mathbb{Z}},i}K(M)}_{\Delta_i}\mathbf{H_0}[\mathbf{U},\mathbf{0}]\right)}$$

for $U_{\widehat{\mathbb{Z}},i}K(M) \subseteq K_i$ is given by

$$\left(\frac{f}{s}\right)_{\zeta_{M,\mathbb{C}}} = \sum_{\lambda \in U^*_\mathbb{Z}} a_{\iota,z',k,\alpha,\rho_i k,\lambda,\sigma}(f).$$

Here

$$[\alpha, k] \in \mathbb{Q}^*\backslash\mathbb{D}_{\mathbf{H}_0} \times {\mathbb{A}^{(\infty)}}^*/K(M)$$

[6]horz. is the tensor product with the localization of $\mathcal{O}_{\mathrm{M}({}^K_\Delta\mathbf{O})_R}$ w.r.t. all functions, which have no component of the fibre above p in its divisor.

with $k \in K(1)$ is determined by $\zeta_M = \exp\bigl(\alpha(k/M)\bigr)$. (It suffices of course to look at the fibre over the $\zeta_{M,\mathbb{C}}$ corresponding to $[\alpha, 1]$). Then, since by assumption these coefficients lie in R, $f/s_{z'}$ is defined over R and hence f itself. The descriptions of spaces of sections in the theorem follow because we investigated at least one boundary point on each connected component. If f was a meromorphic section, we may refine Δ in a way such that for all i its pullback Δ_i contains the cone $\sigma \subset U(\mathbb{R})$ (up to Tate-twist) such that there exists a Fourier expansion in a neighborhood of the corresponding boundary point (see 2.8.6). The statement about the Fourier coefficients of the Galois conjugates was shown in 2.8.2. □

CHAPTER 3

Volumes of Orthogonal Shimura Varieties

3.1. Special cycles

Let $L_{\mathbb{Z}_{(p)}}$ be a lattice with unimodular quadratic form of signature $(l-2, 2)$. Recall (2.4) the associated p-integral Shimura data $\mathbf{S}(L)$ and $\mathbf{O}(L)$. In this section we further investigate the structure of the corresponding integral models $\mathrm{M}({}^K_\Delta\mathbf{S}(L))$, and $\mathrm{M}({}^K_\Delta\mathbf{O}(L))$, and define the special cycles on them.

3.1.1. Proposition. *If* $\dim(L) \geq 3$, *the reflex field* (2.5.2) *of* $\mathbf{S}(L)$ *is* $\mathbb{Q}$. *The compact dual* (2.5.12) *is the zero quadric, i.e.,*

$$\mathrm{M}^\vee(\mathbf{S}(L))(R) = \{\langle z\rangle \in \mathbb{P}(L_R) \mid \langle z, z\rangle = 0\}$$

for any ring R *over* $\mathbb{Z}_{(p)}$.

If $\dim(L) = 2$, *the reflex field of* $\mathbf{S}(L)_+$, *respectively* $\mathbf{S}(L)_-$ (*see* 2.4.4) *is the imaginary quadratic field* (*subfield of* $\mathbb{C}$) *associated with the negative-definite binary quadratic form* L. *The compact dual is a point given by the isotropic vector corresponding to* the *filtration in* $\mathbb{D}_{\mathbf{S}(L)_+}$, *respectively* $\mathbb{D}_{\mathbf{S}(L)_-}$.

Proof. Let D be a negative definite subspace defined over $\mathbb{Q}$. There is an imaginary quadratic field F, unramified at p, and a morphism

$$\mu\colon \mathbb{G}_{m,F} \hookrightarrow \mathrm{GSpin}(L_F),$$

acting with exponent 2 on one isotropic vector $z \in D_F$, with exponent 0 on $\bar{z}$ and with exponent 1 on the orthogonal complement. Over $\mathbb{C}$ it yields the associated morphism u_h (2.5.2), where h is one of the 2 morphisms corresponding to D. The morphism $\overline{u_h}$ is also of the form $u_{h'}$ and hence conjugated to u_h if $l \geq 3$. The conjugacy class of these morphisms is therefore defined over $\mathbb{Q}$. If $l = 2$, $\mathrm{GSpin}(L)$ is a torus and hence the reflex field is F. Furthermore, a morphism u_h defines a filtration $0 \subset \langle z\rangle \subset \langle z\rangle^\perp \subset L$ which is completely determined by $\langle z\rangle$. This determines the compact dual by 2.5.12(2). □

3.1.2. Remark. The Hodge embedding from Lemma 2.4.4 induces a morphism (cf. Main Theorem 2.5.12)

$$\mathrm{M}^\vee(\mathbf{S}(L)) \hookrightarrow \mathrm{M}^\vee(\mathbf{H}(\mathrm{C}^+(L), \langle\cdot,\cdot\rangle_\delta)).$$

Here $\mathrm{M}^\vee(\mathbf{S}(L))$ is the space of isotropic lines in L and $\mathrm{M}^\vee(\mathbf{H}(\mathrm{C}^+(L), \langle\cdot,\cdot\rangle_\delta))$ is the space of Lagrangian saturated sublattices of $\mathrm{C}^+(L)$ with respect to the form $\langle\cdot,\cdot\rangle_\delta$. The proof of Lemma 2.4.4 shows that this map is given (for any R) by sending a $\langle z\rangle$ in L_R, with z assumed to be primitive, to the sublattice $zw\,\mathrm{C}^+(L_R)$, where w is a primitive vector with $\langle z, w\rangle = 1$ (this sublattice actually does not depend on w). One also immediately sees that this is compatible with the Borel embedding because $w = \bar{z}/\langle z, \bar{z}\rangle$ leads to the projector considered in the proof of Lemma 2.4.4.

3.1.3. Lemma. *Let $K \subset \mathrm{SO}(L_{\mathbb{A}^{(\infty)}})$ be a neat admissible compact open subgroup and Δ be a smooth K-admissible rational polyhedral cone decomposition. The (geometric) connected components of $\mathrm{M}({}^K_\Delta\mathbf{O})_{\overline{\mathbb{Q}}}$ and $\mathrm{M}({}^K_\Delta\mathbf{O})_{\overline{\mathbb{F}}_p}$ are in bijection and defined over $\mathrm{Spec}(\mathbb{Z}_{(p)}[\zeta_N])$ where ζ_N is a primitive Nth root of unity and N some integer coprime to p.*

PROOF. This follows directly from 2.5.5, 2.5.9 using [52, Theorem 5.17]. □

3.1.4. Lemma. *Let $L_\mathbb{Z}$ be a lattice with quadratic form of discriminant $D \neq 0$. Let K be a compact open subgroup of $\mathrm{SO}(L_{\mathbb{A}^{(\infty)}})$, and pick a set $\{g_1, \dots, g_n\}$ of representatives of the cosets*

$$\mathrm{SO}(L_\mathbb{Q})\backslash \mathrm{SO}(\mathbb{A}^{(\infty)})/K.$$

We have

$$[\mathrm{SO}(L_\mathbb{Q})\backslash \mathbb{D}_\mathbf{O} \times (\mathrm{SO}(\mathbb{A}^{(\infty)})/K)] = \bigcup_i [\Gamma_i \backslash \mathbb{D}_\mathbf{O}],$$

where $\Gamma_i = \mathrm{SO}(L_\mathbb{Q}) \cap g_i K g_i^{-1}$.

If K is the discriminant kernel of $L_{\widehat{\mathbb{Z}}}$ then for each i there is the lattice $L_\mathbb{Z}^{(i)}$ characterized by $L_{\widehat{\mathbb{Z}}}^{(i)} = g_i L_{\widehat{\mathbb{Z}}}$ (its class is independent of the choice of g_i), and Γ_i is the discriminant kernel of the corresponding $L_\mathbb{Z}^{(i)}$, in particular a subgroup of $\mathrm{SO}(L_\mathbb{Z}^{(i)})$.

PROOF. Analogous to [52, Lemma 5.13]. □

3.1.5. Let $L_{\mathbb{Z}_{(p)}}$ be a lattice with unimodular quadratic form, i.e., inducing an isomorphism $L_{\mathbb{Z}_{(p)}} \cong (L_{\mathbb{Z}_{(p)}})^*$, and $L'_{\mathbb{Z}_{(p)}}$ be a saturated sublattice with unimodular form, of signature $(l-2, 2)$ and $(l-2-m, 2)$, respectively, where $0 < m \leq l-2$.

Recall the embedding 2.4.4 here used modulo $\mathbb{G}_\mathrm{m}$

$$\iota\colon \mathbf{O}(L') \hookrightarrow \mathbf{O}(L),$$

in the case $l - m > 2$ and the 2 embeddings

$$\iota_\pm\colon \mathbf{O}(L')_\pm \hookrightarrow \mathbf{O}(L),$$

in the case $l - m = 2$.

Let an admissible compact open subgroup K be given. For each $g \in \mathrm{SO}(L_{\mathbb{A}^{(\infty,p)}})$ we get a conjugated embedding:

$$(\iota, g)\colon {}^{K'_g}\mathbf{O}(L') \hookrightarrow {}^K\mathbf{O}(L)$$

for $K'_g := \iota^{-1}(gKg^{-1})$.

In addition, we may find smooth (with respect to K, respectively K'_g) complete projective r.p.c.d. Δ, and Δ'_g (2.2.25), such that models exist (2.5.9) and we have, in the end, embeddings of p-ECMSD:

$$(\iota, g)\colon {}^{K'_g}_{\Delta'_g}\mathbf{O}(L') \hookrightarrow {}^K_\Delta\mathbf{O}(L)$$

(respectively with $\pm$ if $l - m = 2$).

Let $M_{\mathbb{Z}_{(p)}}$ be a lattice of dimension m, $m \leq l-2$, with unimodular positive-definite quadratic form and define

$$\mathrm{I}(L, M)(R) := \{\alpha\colon M_R \to L_R \mid \alpha \text{ is an isometry}\}.$$

We have $\mathrm{I}(L,M)(\mathbb{Q}) \neq \varnothing \Leftrightarrow \mathrm{I}(L,M)(\mathbb{A}^{(\infty)}) \neq \varnothing$ by Hasse's principle. If these sets are nonempty we may find even an $x \in \mathrm{I}(L,M)(\mathbb{Z}_{(p)})$.

For a K-invariant admissible[1] Schwartz function $\varphi \in S\big((M^* \otimes L)_{\mathbb{A}^{(\infty)}}\big)$

$$\mathrm{I}(L,M)(\mathbb{A}^{(\infty)}) \cap \operatorname{supp}(\varphi) = \coprod_j K g_j^{-1} x$$

with finitely many $g_j \in \mathrm{SO}(L_{\mathbb{A}^{(\infty,p)}})$. We assume that Δ has been refined such that Δ'_{g_j} exists for all j with the properties claimed above.

3.1.6. Definition (Special cycle). If $\mathrm{I}(L,M)(\mathbb{Q}) \neq \varnothing$, we define the p-integral **special cycle** associated with $M_{\mathbb{Z}_{(p)}}$ and φ on $\mathrm{M}({}^K_\Delta\mathbf{O}(L))$ as

$$\mathrm{Z}(L,M,\varphi;K) := \sum_j \varphi(g_j^{-1}x)\operatorname{im}\big(\mathrm{M}(\iota,g_j)\big),$$

where ι is associated with the embedding $x^\perp \hookrightarrow L_{\mathbb{Z}_{(p)}}$. If $\mathrm{I}(L,M)(\mathbb{Q}) = \varnothing$ we define $\mathrm{Z}(L,M,\varphi;K) := 0$.

Note that if $l-m=2$, $\mathrm{M}({}^{K'_{g_j}}\mathbf{O}(x^\perp))$ is actually equipped with a morphism to $\operatorname{Spec}(\mathcal{O}_\mathbf{X})$ which is a quadratic extension of $\operatorname{Spec}(\mathbb{Z}_{(p)})$. In the definition above, we consider, however, the structural morphism to $\operatorname{Spec}(\mathbb{Z}_{(p)})$. (For the $\mathbb{C}$-points this implies that we have in any case the same special cycles as defined, e.g., in [2] or [35].) The sum is finite and independent of the choice of the g_i and x. The formation is compatible with Hecke operators (2.2.24) in the obvious sense.

If $\mathrm{I}(L,M)(\mathbb{Q}) \neq \varnothing$ and $M_\mathbb{Q}$ is only a $\mathbb{Q}$-vector space with nondegenerate quadratic form and φ any Schwartz function, we write

$$\mathrm{I}(L,M)(\mathbb{A}^{(\infty)}) \cap \operatorname{supp}(\varphi) = \coprod_j K g_j^{-1} x$$

for any $x \in \mathrm{I}(L,M)(\mathbb{Q})$ and $g_j \in \mathrm{SO}(L_{\mathbb{A}^{(\infty)}})$.

3.1.7. Definition (Rational special cycle). We define the **rational special cycle** associated with $M_\mathbb{Q}$ and φ on $\mathrm{M}({}^K_\Delta\mathbf{O}(L))$ as

$$\mathrm{Z}(L,M,\varphi;K) := \sum_j \varphi(g_j^{-1}x)\overline{\operatorname{im}\big(\mathrm{M}(\iota,g_j)\big)}^{\mathrm{Zar}}.$$

Here ι is associated with the embedding $x^\perp \hookrightarrow L_\mathbb{Q}$ and $\mathrm{M}(\iota,g_j)$ is only a morphism of rational Shimura varieties. Here even Δ may be chosen arbitrarily (such that $\mathrm{M}({}^K_\Delta\mathbf{O}(L))$ exists). In that case we understand the Zariski closure of the embedding of the uncompactified rational Shimura variety.

From the extension property (cf. Main Theorem 2.5.5(3)) and the properties of toroidal compactifications (cf. Main Theorem 2.5.9) follows that this definition coincides with the previous one if there exists a lattice $M_{\mathbb{Z}_{(p)}}$ such that the restriction of the quadratic form is unimodular, φ is admissible, and Δ is chosen in such a way that we have embeddings of p-ECMSD.

---

[1] by this we mean that φ is the tensor product of a Schwartz function $\varphi \in S\big((M^* \otimes L)_{\mathbb{A}^{(\infty,p)}}\big)$ with the characteristic function of $(M^* \otimes L)_{\mathbb{Z}_p}$

3.2. The arithmetic of Borcherds products

3.2.1. Let L be a finite-dimensional $\mathbb{Q}$-vector space with nondegenerate quadratic form $Q_L \in \mathrm{Sym}^2(L^*)$ and M be any finite-dimensional $\mathbb{Q}$-vector space. Denote $\mathfrak{M} = M \oplus M^*$ equipped with its standard symplectic form. Consider the following elements of $\mathrm{Sp}(\mathfrak{M})$:

$$g_l(\alpha) := \begin{pmatrix} \alpha & 0 \\ 0 & {}^t\alpha^{-1} \end{pmatrix}$$
$$u(\beta) := \begin{pmatrix} 1 & \beta \\ 0 & 1 \end{pmatrix}$$
$$d(\gamma) := \begin{pmatrix} 0 & -{}^t\gamma^{-1} \\ \gamma & 0 \end{pmatrix}$$

for $\alpha \in \mathrm{Aut}(M)$, $\beta \in \mathrm{Hom}(M^*, M)$ and $\gamma \in \mathrm{Iso}(M, M^*)$. We denote the image of g_l and u by G_l and U respectively, and their product by P, which is a maximal parabolic of $\mathrm{Sp}(\mathfrak{M})$.

Let R be one of $\mathbb{R}, \mathbb{Q}_p, \mathbb{A}^{(\infty)}$ or $\mathbb{A}$ and χ its standard additive character (depending on a choice of $i \in \mathbb{C}$). Let $\mathrm{Sp}'(\mathfrak{M}_R)$ be $\mathrm{Sp}(\mathfrak{M}_R)$ if m is even and $\mathrm{Mp}(\mathfrak{M}_R)$ (metaplectic double cover) if m is odd. Recall the **Weil representation** ω of $\mathrm{Sp}'(\mathfrak{M}_R)$ acting on $S\big((M^* \otimes L)_R\big)$ (Schwartz–Bruhat functions). The elements defined above have lifts to $\mathrm{Sp}'(\mathfrak{M}_R)$, denoted $r''(\cdots)$ acting as

$$\omega\Big(r''\big(g_l(\alpha)\big)\Big)\varphi\colon x^* \mapsto \widetilde{\Upsilon}(\gamma_0\alpha)\widetilde{\Upsilon}(\gamma_0)^{-1}|\alpha|^{l/2}\varphi({}^t\alpha x^*)$$
$$\omega\Big(r''\big(u(\beta)\big)\Big)\varphi\colon x^* \mapsto \chi\big((x^*)^! Q_L \cdot \beta\big)\varphi(x^*)$$
$$\omega\Big(r''\big(d(\gamma)\big)\Big)\varphi\colon x^* \mapsto \widetilde{\Upsilon}(\gamma)|\gamma|^{-l/2}\int_{(M\otimes L^*)_R} \varphi({}^t\gamma x)\chi(-\langle x^*, x\rangle)\,\mathrm{d}x.$$

$\widetilde{\Upsilon}$ is a certain eigth root of unity. For $m=1$ we have just $\widetilde{\Upsilon}(\gamma) = \Upsilon(\gamma \otimes Q_L)$, where Υ is the Weil index [64]. In the third formula $|\gamma|$ is computed with respect to the measure $\mathrm{d}x$ and its dual, and hence the formula is independent of this measure.

If a global lattice $L_{\mathbb{Z}} \subset L_{\mathbb{Q}}$ with $Q_L \in \mathrm{Sym}^2(L_{\mathbb{Z}}^*)$ and a lattice $M_{\mathbb{Z}} \subset M_{\mathbb{Q}}$ are given we often restrict, as in [2], to the subspace of $S\big((M^* \otimes L)_{\mathbb{A}^{(\infty)}}\big)$ given by characteristic functions of cosets in $(L_{\hat{\mathbb{Z}}}^*/L_{\hat{\mathbb{Z}}}) \otimes M_{\hat{\mathbb{Z}}}^*$. We denote the space by $\mathbb{C}[(L_{\mathbb{Z}}^*/L_{\mathbb{Z}}) \otimes M_{\mathbb{Z}}^*]$.

3.2.2. In the $m=1$ case ($M = \mathbb{Z}$) we denote $\mathbb{C}[L_{\mathbb{Z}}^*/L_{\mathbb{Z}}]$ also by $\mathrm{Weil}(L_{\mathbb{Z}}^*/L_{\mathbb{Z}})$, and consider it as a representation of $\mathrm{Sp}_2'(\mathbb{Z})$, $\mathrm{Sp}_2'(\widehat{\mathbb{Z}})$ or a suitable $\mathrm{Sp}_2'(\mathbb{Z}/N\mathbb{Z})$. $\mathrm{Sp}_2'(\mathbb{Z})$ is generated by 2 elements S and T, satisfying $Z = S^2 = (ST)^3$ and ($Z^4 = 1$ or $Z^2 = 1$ according to whether l is odd or even), projecting down to $\left(\begin{smallmatrix} 0 & -1 \\ 1 & 0 \end{smallmatrix}\right)$ and $\left(\begin{smallmatrix} 1 & 1 \\ 0 & 1 \end{smallmatrix}\right)$, respectively, in $\mathrm{Sp}_2(\mathbb{Z})$. They act explicitly as:

$$T\chi_\kappa = \exp\big(2\pi i Q_L(\kappa)\big)\chi_\kappa,$$
$$S\chi_\kappa = \frac{\Upsilon_\infty(Q_L)^{-1}}{\sqrt{D}} \sum_{\delta \in L_{\mathbb{Z}}^*/L_{\mathbb{Z}}} \exp(-2\pi i\langle\delta, \kappa\rangle)\chi_\delta.$$

Note that the occurring correction factor $\widetilde{\Upsilon}_f(1) = \Upsilon_f(Q_L)$ is the product over the local $\Upsilon_p(Q_L)$'s which is, however, the same as $\Upsilon_\infty(Q_L)^{-1}$ and we have $\Upsilon_\infty(Q_L) = \exp\big(2\pi i(p-q)\big)$, if the signature of Q is (p, q). These are the same formulas used in [2]. The Weil representation of $\mathrm{Sp}_2'(\widehat{\mathbb{Z}})$, or $\mathrm{Sp}_2'(\mathbb{Z})$, always factors through $\mathrm{Sp}_2'(\mathbb{Z}/N\mathbb{Z})$, where N is related to the exponent of $L_{\mathbb{Z}}^*/L_{\mathbb{Z}}$.

If l is *even*, the Borel subgroup $\left(\begin{smallmatrix} * & * \\ & * \end{smallmatrix}\right) \subset \mathrm{Sp}_2(\mathbb{Z}/N\mathbb{Z})$ acts on $\chi_{L_{\hat{\mathbb{Z}}}}$ by the *character*

$$\begin{pmatrix} \alpha & \beta \\ & \delta \end{pmatrix} \mapsto \Upsilon(\alpha Q_L)\Upsilon(Q_L)^{-1}$$

valued in μ_2.

By construction of the Weil representation, we have

$$\mathrm{Weil}(L_{\mathbb{Z}}^*/L_{\mathbb{Z}}) = \bigotimes_{p|D} \mathrm{Weil}(L_{\mathbb{Z}_p}^*/L_{\mathbb{Z}_p}), \tag{3.1}$$

where $\mathrm{Weil}(L_{\mathbb{Z}_p}^*/L_{\mathbb{Z}_p})$ is actually a representation of $\mathrm{Sp}_2'(\mathbb{Z}_p)$. If $L_{\mathbb{Z}_p}^*/L_{\mathbb{Z}_p}$ is cyclic of prime order or of order 4, we have

$$\mathrm{Weil}(L_{\mathbb{Z}_p}^*/L_{\mathbb{Z}_p}) = \mathrm{Weil}(L_{\mathbb{Z}_p}^*/L_{\mathbb{Z}_p})^+ \oplus \mathrm{Weil}(L_{\mathbb{Z}_p}^*/L_{\mathbb{Z}_p})^-$$

(decompositions into odd and even functions), and both summands are *irreducible.* (If $p=2$ and the order is 2, $\mathrm{Weil}(L_{\mathbb{Z}_p}^*/L_{\mathbb{Z}_p})^- = 0$.) Together with (3.1) this yields the decomposition of the Weil representation of L into irreducible representations, if D is square-free, respectively $\frac{D}{2}$ square-free and $L_{\mathbb{Z}_2}^*/L_{\mathbb{Z}_2}$ cyclic of order 4.

3.2.3. Let $L_{\mathbb{Q}}$ be a nondegenerate quadratic space of signature (p,q). Consider a 1-dimensional space $M_{\mathbb{Q}} = \mathbb{Q}$. We canonically identify $M^* \otimes L = L$.

Let N be a maximal negative definite subspace of $L_{\mathbb{R}}$. We have the Gaussian $\varphi_\infty^0 \in S(L_{\mathbb{R}}) \otimes \mathcal{C}(\mathbb{D}_{\mathbf{O}})$, defined by

$$\varphi_\infty^0(x,N) := \mathrm{e}^{-\pi\langle x, \alpha_N x\rangle},$$

where $\alpha_N \in \mathrm{End}(L_{\mathbb{R}})$ acts as -1 on N and as $+1$ on the orthogonal complement. It satisfies

$$\varphi_\infty^0(hx, hN) = \varphi_\infty^0(x,N) \quad \forall h \in \mathrm{SO}(L_{\mathbb{R}}).$$

For any $\varphi \in S(L_{\mathbb{A}^{(\infty)}})$, define the theta function

$$\Theta(g', N, h; \varphi) = \sum_{v \in L_{\mathbb{Q}}} \sum \omega(g')(\varphi_\infty(\cdot, N) \otimes \omega(h)\varphi)(x)$$

as a function of $g' \in \mathrm{Sp}_2'(\mathbb{A}), N \in \mathbb{D}_{\mathbf{O}}, h \in \mathrm{SO}(L_{\mathbb{A}^{(\infty)}})$. Here ω is the Weil representation (3.2.1).

We have (cf. [63]) for any $\gamma \in \mathrm{SO}(L_{\mathbb{Q}})$ and $\gamma' \in \mathrm{Sp}_2'(\mathbb{Q})$ (canonical lift of $\mathrm{Sp}_2(\mathbb{Q})$ to $\mathrm{Sp}_2'(\mathbb{A})$),

$$\Theta(\gamma' g', \gamma N, \gamma h; \varphi) = \Theta(g', N, h; \varphi)$$

and for $g_1' \in \mathrm{Sp}_2'(\mathbb{A}^{(\infty)})$ and $h_1 \in \mathrm{SO}(L_{\mathbb{A}^{(\infty)}})$:

$$\Theta(g' g_1', N, hh_1; \varphi) = \Theta(g', N, h; \omega(g_1')\omega(h_1)\varphi).$$

By these invariance properties, we may consider Θ as a function

$$\mathrm{Sp}_2'(\mathbb{Q}) \backslash \mathrm{Sp}_2'(\mathbb{A}^{(\infty)}) \times \mathrm{M}({}^K\mathbf{O}(L))(\mathbb{C}) \to \left(S(L_{\mathbb{A}^{(\infty)}})^K\right)^*.$$

We have (cf. [35, p. 10]):

$$\omega(k_\infty')\varphi_\infty^0(\cdot, N) = \chi_\ell(k_\infty')\varphi_\infty^0(\cdot, N)$$

for $\ell = (p-q)/2$ and $k_\infty' \in K_\infty'$ (inverse image of $K_\infty = \mathrm{SO}(2)$ in $\mathrm{Sp}_2'(\mathbb{R})$). It follows that

$$\Theta(g' k_\infty' k', N, h) = \chi_\ell(k_\infty')({}^t\omega(k'))^{-1}\Theta(g', N, h)$$

for all $k'_\infty \in K'_\infty$ and $k' \in K'$ (inverse image of $\mathrm{Sp}_2(\widehat{\mathbb{Z}})$ in $\mathrm{Sp}'_2(\mathbb{A}^{(\infty)})$).

If ℓ is half-integral χ_ℓ is the character of K'_∞ which is uniquely characterized by the fact that χ_ℓ^2 descends to K_∞ and is the (2ℓ)th power of the standard character $\mathrm{SO}_2(\mathbb{R}) \to \mathbb{C}^*$ (with respect to the same choice of $i \in \mathbb{C}$ as in the definition of the Weil representation).

Now consider an automorphic function

$$F\colon \mathrm{Sp}_2(\mathbb{Q})\backslash \mathrm{Sp}'_2(\mathbb{A}) \to S(L_{\mathbb{A}^{(\infty)}}),$$

satisfying

$$F(g'k'_\infty k') = \chi_{-\ell}(k'_\infty)\omega(k')^{-1}F(g')$$

and invariant under some compact open subgroup $K \subset \mathrm{SO}(L_{\mathbb{A}^{(\infty)}})$.

Define the **Borcherds lift** of F by

$$\Phi(N,h;F) := \int^\bullet_{\mathrm{Sp}'_2(\mathbb{Q})\backslash \mathrm{Sp}'_2(\mathbb{A})} \Theta\big(g',N,h;F(g')\big)\,\mathrm{d}g',$$

where $\bullet$ denotes Borcherds' regularization [2], described in [35] in the adelic language. $\mathrm{d}g'$ is the Tamagawa measure. Note that this definition does not depend on K. By definition, we have

$$\Phi(N,h\rho;F) = \Phi(N,h;\omega(\rho)F). \tag{3.2}$$

It compares to the language of Borcherds as follows: If $K \subset \mathrm{SO}(\mathbb{A}^{(\infty)})$ is the discriminant kernel with respect to a lattice $L_\mathbb{Z}$, we may consider Θ as valued in $\mathbb{C}[L_\mathbb{Z}^*/L_\mathbb{Z}]^*$ and F, valued in $\mathbb{C}[L_\mathbb{Z}^*/L_\mathbb{Z}]$, relates to a usual modular form (if ℓ is odd, of half integer weight) by

$$y^{\ell/2}F(g'_\tau) = f(\tau) \quad \tau = x+iy \in \mathbb{H},$$

where $g_\tau = \left(\begin{smallmatrix} y^{1/2} & y^{-1/2}x \\ 0 & y^{-1/2}\end{smallmatrix}\right)$.

3.2.4. Lemma. *We have*

$$\Phi(\iota(N),1;F) = \Phi_{\mathrm{Bor}}(\iota,L_\mathbb{Z},N,f)$$

for $N \in \mathrm{Grass}^-(L_\mathbb{R})$ *and according to* $\iota\colon \mathrm{Grass}^-(L_\mathbb{R}) \hookrightarrow \mathbb{D}_\mathbf{O}$. ι *is a parameter in Borcherds' theory (cf.* [2, p. 46]*).*

PROOF. In 3.2.2, we saw that the Weil representation composed with the inclusion $\mathrm{Sp}'_2(\mathbb{Z}) \to \mathrm{Sp}'_2(\widehat{\mathbb{Z}})$ acts on the subspace $\mathbb{C}[L_\mathbb{Z}^*/L_\mathbb{Z}] \subset S(L_{\mathbb{A}^{(\infty)}})$ by the same formulas used in [2]. □

We call f, respectively F, nearly holomorphic if f is holomorphic on $\mathbb{H}$ and has a Fourier expansion

$$f(\tau) = \sum_{k\in\mathbb{Q}} a_k \exp(2\pi i k\tau)$$

with $a_k \in \mathbb{C}[L_\mathbb{Z}^*/L_\mathbb{Z}]$ and $a_k = 0$ unless $k \gg -\infty$.

3.2.5. Definition. Let $\mathcal{E}$ be the restriction of the tautological line bundle on $\mathbb{P}(L)$ to the zero-quadric $\mathrm{M}^\vee\big(\mathbf{O}(L)\big)$. It carries an $\mathrm{SO}(L)$ action. $\mathcal{E}_\mathbb{C}|_{h(\mathbb{D}_\mathbf{O})}$ carries a natural Hermitian metric $h_\mathcal{E}$ which we normalize as follows:

$$h_\mathcal{E}(v) := -\tfrac{1}{2}\mathrm{e}^{-C}\langle v,\overline{v}\rangle,$$

where $C = \gamma + \log(2\pi)$ and $\gamma = -\Gamma'(1)$ is Euler's constant.

Denote $\overline{\mathcal{E}} = (\mathcal{E}, h_E)$.

3.2.6. Assume from now on that the signature of $L_{\mathbb{Q}}$ is $(l-2,2)$. Borcherds proved in [2, Theorem 13.3] that $\Phi(F)$ is the logarithm of the Hermitian norm (with respect to the metric $\Xi^* h_{\mathcal{E}}$) of a modular form (i.e., a section in a tensor power of $\Xi^*\mathcal{E}$) on $\mathrm{M}({}^K_\Delta\mathbf{O}(L))$. The task of the rest of this section is a translation of its famous product expansion into the adelic language. In the next paragraphs we will not consider any fixed integral structure. All spaces will be defined over $\mathbb{Q}$ and we will write L, I, U, etc. instead of $L_{\mathbb{Q}}, I_{\mathbb{Q}}, U_{\mathbb{Q}}$.

3.2.7. Let I be an isotropic line in L and $z' \in L$ a vector with $\langle z', I\rangle \neq 0$. We fix a measure $\mathrm{d}x$ on $\mathbb{A}^{(\infty)}$ which has the property, that the measure of the set

$$\{\langle \delta, z'\rangle \mid \delta \in I_{\mathbb{A}^{(\infty)}}, a_k(x+\delta) = a_k(x) \text{ for all } k \in \mathbb{Q}\}$$

is an integer for all $x \in L_{\mathbb{A}^{(\infty)}}$. Since f is a finite linear combination of input modular forms considered by Borcherds, whose Fourier coefficients lies in the subspace $\mathbb{C}[L_{\mathbb{Z}}^*/L_{\mathbb{Z}}]$ for certain lattices $L_{\mathbb{Z}}$, this requirement does indeed make sense. We let $\chi_\alpha\colon \mathbb{A}^{(\infty)} \to \mathbb{C}^*$, $\alpha \in \mathrm{Iso}(\mathbb{Z}, \mathbb{Z}(1))$, be the corresponding additive character, i.e., such that $\mathrm{d}x$ is self-dual with respect to it, and such that $\chi_\alpha|_{\mathbb{Q}} = \exp(\alpha(q\cdot))$ for a $q > 0$.

We let $U := (I^\perp/I) \otimes I \subseteq \Lambda^2 L$. It acts (Lie algebra action, see proof of Proposition 2.4.5) on L, and we have a boundary morphism (cf. 2.4.5):

$$\mathbf{H}_0[U_{\mathbb{Q}}, 0] \Longrightarrow \mathbf{O}(L_{\mathbb{Q}}).$$

We may identify U^* with $(I^\perp/I) \otimes I^*$ by means of Q_L.

z' induces the following quadratic form on U:

$$Q_{z'}(k \wedge z_1) = Q_L\big((k \wedge z_1) \circ z'\big) = (\langle z_1, z'\rangle)^2 Q_L(k).$$

Here "$\circ$" denotes the Lie algebra action of U on L. Its inverse on U^* is:

$$Q_{z'}(k \wedge z_1') = \left(\frac{\langle z, z_1'\rangle}{\langle z, z'\rangle}\right)^2 Q_L(k)$$

for an arbitrary $z \in I$.

We may also embed U^* in $I^\perp$ by associating

$$\lambda = k \wedge z_1' \longmapsto \lambda_{z'} := \frac{\langle z, z_1'\rangle}{\langle z, z'\rangle}\tilde{k}, \tag{3.3}$$

where $\tilde{k} \in I^\perp$ is the unique lift of k with $\langle z', \tilde{k}\rangle = 0$. The embedding may be seen as the composition

$$U^* \xrightarrow{\langle\cdot,\cdot\rangle_{z'}} U \xrightarrow{\circ z'} \langle I, z'\rangle^\perp.$$

The pullback of Q_L is the quadratic form $Q_{z'}$.

3.2.8. Definition/Lemma (Weyl vector). *If $\sigma \subset U_{\mathbb{R}}$ is a sufficiently small cone, for example a Weyl chamber [2, §10], the* **Weyl vector** *$\rho(F, I, \sigma) \in U_{\mathbb{Q}}^*$ is characterized by*

$$8\pi\rho(F, I, \sigma)(\nu) = \sqrt{Q_{z'}(\nu)}\Phi(\langle\nu\rangle, 1, F^{U,z'}) \quad \text{for } \nu \in \sigma \subset U_{\mathbb{R}},$$

where $F^{U,z'}$, as a $S(U_{\mathbb{A}^{(\infty)}})$-valued function, is defined as:

$$F^{U,z'}(g')(w) := C^{-1}\int_{\delta \in I_{\mathbb{A}^{(\infty)}}} F(g')(w \circ z' + \delta)\, \mathrm{d}\langle z', \delta\rangle$$

(any invariant measure). The Weyl vector does not depend on the choice of z'.

PROOF. Compare [2, p. 40] for this characterization of the Weyl vector. First it is independent of z': Scaling z' to $\beta z'$ changes the $w \circ z'$ to $\beta w \circ z'$. $F^{U,z'}(w,z')$ changes to $\beta^{-1}F^{U,z'}(\beta w, g')$. Since the argument of the quadratic form on $U_\mathbb{Q}$ and the argument in $K_{\mathbb{A}^{(\infty)}}$ of $F^{U,z'}$ are both scaled by β, $\Phi(\cdots)$ is only scaled by β^{-1} coming from scaling $F^{U,z'}$ hence the Weyl vector remains the same. (Observe that σ is invariant under scaling, too.) Replacing z' by $\exp(w)z'$ for $w \in U$ does neither change $F^{U,z'}$ nor $Q_{z'}$. □

3.2.9. Definition/Lemma (Adelic Borcherds product). *Let F be a modular form as above, I, z', α be chosen and $\sigma \subset U_\mathbb{R}$ be a sufficiently small cone. The associated* **adelic Borcherds product** *is the formal function*

$$\mathrm{BP}(F,I,z',\alpha,\sigma) := C^{-a_0(0)/2} \prod_{\substack{\delta\in I_{\mathbb{A}^{(\infty)}} \\ \chi_\alpha(\langle z',\delta\rangle)\neq 1}} (1-\chi_\alpha(\langle z',\delta\rangle))^{(a_0(\delta)/2)\,\mathrm{d}\langle z',\delta\rangle}$$

$$\cdot \prod_{\substack{\lambda\in U_\mathbb{Q}^* \\ \langle\lambda,\sigma\rangle>0}} \prod_{\delta\in I_{\mathbb{A}^{(\infty)}}} \left(1-\chi_\alpha(\langle z',\delta\rangle)\left[\frac{\lambda}{C}\right]\right)^{a_{Q_{z'}(\lambda)}(\lambda_{z'}+\delta)\,\mathrm{d}\langle z',\delta\rangle}$$

in $\mathbb{C}[\![U_\mathbb{Z}^* \cap \sigma^\vee]\!][U_\mathbb{Z}^*]$ for a suitable lattice $U_\mathbb{Z} \subseteq U_\mathbb{Q}$. Here C is the conductor, i.e., the volume of $\widehat{\mathbb{Z}}$ with respect to $\mathrm{d}x$.

$\mathrm{BP}(F,I,z',\alpha,\sigma)$ is independent of the chosen measure.

PROOF OF THE INDEPENDENCE. Rewrite the first integral as

$$\left.\frac{\prod_{\delta\in I_{\mathbb{A}^{(\infty)}}}(1-\chi_\alpha(\langle z',\delta\rangle)[C^{-1}\lambda])^{(a_0(\delta)/2)\,\mathrm{d}\langle z',\delta\rangle}}{(1-[C^{-1}\lambda])^{(a_0(0)/2)}}\right|_{\lambda=0}.$$

Now consider the multiplicative integral

$$\prod_{\delta\in(z')^{-1}(C\hat{\mathbb{Z}})+a} (1-\chi_\alpha(\langle z',\delta\rangle)[C^{-1}\lambda])^{\mathrm{d}\langle z',\delta\rangle} = (1-\chi_\alpha(\langle z',a\rangle)[C^{-1}\lambda]). \tag{3.4}$$

All multiplicative integrals occurring in $\mathrm{BP}(F,I,z',\alpha,\sigma)$ split into a product of these integrals by hypothesis on the measure. Now changing the measure to $N\,\mathrm{d}\delta$, we get the expression

$$\prod_{\delta\in(z')^{-1}(C\hat{\mathbb{Z}})} (1-\chi_\alpha(N^{-1}\langle z',a\rangle)\chi_\alpha(N^{-1}\langle z',\delta\rangle)[(NC)^{-1}\lambda])^{\mathrm{d}N^{-1}\langle z',\delta\rangle}$$
$$= \prod_{i\in\mathbb{Z}/N\mathbb{Z}} (1-\chi_\alpha(N^{-1}\langle z',a\rangle)\zeta^i[(NC)^{-1}\lambda])$$
$$=(1-\chi_\alpha(\langle z',a\rangle)[C^{-1}\lambda])$$

here ζ is some primitive Nth root of unity in $\mathbb{C}$. This shows invariance of the multiplicative integral (3.4). We have to check what happens to

$$\frac{1}{(1-[\lambda C^{-1}])^{a_0(0)/2}}.$$

It changes to $1/(1-[\lambda(NC)^{-1}])^{a_0(0)/2}$ hence gets multiplied by

$$(1+[\lambda(CN)^{-1}]+[2\lambda(CN)^{-1}]+\cdots+[(N-1)\lambda(CN)^{-1}])^{a_0(0)/2},$$

which for $\lambda = 0$ is equal to $N^{a_0(0)/2}$. $C^{-a_0(0)/2}$ gets $(NC)^{-a_0(0)/2}$. Therefore the whole expression $\mathrm{BP}(F, I, z', \alpha, \sigma)$ is invariant. □

We start by observing the following lemmas about the adelic Borcherds products:

3.2.10. Lemma. *$\mathrm{BP}(F, I, z', \alpha, \sigma)$ multiplied by the trivialization $(s_{z'})^{\otimes a_0(0)/2}$ (2.8.2) is invariant under $z' \mapsto \beta z'$ for $\beta \in \mathbb{Q}_{>0}$, i.e., we have*

$$\mathrm{BP}(F, I, \beta z', \alpha, \sigma) = \beta^{-a_0(0)/2}\, \mathrm{BP}(F, I, z', \alpha, \sigma).$$

PROOF. Changing z' in the various $\langle z', \delta \rangle$ has the same effect as scaling the measure, hence (by the last lemma) as multiplying C by β. The expression $\lambda_{z'}$ gets divided by β (3.3), $Q_{z'}(\lambda)$ gets divided by β^2. Substituting $\beta^2\lambda$ for λ, we get the claimed expression. □

3.2.11. Lemma. *For $w = k \wedge z_1 \in U_{\mathbb{Q}}$, $\mathrm{BP}(F, I, \exp(w)z', \alpha, \sigma)$ is equal to $\mathrm{BP}(F, I, z', \alpha, \sigma)$ transformed by the homomorphism $[\lambda] \mapsto \exp\Big(\alpha\big(\lambda(w)\big)\Big)[\lambda]$.*

PROOF. Let $w = k \wedge z_1$ and

$$z'' := \exp(w)z' = z' + \langle z_1, z' \rangle k - \big(\langle k, z' \rangle + \langle z_1, z' \rangle Q(k)\big) z_1.$$

First $Q_{z'} = Q_{z''}$ because $\langle z, z' \rangle = \langle z, z'' \rangle$.

We have (see (3.3) for the definition):

$$\lambda_{z''} + \delta = \lambda_{z'} + \delta - \frac{\lambda(w)}{\langle z_1, z' \rangle} z_1.$$

If we substitute $\delta \mapsto \delta + (\lambda(w)/\langle z_1, z' \rangle) z_1$, we get the same by applying $[\lambda C^{-1}] \mapsto \chi_\alpha(\lambda w)[\lambda C^{-1}]$, i.e., by $[\lambda] \mapsto \exp\big(\alpha(\lambda w)\big)[\lambda]$. □

3.2.12. We recall some definitions from [26]. Let $L_{\mathbb{Z}}$ be a lattice of dimension l with nondegenerate quadratic form in $\mathrm{Sym}^2(L_{\mathbb{Z}}^*)$. Let $M_{\mathbb{Z}}$ be a lattice of dimension m with quadratic form, not necessarily integer valued. We assume $l - m \geq 1$ throughout. Let κ be a coset in $(L_{\mathbb{Z}}^*/L_{\mathbb{Z}}) \otimes M_{\mathbb{Z}}^*$.

For each prime p and $s \in \mathbb{N}_0$, we define [26, Definition 5.9]:

$$\begin{aligned}
\lambda_p(L_{\mathbb{Z}}; s) &:= \frac{\mathrm{vol}\big(\mathrm{SO}'(L_{\mathbb{Z}_p} \oplus H^s_{\mathbb{Z}_p})\big)}{\prod_{i=1}^s (1 - p^{-2i})}, \\
\mu_p(L_{\mathbb{Z}}, M_{\mathbb{Z}}, \kappa; s) &:= \mathrm{vol}\big(\mathrm{I}(L \oplus H^s, M)(\mathbb{Q}_p) \cap (\kappa \oplus (H^s \otimes M^*)_{\mathbb{Z}_p})\big) \\
&= |d(L_{\mathbb{Z}_p})|^{m/2} |d(M_{\mathbb{Z}_p})|^{-s+(1+m-l)/2} \beta_p(L_{\mathbb{Z}_p}, M_{\mathbb{Z}_p}, \kappa; s).
\end{aligned}$$

The observation [26, 5.11] is that, for $p \neq 2$, these functions are actually polynomials in p^{-s} for sufficiently large s. In the following it is understood that we replaced them by the corresponding polynomial and consider them as functions of $s \in \mathbb{C}$. $\lambda_2(L_{\mathbb{Z}}; s)$ is replaced by $\lambda_2(L_{\mathbb{Z}}; 0)$.

We define furthermore [26, Definition 6.1]:

$$\begin{aligned}
\lambda_\infty(L_{\mathbb{Z}}; s) &:= \Gamma_{l-1,l}(s), \\
\mu_\infty(L_{\mathbb{Z}}, M_{\mathbb{Z}}, \kappa; s) &:= \Gamma_{m,l}(s).
\end{aligned}$$

and the following Euler products [26, 10.2, 10.3]:

$$\lambda(L_{\mathbb{Z}}; s) := \prod_{\nu} \lambda_{\nu}(L_{\mathbb{Z}}; s),$$

$$\mu(L_{\mathbb{Z}}, M_{\mathbb{Z}}, \kappa; s) := \prod_{\nu} \mu_{\nu}(L_{\mathbb{Z}}, M_{\mathbb{Z}}, \kappa; s).$$

The μ functions are related to Whittaker integrals occurring as Euler factors of Fourier coefficients of Eisenstein series. The coefficient μ_p is equal to the corresponding p-factor ([26, Theorem 4.5]), whereas the infinite factor is only indirectly related to μ_∞ ([26, Theorem 4.7]). In the case $m = 1$ it occurs as a limit of the Archimedean factor up to a factor $-|2d(M_{\mathbb{Z}})|^{s/2}$, where $d(M_{\mathbb{Z}})$ is the discriminant of the associated bilinear form (a number in this case) on $M_{\mathbb{Z}}$. More precisely, we have (cf. [26, Section 10])

$$\lim_{\alpha \to \infty} |\alpha|^{-l/2} e^{2\pi\alpha^2 q} E_q(\Psi_\infty \Phi(\chi_\kappa); g_l(\alpha), s - s_0) = |4q|_\infty^{-s/2} \mu(L_{\mathbb{Z}_p}, \langle q \rangle, \kappa; s) \tag{3.5}$$

(the limit corresponds to taking $\tau \to i\infty$ in the classical picture). It is the derivative in s of this quantity is actually is related to the integral of a Borcherds form (cf. Theorem 3.2.14 below). Therefore it is convenient to define also for $m > 1$:

$$\widetilde{\mu}(L_{\mathbb{Z}}, M_{\mathbb{Z}}, \kappa; s) = |2d(M_{\mathbb{Z}})|_\infty^{-s/2} \mu(L_{\mathbb{Z}}, M_{\mathbb{Z}}, \kappa; s).$$

In (3.5) $E_q(\Psi; g, s)$ is the qth Fourier coefficient of the Eisenstein series defined as:

$$\mathrm{Sp}_2'(\mathbb{Q}) \backslash \mathrm{Sp}_2'(\mathbb{A}) \times \mathbb{C} \longrightarrow \mathbb{C}$$

$$(g, s) \longmapsto E(\Psi; g, s) := \sum_{\gamma \in P(\mathbb{Q}) \backslash \mathrm{Sp}(\mathfrak{M}_{\mathbb{Q}})} \Psi(s)(\gamma g).$$

Here $\Psi(s) := \Psi_{\infty, l/2}(s) \Phi(\chi_\kappa, s)$; where $\Phi(\chi_\kappa; s)$ is the natural continuation in $s \in \mathbb{C}$ of

$$\Phi(\chi_\kappa; s_0) \in \mathrm{I}_{\mathbb{A}^{(\infty)}}(s_0, \xi'_{\mathbb{A}^{(\infty)}})$$

given by $\Phi(\chi_\kappa; s_0)(g) = (\omega(g)\chi_\kappa)(0)$ where ω is the Weil-representation and ξ' is (essentially) the associated quadratic character; and where

$$\Phi_{\infty, l/2}(s) \in \mathrm{I}_{\mathbb{R}}(s, \xi'_{\mathbb{R}}).$$

is the standard section of weight $\frac{l}{2}$. See [26, Section 4] for details and also [11] for a more classical definition of this Eisenstein series. Also for $m > 1$, μ_p always occurs as the p-factor of the Fourier coefficient of the analogously defined Eisenstein series of higher genus. The relation between μ_∞ and $E_{Q,\infty}$, however, is not so clear.

We define also:

$$\widetilde{\lambda}(L_{\mathbb{Z}}; s) := |d(L_{\mathbb{Z}})|_\infty^{s/2} \lambda(L_{\mathbb{Z}}; s).$$

The functions μ_p and λ_p can be computed explicitly. See [26, Theorem 8.1, 8.2]. For explicit computations of $\widetilde{\lambda}$ from this we refer the reader to Appendix B. We will continue the discussion of these functions and their relations in 3.5.1.

3.2.13. Using the q-expansion principle 2.8.7 we will show the following theorem on integral properties of Borcherds products. It is a strengthening of the main results of Borcherds [2, Theorem 13.3] and Kudla [35, Theorem 2.12] (respectively Bruinier–Kühn [11, Theorem 4.11]).

Let $p \neq 2$ be a prime.[2] Let $L_{\mathbb{Q}}$ again be a quadratic space of signature $(l-2,2)$ and let $L_{\mathbb{Z}_{(p)}} \subset L_{\mathbb{Q}}$ be a lattice such that the quadratic form is unimodular. For this to exist it is necessary that p does not divide a minimal discriminant of $L_{\mathbb{Q}}$.

Assume that f has Fourier coefficients of the form $a_q = \chi_{L_{\mathbb{Z}_p}} \otimes a_q^{(p)}$, $a_q^{(p)}$ as in 3.2.3 with values in $S(L_{\mathbb{A}^{(\infty,p)}})$. In this case, f is invariant under a (w.l.o.g.) *admissible* compact open subgroup $K \subseteq \mathrm{SO}(\mathbb{A}^{(\infty)})$.

3.2.14. Theorem. *There is an $N \in \mathbb{N}$, coprime to p, such that for $R = \mathbb{Z}_{(p)}[\zeta_{N,\mathbb{C}}] \subset \mathbb{C}$, and up to multiplication of F by a large constant $\in \mathbb{Z}$, we have*

(1) *For any r.p.c.d. Δ, there is a section*

$$\Psi(F) \in H^0(\mathrm{M}({}^K_\Delta\mathbf{O})_R, (\Xi^*\mathcal{E})_{\text{horz.}}^{\otimes a_0(0)/2})$$

such that on $\mathrm{M}({}^K\mathbf{O})(\mathbb{C})$, we have (see 3.2.5 for the definition of $h_{\mathcal{E}}$)

$$\Phi(\nu, h; F) = -2\log \Xi^* h_{\mathcal{E}}\big(\Psi_{\mathbb{C}}(\nu, h; F)\big).$$

There is a locally constant, invertible function Λ on $\mathrm{M}({}^K_\Delta\mathbf{O})_R$ such that

$$\Lambda\Psi(F) \in H^0(\mathrm{M}({}^K_\Delta\mathbf{O}), (\Xi^*\mathcal{E})_{\text{horz.}}^{\otimes a_0(0)/2})$$

(i.e., $\Lambda\Psi(F)$ is defined over $\mathbb{Z}_{(p)}$).

(2)

$$\mathrm{div}(\Psi(F)^2) = \sum_{q<0} Z(L_{\mathbb{Z}}, \langle -q \rangle, a_q; K) + D_\infty$$

(cf. 3.1.6), where the support of D_∞ is contained in the support of the exceptional divisor.

(3)

$$\int_{\mathrm{M}({}^K\mathbf{O})_{\mathbb{C}}} \log \Xi^* h_{\mathcal{E}}\big(\Lambda\Psi_{\mathbb{C}}(\nu, h; F)\big)\, \mathrm{c}_1(\Xi^*\mathcal{E}, \Xi^*h)^{l-2}$$
$$\equiv \sum_{q<0} \mathrm{vol}_{\mathcal{E}}\big(\mathrm{M}({}^K\mathbf{O})\big) \cdot \frac{\mathrm{d}}{\mathrm{d}s} \widetilde{\mu}(L_{\mathbb{Z}}, \langle -q \rangle, a_q; s)\Big|_{s=0}$$

in $\mathbb{R}^{(p)}$ (see 3.2.12 for the definition of $\widetilde{\mu}$).

(4) *For any saturated isotropic line $I \subset L_{\mathbb{Z}_{(p)}}$ and primitive isotropic vector z' in $L_{\mathbb{Z}_{(p)}}$ with $\langle I, z' \rangle \neq 0$ consider the associated boundary morphism (2.4.5)*

$$\iota\colon \mathbf{H}_0[(I^\perp/I) \otimes I, 0] \to \mathbf{O}(L).$$

Using the trivialization $(s_{z'})^{\otimes a_0(0)/2}$ of 2.8.2, we get a Fourier expansion defined in 2.8.5 and 2.8.6 which is valid on a set of the form

$$\{\alpha\} \times \big(x + U_{\mathbb{R}} + \alpha(\sigma)\big) \times \{1\} \subset \mathbb{D}_{\mathbf{H}_0} \times U_{\mathbb{C}} \times U_{\mathbb{A}^{(\infty)}},$$

where $\sigma \subset U_{\mathbb{R}}$ is some rational polyhedral cone. The corresponding formal function

$$\sum_{\lambda \in U_{\mathbb{Q}}^*} a_{\iota, z', \lambda, \alpha, 1, \sigma}\big(\Psi(F)\big)[\lambda]$$

is, up to multiplying $\Psi(F)$ by a constant of absolute value 1, given by:

$$[\rho(F, I, \sigma)] \cdot \mathrm{BP}(F, I, z', \alpha, \sigma). \tag{3.6}$$

[2] $p \neq 2$ in order to be able to apply the Main Theorem 2.5.9

Remark. The expansion in (4) suffices to know every other expansion, i.e., for $\rho \neq 1$, by property (3.2).

PROOF. First observe that F is a finite linear combination of functions valued in $\mathbb{C}[L_\mathbb{Z}/L_\mathbb{Z}^*]$ for various saturated lattices $L_\mathbb{Z}$ of $L_{\mathbb{Z}_{(p)}}$. Hence we may assume that F is of this kind, observing that all properties are stable under taking finite linear combinations, respectively change of K. Hence w.l.o.g. F corresponds to an f considered in [2].

Recall the normalization of the Hermitian metric $h_\mathcal{E}$ (3.2.5). Consider the boundary morphism as in 4. associated with some I and a $\langle z' \rangle$:

$$\iota \colon \mathbf{H}_0[(I^\perp/I) \otimes I, 0] \to \mathbf{O}(L)$$

$\Psi/s_{z'}$ may be seen as a function on $L_{0,\mathbb{C}} = \langle z', I \rangle_\mathbb{C}^\perp$. We denote the corresponding variable by Z. We have

$$h_\mathcal{E}(\Psi) = \left|\frac{\Psi}{s}\right|^2 h_\mathcal{E}(s),$$

and

$$h_\mathcal{E}(s) = \left((2\pi)^{-1} \mathrm{e}^{-\gamma} \langle \Im(Z), \Im(Z) \rangle (2\pi)^2\right)^{a_0(0)/2}.$$

Hence we get

$$\log h_\mathcal{E}(\Psi) = \log\left|\frac{\Psi}{s}\right|^2 + \frac{a_0(0)}{2}(\log(\langle \Im(Z), \Im(Z) \rangle) + \log(2\pi) - \gamma)$$

and (1) follows from [2, Theorem 13.3], (3) is essentially [35, Theorem 2.12(ii)], using [26, Theorem 4.7 and Section 10.2]]. See [11, Theorem 4.11] for an independent proof (use [26, Lemma 7.3] for the comparison). Observe that the function Ψ differs from the one in [35] or [2] by $(2\pi i)^{a_0(0)/2}$. The constant of the Hermitian metric has been adjusted, such that (3) holds without an additive constant. Compare also to formula (3.43) of the proof of Main Theorem 3.5.5, where the same constant appears naturally, too.

That some multiple of F exists, such that *all* coefficients are integral and even is due to the classical q-expansion principle. Further multiplication with the order of the character of $\mathrm{SO}(L_\mathbb{Q})$ occurring in [2] (cf. [6, Theorem 3.22] for its finiteness) we get a section of the *complex* bundle $\Xi^* \mathcal{E}_\mathbb{C}$.

It suffices to prove (4) for any choice of z' and $\alpha \colon 1 \mapsto 2\pi i$ (the same choice as in the construction of the Borcherds lift). For note that change of z' by β' changes BP by $\beta^{-a_0(0)/2}$ (3.2.10) and $s_{z'}^{\otimes a_0(0)/2}$ by $\beta^{a_0(0)/2}$. Changing z' by $\exp(w)z'$ has the effect of translating by w in the $U_\mathbb{C}$-domain. Hence a formal function gets transformed by $[\lambda] \mapsto \exp(2\pi i \lambda(w))[\lambda]$. This is the same effect as observed by BP (3.2.11). The Weyl vector is independent of z', too, so the above transformation has the effect of multiplication by a constant of absolute value 1.

Choose $z = z_{\mathrm{Bor}} \in L_\mathbb{Z} \cap I$ and $z'_{\mathrm{Bor}} \in L_\mathbb{Z}^*$ as in [2], i.e., z_{Bor} primitive and isotropic and $\langle z_{\mathrm{Bor}}, z'_{\mathrm{Bor}} \rangle = 1$.

We may now show (4) for $z' := z'_{\mathrm{Bor}} - Q(z'_{\mathrm{Bor}}) z_{\mathrm{Bor}}$, which is now isotropic, but satisfies still $\langle z_{\mathrm{Bor}}, z' \rangle = 1$. z' may not lie in $L_\mathbb{Z}^*$, however.

As measure we may choose the standard invariant measure on $\mathbb{A}^{(\infty)}$, giving $\widehat{\mathbb{Z}}$ the volume one. The trivialization now coincides (up to the factor $(2\pi i)^{a_0(0)/2}$ that we incorporated in the characterization of $\Psi(F)$) with the one given in [2], up to identification of $I^\perp/I$ with $I \otimes (I^\perp/I)$ via $k \mapsto k \otimes z_{\mathrm{Bor}}$. This interchanges Q and $Q_{z'}$.

Recall from [2, Theorem 13.3(v)] that the Fourier expansion with respect to this particular choice is (up to a constant of absolute value 1) given by

$$\prod_{\delta \in \mathbb{Z}/N\mathbb{Z}} (1 - e(\delta/N))^{a_{\delta z/N}(0)/2} \cdot [\rho(U, \sigma, F_U)] \cdot \prod_{\substack{\lambda \in (\langle z\rangle^\perp/\langle z\rangle)^* \\ \langle\lambda,\sigma\rangle>0}} \prod_{\substack{\delta \in (L_\mathbb{Z})^*/L_\mathbb{Z} \\ \delta|_{z^\perp}=\lambda}} (1 - e(\langle\delta, z'\rangle)[\lambda])^{a_{\lambda^2/2}(\delta)}. \tag{3.7}$$

The first product in BP may be rewritten as

$$\prod_{\delta \in \frac{1}{N}\mathbb{Z}/\mathbb{Z}} (1 - e(\delta))^{a_0(\delta z)/2},$$

and coincides with the product in Borcherds' expression. The Weyl vector is the same, too.

3.2.15. Lemma. *If* $\langle z, z'\rangle = 1$, *there is a bijection between*

$$\{\lambda \in U_\mathbb{Q}^*, \delta \in \mathbb{Q}/\mathbb{Z} = \mathbb{A}^{(\infty)}/\widehat{\mathbb{Z}} \mid \lambda_{z'} + \delta z \in L_\mathbb{Z}^*\}.$$

and

$$\{k_{\mathrm{Bor}} \in (\langle z\rangle_\mathbb{Z}^\perp/\langle z\rangle_\mathbb{Z})^*, \delta_{\mathrm{Bor}} \in L_\mathbb{Z}^*/L_\mathbb{Z}, \delta_{\mathrm{Bor}}|_{z^\perp} = k_{\mathrm{Bor}}\}.$$

PROOF. The map is given by sending (λ, δ) to $(k_{\mathrm{Bor}}, \delta_{\mathrm{Bor}}) = (\lambda z, \lambda_{z'} + \delta z)$. For $v \in \langle z\rangle_\mathbb{Z}^\perp$, we have

$$\langle k_{\mathrm{Bor}}, v\rangle = \langle \lambda z, v\rangle = \langle \lambda_{z'}, v\rangle = \langle \lambda_{z'} + \delta z, v\rangle \in \mathbb{Z},$$

hence $k_{\mathrm{Bor}} \in (\langle z\rangle_\mathbb{Z}^\perp/\langle z\rangle_\mathbb{Z})^*$.

In the other direction, we map $(k_{\mathrm{Bor}}, \delta_{\mathrm{Bor}})$ to $(\lambda, \delta) = (k_{\mathrm{Bor}} \otimes z', \langle\delta_{\mathrm{Bor}}, z'\rangle)$. These maps are inverse to each other. One direction is clear; for the other we have to show:

$$\delta_{\mathrm{Bor}} \equiv \tilde{k}_{\mathrm{Bor}} + \langle\delta_{\mathrm{Bor}}, z'\rangle z \pmod{L_\mathbb{Z}}),$$

where $\tilde{k}_{\mathrm{Bor}}$ is the representative in $\langle z\rangle_\mathbb{Q}^\perp$ that satisfies $\langle\tilde{k}_{\mathrm{Bor}}, z'\rangle = 0$. This follows from the condition $\delta_{\mathrm{Bor}}|_{z^\perp} = k_{\mathrm{Bor}}$. □

The lemma shows that the second product occurring in Borcherds' Fourier expansion is equal to the second product appearing in BP. This finishes the proof of (4).

We now show the integrality properties stated in (1) and (2).

We can multiply $\Psi(F)$ by an appropriate complex number of absolute value 1 on every connected component and hence may assume that *one* Fourier expansion for each connected component is given *precisely* by

$$[\rho(\omega(\rho)F, I, \sigma)]\,\mathrm{BP}(\omega(\rho)F, I, z', \alpha, \sigma). \tag{3.8}$$

We get a finite number of different expansions with coefficients in some ring $R = \mathbb{Z}_{(p)}[\zeta_{N,\mathbb{C}}]$, $p \nmid N$ which we may assume to contain the ring described by 2.8.7. By multiplying F again by a large integer, we may also assume that all Weyl vectors involved are integral.

Therefore, by 2.8.7, $\Psi(F)$ extends to an (integral) section of

$$H^0(\mathrm{M}({}^K_\Delta\mathbf{O})_R, \Xi^*(\mathcal{E})_{\mathrm{horz.}}^{a_0(0)/2})$$

for any Δ. Furthermore, the expansion (3.8) is invertible integrally, hence we may infer that $\Psi(F)^{-1}$ is a section[3] of

$$H^0(\mathrm{M}({}^K_\Delta\mathbf{O})_R, \Xi^*(\mathcal{E})_{\text{horz.}}^{-a_0(0)/2}).$$

It follows that $\Psi(F)$ cannot have any connected (= irreducible) component of the fibre above p in its divisor.

To prove that $\Psi(F)$ is actually defined over $\mathbb{Z}_{(p)}$, i.e., lies in

$$H^0(\mathrm{M}({}^K_\Delta\mathbf{O}), \Xi^*(\mathcal{E})_{\text{horz.}}^{a_0(0)/2}),$$

we have to check the requested invariance property of the Fourier coefficients.

We compute

$$a_{\iota,z',\lambda,\alpha,\rho,\sigma}\big({}^\tau\Psi(F)\big) = {}^\tau a_{\iota,z',\lambda,\alpha,\rho\iota(k_\tau)^{-1},\sigma}\big(\Psi(F)\big)$$

for all $\tau \in \mathrm{Gal}(\mathbb{Q}(\zeta_N) \mid \mathbb{Q})$.

By the transformation property (3.2), we are reduced to compute

$$a_{\iota,z',\lambda,\alpha,1,\sigma}\big({}^\tau\Psi(F')\big) = {}^\tau a_{\iota,z',\lambda,\alpha,1,\sigma}\big(\Psi(\omega(\iota(k_\tau))^{-1}F'\big)$$

for $F' = \omega(\rho)F$. We will, however, keep the letter F.

We have seen that, up to a constant of absolute value 1, the coefficients

$$a_{\iota,z',\lambda,\alpha,1,\sigma}\big(\Psi(\omega(\iota(k_\tau))^{-1}F\big)$$

are the ones forming $\mathrm{BP}\big(\omega(\iota(k_\tau))^{-1}F, I, z', \alpha, \sigma\big)$. We have by definition

$$a_m\big(\omega(\iota(k_\tau))^{-1}F\big)(\rho) = a_m(F)(\iota(k_\tau)\rho).$$

However $\iota(k_\tau)$ acts trivial on $\lambda_{z'}$ because the latter is perpendicular to $I \oplus \langle z'\rangle$, and it acts by multiplication by k_τ on δ. Substituting $k_\tau^{-1}\delta$ for δ and using invariance of the measure ($k_\tau \in \widehat{\mathbb{Z}}^*!$), we get the same expression, but all $\chi_\alpha(\langle z', \delta\rangle)$ changed to $\chi_\alpha(\langle z', k_\tau^{-1}\delta\rangle) = \chi_\alpha(k_\tau^{-1}\langle z', \delta\rangle) = \chi_\alpha(\langle z', \delta\rangle)^{\tau^{-1}}$. This shows:

$$a_{\iota,z',\lambda,\alpha,1,\sigma}\Big(\Psi\big(\omega(\iota(k_\tau))^{-1}F\big)\Big) = \Lambda_\tau{}^{\tau^{-1}} a_{\iota,z',\lambda,\alpha,1,\sigma}\big(\Psi(F)\big).$$

with $|\Lambda_\tau| = 1$ and $\Lambda_\tau \in R^*$ (because $\Psi(F)$ is defined over R and invertible in the sense above).

Therefore

$${}^\tau\Psi(F) = ({}^\tau\Lambda_\tau)\Psi(F)$$

for all $\tau \in \mathrm{Gal}(\mathbb{Q}(\zeta_N)|\mathbb{Q})$ where Λ_τ locally constant invertible on $\mathrm{M}({}^K\mathbf{O})_R$ with $|\Lambda_\tau|_\infty \equiv 1$ constant. From Hilbert 90 follows that there exists a Λ, locally constant invertible on $\mathrm{M}({}^K\mathbf{O})_R$, such that

$$({}^\tau\Lambda)/\Lambda = {}^\tau\Lambda_\tau. \tag{3.9}$$

i.e.,

$${}^\tau\big(\Lambda\Psi(F)\big) = \Lambda\Psi(F),$$

hence $\Lambda\Psi(F)$ is defined over $\mathbb{Z}_{(p)}$.

It remains to show (3). Without the Λ this is proven in [35] or [11] as an identity in $\mathbb{R}$, cf. beginning of this proof. The evaluation of

$$\int \log(\Xi^*h_{\mathcal{E}})\big(\Lambda\Psi_{\mathbb{C}}(\nu, h; F)\big)\, \mathrm{c}_1\big(\Xi^*(\overline{\mathcal{E}})\big)^{l-2}$$

---

[3] Up to the problem of constants, this is $\Psi(-F)$.

differs from the same expression for $\Lambda = 1$ by

$$\sum_{\alpha \in \pi_0(\mathrm{M}({}^K\mathbf{O})_R)} \mathrm{vol}(\alpha) \log(|\Lambda(\alpha)|_\infty).$$

Equation (3.9) together with $|\Lambda_\tau| = 1$ implies $|\Lambda(\tau^{-1}\alpha)^\tau|_\infty = |\Lambda(\alpha)|_\infty$ hence we may write the sum as

$$\begin{aligned}\frac{1}{\phi(N)} & \sum_{\substack{\tau \in \mathrm{Gal}(\mathbb{Q}(\zeta_N)|\mathbb{Q}) \\ \alpha \in \pi_0(\mathrm{M}({}^K\mathbf{O})_R)}} \mathrm{vol}(\alpha) \log(|\Lambda(\tau^{-1}\alpha)^\tau|_\infty) \\ &= \frac{1}{\phi(N)} \sum_{\substack{\tau \in \mathrm{Gal}(\mathbb{Q}(\zeta_N)|\mathbb{Q}) \\ \alpha \in \pi_0(\mathrm{M}({}^K\mathbf{O})_R)}} \mathrm{vol}(\tau\alpha) \log(|\Lambda(\alpha)^\tau|_\infty) \\ &= \frac{1}{\phi(N)} \sum_{\alpha \in \pi_0(\mathrm{M}({}^K\mathbf{O})_R)} \mathrm{vol}(\alpha) \log\left(\left|N_{\mathbb{Q}(\zeta_N)|\mathbb{Q}}(\Lambda(\alpha))\right|_\infty\right) \equiv 0 \quad \text{in } \mathbb{R}^{(p)}.\end{aligned}$$

Here we used $\mathrm{vol}(\tau\alpha) = \mathrm{vol}(\alpha)$ which is true because $\mathrm{c}_1(\Xi^*(\mathcal{E}))$ is Galois invariant. This finishes the proof of (3). □

Later we will need the following general facts about the line bundle of orthogonal modular forms.

3.2.16. Lemma. *With respect to the Hodge embedding* (2.4.4)

$$\mathbf{S}(L) \hookrightarrow \mathbf{H}(\mathrm{C}^+(L), \langle , \rangle_\delta),$$

the pullback of the canonical bundle $\Lambda^g \mathcal{L}$ via the 'dual' embedding

$$\mathrm{M}^\vee(\mathbf{S}(L)) \longrightarrow \mathrm{M}^\vee(\mathbf{H}(\mathrm{C}^+(L), \langle , \rangle_\delta))$$

is, up to a bundle coming from a character $\mathrm{GSpin}(L) \to \mathbb{G}_\mathrm{m}$, *and up to a tensor-power, the bundle* $\mathcal{E}$ (3.2.5).

PROOF. Recall that the induced map on compact duals (3.1.2) is given by $\langle z \rangle \mapsto zw\,\mathrm{C}^+(L)$ where z is assumed to be primitive in $L_{\mathbb{Z}_{(p)}}$ and w is *any* primitive vector with $\langle w, z \rangle = 1$. The fibre of $\mathcal{L}$ at the image of $\langle z \rangle$ is given by the space $zw\,\mathrm{C}^+(L)$ and the fibre of $\mathcal{E}$ is given by $\langle z \rangle$. By an explicit determination of the action of the stabilizer group in both cases the result follows. □

3.2.17. Corollary. *The pullback of the line bundle of Siegel modular forms $\Xi^*(\Lambda^g\mathcal{L})^{\otimes n}$ of some positive weight n is equal to some positive power of the bundle of orthogonal modular forms $\Xi^*\mathcal{E}^{\otimes n'}$. In particular $(\Xi^*\mathcal{E})$ is ample on every* $\mathrm{M}({}^K\mathbf{S}(L))$ *and* $\mathrm{M}({}^K\mathbf{O}(L))$ *and semiample on* $\mathrm{M}({}^K_\Delta\mathbf{O}(L))$.

PROOF. Follows from 3.2.16 and 2.6.5. The bundle of Siegel modular forms $\Xi^*(\Lambda^g\mathcal{L})$ is ample [19]. The ampleness of $\Xi^*\mathcal{E}$ follows from the fact that the Hodge embedding induces an embedding (up to a normalization) of the corresponding Shimura varieties (2.5.5) for suitable levels and the fact that all $\mathrm{M}({}^K\mathbf{S}(L))$ and $\mathrm{M}({}^K\mathbf{O}(L))$ are related by finite etale maps that pullback the $\Xi^*\mathcal{E}$'s into each other. The assertion on base points is true because a suitable power of $\Xi^*(\Lambda^g\mathcal{L})$ defines a morphism to the minimal compactification on every $\mathrm{M}({}^{K'}_\Delta\mathbf{H}_g)$ [19, Theorem 2.3(1)]. □

We will compare the measure given by the volume form $c_1(\Xi^*\overline{\mathcal{E}})^{l-2}$, the highest power of the Chern form of the canonical Hermitian line bundle defined above with the quotient μ of the canonical volume form [26, 2.4] on $\mathrm{SO}(L_\mathbb{R})$ by the one giving K_∞ = stabilizer of some positive definite subspace N in $L_\mathbb{R}$ the volume one.

3.2.18. Lemma.

$$2\lambda_\infty^{-1}(L;0)\mu = \big(c_1(\Xi^*\mathcal{E},\Xi^*h)\big)^{l-2}.$$

PROOF. Choose an orthogonal basis $\{e_1,\dots,e_l\}$ of $L_\mathbb{R}$ with $\langle e_j,e_j\rangle=-1$, $j=1,2$ and $\langle e_j,e_j\rangle=1$, $j>2$. Set $z=\frac{1}{2}(e_3-e_2)$, $z'=\frac{1}{2}(e_3+e_2)$. We have $\langle z,z\rangle=0$ and $\langle z,z'\rangle=\frac{1}{2}$. We let $\xi:=ie_1+e_2$. We have $\langle\xi\rangle\in\mathbb{D}_\mathbf{O}\subset\mathbb{P}^1(L_\mathbb{C})$. It induces an isomorphism $\mathrm{SO}(L_\mathbb{R})/K_\infty\cong\mathbb{D}_\mathbf{O}$. The tangent space at $\langle\xi\rangle$ in $\mathbb{D}_\mathbf{O}\subset\mathrm{M}^\vee\big(\mathbf{O}(L)\big)(\mathbb{C})$ is canonically identified with

$$\mathrm{Hom}(\langle\xi\rangle_\mathbb{C},\langle\xi\rangle_\mathbb{C}^\perp/\langle\xi\rangle).$$

The induced map

$$\mathrm{Lie}\big(\mathrm{SO}(L_\mathbb{R})\big)\to\mathrm{Hom}(\langle\xi\rangle_\mathbb{C},\langle\xi\rangle_\mathbb{C}^\perp/\langle\xi\rangle_\mathbb{C})$$

is given by

$$\Lambda\mapsto\{\xi\mapsto\Lambda\xi \bmod \langle\xi\rangle_\mathbb{C}\}.$$

On $\mathrm{Lie}\big(\mathrm{SO}(L_\mathbb{R})\big)$ we have the basis $\{L_{jk}\}_{j<k}$ with

$$L_{jk}:=e_j\wedge e_k,$$

whose associated volume form is the canonical volume form in the sense [26, 2.4] (see [26, Lemma 3.6]). The basis elements L_{12}, L_{jk}, $j,k\geq 3$ are mapped to 0. Its associated volume form is likewise the canonical on K_∞ [26, Lemma 3.6]. The quotient volume form on $\mathbb{D}_\mathbf{O}$ hence is associated with

$$\bigwedge_{j=3}^{l}\tilde{L}_{2j}^*\wedge\bigwedge_{j=3}^{l}\tilde{L}_{1j}^*\in\bigwedge^{2(l-2)}\mathrm{Hom}(\langle\xi\rangle_\mathbb{C},\langle\xi\rangle_\mathbb{C}^\perp/\langle\xi\rangle_\mathbb{C})^*,$$

where $\tilde{L}_{1j}=\{\xi\mapsto ie_j\}$ and $\tilde{L}_{2j}=\{\xi\mapsto e_j\}$ (wedge product over $\mathbb{R}$!). Hence we have that the volume form associated with μ is (at $\langle\xi\rangle_\mathbb{C}$) given by

$$\frac{1}{\mathrm{vol}(K_\infty)}\bigwedge_{j=3}^{l}\tilde{L}_{2j}^*\wedge\bigwedge_{j=3}^{l}\tilde{L}_{1j}^*,$$

where the volume of K_∞ is computed with respect to the canonical volume form. Observe that $K_\infty=\mathrm{SO}(N)\times\mathrm{SO}(N^\perp)$ because K_∞ is not allowed to change the orientation of N (the $\mathbb{C}$-oriented negative definite plane corresponding to $\langle\xi\rangle$). Hence the volume of K_∞ is $2\pi\cdot\frac{1}{2}\prod_{j=1}^{l-2}2\pi^{j/2}/\Gamma(j/2)$ [26, Lemma 3.7(2)].

Now let $L_0:=\langle e_1,e_4,\dots,e_l\rangle=\langle z,z'\rangle^\perp$. For $Z\in L_{0,\mathbb{C}}$ we write $Z=X+iY$ for the decomposition into real and imaginary part.

Under the parametrization

$$\text{(3.10)}\qquad \begin{aligned} L_{0,\mathbb{C}} &\longrightarrow \mathrm{M}^\vee\big(\mathbf{O}(L)\big)(\mathbb{C})\\ Z &\longmapsto \langle Z+z'-\langle Z,Z\rangle z\rangle \end{aligned}$$

ie_1 is mapped to ξ and we have:

$$c_1(\Xi^*\mathcal{E},\Xi^*h_\mathcal{E})=-\mathrm{d}\,\mathrm{d}^c\log Y^2.$$

We calculate

$$-\partial\bar{\partial}\log(Y^2) = \sum_{j,k\in I} \frac{1}{4}\frac{\partial^2}{\partial y_j \partial y_k}\log(Y^2)\,\mathrm{d}z_j \wedge \mathrm{d}\bar{z}_k$$
$$= \sum_{j,k\in I} -\frac{\delta_j\delta_k y_j y_k}{Y^4}\,\mathrm{d}z_j \wedge \mathrm{d}\bar{z}_k + \sum_{j\in I}\frac{1}{2}\frac{\delta_j}{Y^2}\,\mathrm{d}z_j \wedge \mathrm{d}\bar{z}_j$$

where $I = \{1, 4, \dots, l\}$ and $\delta_1 = -1$ and $\delta_j = 1$ otherwise.

At the point ie_1 this yields:

$$(-\,\mathrm{d}\,\mathrm{d}^c \log Y^2)(ie_1) = \left(\frac{1}{2\pi}\right)\sum_{j\in I}\delta_j\,\mathrm{d}x_j \wedge \mathrm{d}y_j$$

Therefore the volume form is equal to:

$$(-\,\mathrm{d}\,\mathrm{d}^c \log Y^2)^{l-2}(ie_1) = -\left(\frac{1}{2\pi}\right)^{l-2}(l-2)!\,\mathrm{d}x_1 \wedge \mathrm{d}y_1 \wedge x_4 \wedge \mathrm{d}y_4 \wedge \cdots \wedge \mathrm{d}x_l \wedge \mathrm{d}y_l$$

(observe $\mathrm{d}\,\mathrm{d}^c = -(1/2\pi i)\partial\bar{\partial}$ and $\mathrm{d}z_j \wedge \mathrm{d}\bar{z}_j = -2i\,\mathrm{d}x_j \wedge \mathrm{d}y_j$).

The tangent map T of the parametrization (3.10) at the point ie_1 is

$$Z \mapsto \{\xi \mapsto Z - 2i\langle Z, e_1\rangle z \bmod \langle\xi\rangle_{\mathbb{C}}\},$$

and therefore we have: $T(ie_1) = -\tilde{L}_{13}$, $T(e_1) = \tilde{L}_{23}$ and $T(ie_j) = \tilde{L}_{1j}$, $T(e_j) = \tilde{L}_{2j}$ for $j \geq 4$.

Hence we get

$$\mathrm{c}_1(\Xi^*\mathcal{E}, \Xi^* h_{\mathcal{E}})^{l-2}(\xi) = (l-2)!\left(\frac{1}{2\pi}\right)^{l-2}\bigwedge_{j=3}^{l}\tilde{L}_{2j}^* \wedge \bigwedge_{j=3}^{l}\tilde{L}_{1j}^*$$
$$= \Gamma\left(\frac{l}{2}\right)\Gamma\left(\frac{l}{2}-\frac{1}{2}\right)\pi^{3/2-l}\bigwedge_{j=3}^{l}\tilde{L}_{2j}^* \wedge \bigwedge_{j=3}^{l}\tilde{L}_{1j}^*.$$

Now we proceed to calculate the proportionality:

$$\frac{\mathrm{c}_1(\Xi^*\mathcal{E}, \Xi^* h_{\mathcal{E}})^{l-2}(\xi)}{\mu(\xi)} = \Gamma\left(\frac{l}{2}\right)\Gamma\left(\frac{l}{2}-\frac{1}{2}\right)\pi^{3/2-l}(2\pi)^{-1}\cdot 2\prod_{j=1}^{l-2}\frac{\Gamma(j/2)}{2\pi^{j/2}}$$
$$= 2\prod_{j=2}^{l}\frac{1}{2}\frac{\Gamma(j/2)}{\pi^{j/2}} = 2\lambda_\infty^{-1}(L;0).$$

□

3.3. The arithmetic of Borcherds input forms

3.3.1. In this section, we present and refine some results of [3]. We begin by recalling the notation of [loc. cit.]. Let $\Gamma \subset \mathrm{Sp}_2'(\mathbb{R})$ be a subgroup, commensurable with $\mathrm{Sp}_2'(\mathbb{Z})$. Let V be a $\mathbb{C}$-representation of Γ. Here Sp_2', as usual, is either $\mathrm{Sp}_2 = \mathrm{SL}_2$ or Mp_2. Let k be a weight, half-integral, if $\mathrm{Sp}' = \mathrm{Mp}$ and an integer, if $\mathrm{Sp}' = \mathrm{Sp}$.

Let

$$\mathrm{ModForm}(\Gamma, V, k)$$

be the space of (vector valued) modular forms of weight k for Γ for the representation V, meromorphic at the cusps. Let

$$\mathrm{HolModForm}(\Gamma, V, k)$$

be the space of modular forms of weight k for Γ of representation ρ which are holomorphic at the cusps.

3.3.2. Denote

$$\begin{aligned}\mathrm{PowSer}(\Gamma) &= \bigoplus_{\kappa} \mathbb{C}[\![q_\kappa]\!],\\ \mathrm{Laur}(\Gamma) &= \bigoplus_{\kappa} \mathbb{C}[\![q_\kappa]\!][q_\kappa^{-1}],\\ \mathrm{Sing}(\Gamma) &= \bigoplus_{\kappa} \mathbb{C}[\![q_\kappa]\!][q_\kappa^{-1}]/q_\kappa\mathbb{C}[\![q_\kappa]\!],\end{aligned}$$

where κ runs through all cusps for Γ, and q_κ is a uniformizer at the the cusp κ. $\mathrm{Sp}_2'(\mathbb{Z})$ acts on these sets.

Let $\Gamma' \subset \Gamma$ be a subgroup of finite index which acts trivial on V. We define

$$\mathrm{PowSer}(\Gamma, V) = (\mathrm{PowSer}(\Gamma') \otimes V)^\Gamma,$$

where Γ acts on both factors.

Similarly we define $\mathrm{Laur}(\Gamma, V)$ and $\mathrm{Sing}(\Gamma, V)$. We have still (abstractly) a 1:1 correspondence between $f \in \mathrm{PowSer}(\Gamma, V)$ (say) and a collection f_κ for each cusp of Γ (not Γ'!) where

$$f_\kappa = \sum_{m \in \mathbb{Q}_{\geq 0}} a_m q_\kappa^m, \tag{3.11}$$

and a_m lies in a vector subspace V_m of V depending on m.

For $\Gamma = \mathrm{Sp}_2'(\mathbb{Z})$ and if κ is the cusp $i\infty$ we have explicitly

$$V_m = \left\{ v \in V \;\middle|\; \begin{pmatrix} 1 & 1 \\ 0 & 1 \end{pmatrix} v = \mathrm{e}^{2\pi i m} v \right\}, \tag{3.12}$$

and in particular V_m is zero if $m \notin (1/N)\mathbb{Z}$ for some N, depending on Γ'.

There are maps

$$\begin{aligned}\lambda\colon \mathrm{HolModForm}(\Gamma, V^*, k) &\longrightarrow \mathrm{PowSer}(\Gamma, V^*),\\ \lambda\colon \mathrm{ModForm}(\Gamma, V, 2-k) &\longrightarrow \mathrm{Sing}(\Gamma, V)\end{aligned}$$

which in the representation (3.11) for the right-hand side, this corresponds to taking Fourier series.

There is a nondegenerate pairing between $\mathrm{PowSer}(\Gamma, V)$ and $\mathrm{Sing}(\Gamma, V^*)$ given by

$$\langle f, \phi \rangle = \sum_\kappa \mathrm{res}(\langle f_\kappa, \phi_\kappa \rangle q_\kappa^{-1} \, \mathrm{d}q_\kappa).$$

This expression is invariant under the operation of $\mathrm{Sp}_2'(\mathbb{Z})$, and hence gives a well-defined pairing between the spaces in question.

For example, if $\Gamma = \mathrm{Sp}_2'(\mathbb{Z})$ and standard parameter q, up to a scalar, this pairing is the same as

$$\langle f, \phi \rangle = \sum_m \langle a_{-m}, b_m \rangle,$$

for $f = \sum_m a_m q^m$ and $\phi = \sum_m b_m q^m$.

In [loc. cit., Theorem 3.1] it is sketched that one can deduce from Serre duality that

$$\lambda\big(\mathrm{ModForm}(\mathrm{Sp}_2'(\mathbb{Z}), V, 2-k)\big) = \big(\lambda(\mathrm{HolModForm}(\mathrm{Sp}_2'(\mathbb{Z}), V^*, k)\big)^\perp. \tag{3.13}$$

If there is a lattice $V_{\mathbb{Z}} \subset V$ given in [loc. cit., Lemma 4.3] this is refined to

$$\lambda(\mathrm{ModForm}(\mathrm{Sp}_2'(\mathbb{Z}), V, 2-k)_{\mathbb{Z}}) \otimes \mathbb{C} = \left(\mathrm{Gal}(\overline{\mathbb{Q}} \mid \mathbb{Q})\lambda(\mathrm{HolModForm}(\mathrm{Sp}_2'(\mathbb{Z}), V^*, k)\right)^{\perp}.$$

It is also shown that the last space has finite index in $\mathrm{Sing}(\mathrm{Sp}_2'(\mathbb{Z}), V)$. Here $\mathrm{ModForm}(\mathrm{Sp}_2'(\mathbb{Z}), V, 2-k)_{\mathbb{Z}}$ is the space of modular forms such that all Fourier coefficients are in $V_{\mathbb{Z}}$.

Let $L_{\mathbb{Z}}$ be a $\mathbb{Z}$-lattice of dimension l with nondegenerate quadratic form $Q_L \in \mathrm{Sym}^2(L_{\mathbb{Z}}^*)$. We let $\mathrm{Weil}(L_{\mathbb{Z}}^*/L_{\mathbb{Z}}) = \mathbb{C}[L_{\mathbb{Z}}^*/L_{\mathbb{Z}}]$ be the Weil representation of $\mathrm{Sp}_2'(\mathbb{Z})$ and $2k \equiv l \bmod 2$. If $V^* \subseteq \mathrm{Weil}(L_{\mathbb{Z}}^*/L_{\mathbb{Z}})^*$ is a subrepresentation which, as a linear subspace, is defined over $\mathbb{Q}$ with respect to the natural $\mathbb{Q}$-structure $\mathbb{Q}[L_{\mathbb{Z}}^*/L_{\mathbb{Z}}]$, Proposition 3.3.11 below states that

$$\lambda\big(\mathrm{HolModForm}(\mathrm{Sp}_2'(\mathbb{Z}), V^*, k)\big)$$

is actually Galois stable.

3.3.3. We start by a general discussion about the arithmetic of 'vector valued modular forms' for GL_2 here. We denote by $\mathbb{Q}^{\mathrm{cycl}}$ the maximal Abelian extension $\mathbb{Q}^{\mathrm{ab}}$ of $\mathbb{Q}$ but equipped with a fixed compatible system of primitive Nth roots of unity $\{\zeta_N\}$. Define

$$\begin{aligned} \tau\colon\ \mathrm{GL}_2(\widehat{\mathbb{Z}}) &\longrightarrow \mathrm{Gal}(\mathbb{Q}^{\mathrm{cycl}}/\mathbb{Q}) \\ k &\longmapsto \tau_k := \mathrm{rec} \circ \det(k) \end{aligned}$$

which is characterized by ${}^{\tau_k}\zeta_N = (\zeta_N)^{\det(k)}$.

3.3.4. Proposition. *For each $n \in \mathbb{Z}$ and an open subgroup $K \subseteq \mathrm{GL}_2(\widehat{\mathbb{Z}})$, we have a natural exact functor from the category of τ-semilinear representations of K (A.3.1) to vector bundles over $\mathrm{M}({}^{K}\mathbf{H}_1)_{\mathbb{Q}}$ denoted*

$$V \mapsto \Xi^*\mathcal{L}_V^{\otimes n}.$$

For $K_1 \subseteq K_2 \subseteq \mathrm{GL}_2(\widehat{\mathbb{Z}})$ open subgroups, we have a natural isomorphism:

$$H^0(\mathrm{M}({}^{K_1}\mathbf{H}_1)_{\mathbb{Q}}, \Xi^*\mathcal{L}_V^{\otimes n}) \cong H^0(\mathrm{M}({}^{K_2}\mathbf{H}_1)_{\mathbb{Q}}, \Xi^*\mathcal{L}^{\otimes n}_{\mathrm{ind}_{K_1}^{K_2} V}).$$

PROOF. The functor is defined as follows: Let N be such that the τ-semilinear representation (V, ω) has a $\mathbb{Q}[\zeta_N]$-structure V_N and $K(N)$ acts just via τ on $V \cong V_N \otimes_{\mathbb{Q}(\zeta_N)} \mathbb{Q}^{\mathrm{cycl}}$.

We have the following obvious commutative diagram of extended Shimura data:

$$\begin{array}{ccc} {}^{K(N)}\mathbf{H}_1 & \xrightarrow{(\mathrm{id},k)} & {}^{K(N)}\mathbf{H}_1 \\ \downarrow{\scriptstyle (\det,1)} & & \downarrow{\scriptstyle (\det,1)} \\ {}^{K(N,\mathbb{G}_{\mathrm{m}})}\mathbf{H}_0 & \xrightarrow{(\mathrm{id},\det(k))} & {}^{K(N,\mathbb{G}_{\mathrm{m}})}\mathbf{H}_0. \end{array}$$

Applying the canonical model functor M to this diagram, we see, that $\mathrm{M}(\det, 1)$ is compatible with the (Hecke) action of $\mathrm{GL}(\mathbb{Z}/N\mathbb{Z})$.

From the explicit description of the canonical model (cf., e.g., 2.7.13), we know that

$$\mathrm{M}({}^{K(N,\mathbb{G}_{\mathrm{m}})}\mathbf{H}_0)_{\mathbb{Q}} \cong \mathrm{Spec}\big(\mathbb{Q}(\zeta_N)\big),$$

in such a way that the Hecke action on the left-hand side becomes the Galois action on the right-hand side under the natural identification:

$$(\mathbb{Z}/N\mathbb{Z})^* \to \mathrm{Gal}(\mathbb{Q}(\zeta_N)/\mathbb{Q}).$$

The τ-semilinear representation V thus gives rise to a vector bundle $\mathcal{V}$ with *right* $K/K(N)$-action on

$$\mathrm{Spec}\big(\mathbb{Q}(\zeta_N)\big)$$

which is compatible with the Hecke action (acting via $\tau = \mathrm{rec} \circ \det$). In detail: we let $k \in K$ act on V as $v \mapsto \omega(k)^{-1}v$ (right action). This action is compatible with the action on $\mathrm{Spec}\big(\mathbb{Q}(\zeta_N)\big)$ given by letting $k \in K$ act on a function $\alpha \in \mathbb{Q}(\zeta_N)$ as $\tau_k(\alpha)$, (i.e., the action $k \mapsto \mathrm{Spec}(\tau_k)$ on $\mathrm{Spec}(\mathbb{Q}(\zeta_N))$, if we have

$$\omega(k)(\alpha v) = \tau_k(\alpha)\omega(k)v.$$

This is precisely the condition of ω being τ-semilinear.

We define $\Xi^*\mathcal{L}_V^{\otimes n}$ as a quotient of the bundle on $\mathrm{M}({}^{K(N)}\mathbf{H}_1)_{\mathbb{Q}}$

$$\Xi^*\mathcal{L}^{\otimes n} \otimes \mathrm{M}(\det, 1)^*\mathcal{V},$$

by the action of $K/K(N)$ on both factors. Since the action is compatible with the Hecke-action on $\mathrm{M}({}^{K(N)}\mathbf{H}_1)_{\mathbb{Q}}$, it descends along the quotient $\mathrm{M}({}^{K(N)}\mathbf{H}_1)_{\mathbb{Q}} \to \mathrm{M}({}^{K}\mathbf{H}_1)_{\mathbb{Q}}$. Up to isomorphism, this bundle does not depend on the N and V_N chosen.

The induction formula is a variant of Shapiro's Lemma. □

3.3.5. We will now investigate the q-expansion of sections in

$$H^0(\mathrm{M}({}^{K_1}\mathbf{H}_1)_{\mathbb{Q}}, \Xi^*\mathcal{L}_V^{\otimes n}).$$

We fix the boundary component $\mathbf{B}$ (corresponding to $i\infty$ in the classical picture) with

$$P_{\mathbf{B}} = \begin{pmatrix} * & * \\ 0 & 1 \end{pmatrix} \qquad \mathbb{D}_{\mathbf{B}} = \mathbf{H}_0 \times \mathbb{C}$$

with its standard splitting

$$P_{\mathbf{B}} = \begin{pmatrix} 1 & * \\ 0 & 1 \end{pmatrix} \rtimes \begin{pmatrix} * & 0 \\ 0 & 1 \end{pmatrix}$$

(corresponding to $0 \in \mathbb{C}$ in the classical picture). It gives rise to a boundary morphism of rational mixed Shimura data

$$\mathbf{H}_0[0, \mathbb{Q}] \Longrightarrow \mathbf{H}_1$$

with respect to which each $f \in H^0(\mathrm{M}({}^{K_1}\mathbf{H}_1)_{\mathbb{C}}, \Xi^*\mathcal{L}_V^{\otimes n})$ has a q-expansion, depending on some fixed rational trivializing section $s \in H^0(\mathrm{M}({}^{K}\mathbf{H}_0[0, \mathbb{Q}])_{\mathbb{Q}}, \Xi^*\mathcal{L})$ as in 2.8.2:

$$g(z, \rho) := \frac{f}{s}(z, \rho) = \sum_{k \in (1/N)\mathbb{Z}} a_{k,\alpha,\rho}(f) \exp\big(\alpha(kz)\big),$$

where $a_{k,\alpha,\rho} \in (V_{\mathbb{C}})_{k,\alpha}$, where

$$(V_{\mathbb{C}})_{k,\alpha} = \left\{ v \in V_{\mathbb{C}} \;\middle|\; \begin{pmatrix} 1 & 1 \\ 0 & 1 \end{pmatrix} v = \exp\big(\alpha(k)\big)v \right\},$$

and α is determined by the connected component of $\mathbf{H}^{\pm}$ that contains z. They satisfy (2.8.2)

$$a_{k,\alpha,\rho}({}^{\tau}f) = {}^{\tau}a_{k,\alpha,\rho\left(\begin{smallmatrix} k_\tau^{-1} & \\ & 1 \end{smallmatrix}\right)}(f).$$

Now choose a $\mathbb{Q}$-structure $V_{\mathbb{Q}}$ on V such that ω gives rise to a cocycle $\omega' \in H^1\big(K, \mathrm{GL}_{\mathbb{Q}^{\mathrm{cycl}}}(V)\big)$ where K acts via τ (cf. A.3.2). Any holomorphic function

$$g \colon \mathbb{H}^{\pm} \times \big(\mathrm{GL}_2(\mathbb{A}^{(\infty)})\big) \to V_{\mathbb{C}}$$

which satisfies

$$g(\gamma z, \gamma \rho) = j(\gamma, z)^n g(z, \rho),$$

where j is the usual automorphy factor, and

$$g(z, \rho k) = \omega'(k^{-1})^{(\exp(\alpha(\det(\rho k)/N)))} g(z, \rho) \tag{3.14}$$

for $\rho \in \mathrm{GL}_2(\widehat{\mathbb{Z}})$, where $\omega'(k^{-1})^{(\exp(\alpha(\det(\rho k)/N)))}$ means that $\omega'(k^{-1})$ acts on $V_{\mathbb{C}}$ by mapping ζ_N to $\exp\big(\alpha(\det(\rho k)/N)\big)$, gives rise to a section

$$f = gs \in H^0(\mathrm{M}({}^{K(1)}\mathbf{H}_1)_{\mathbb{C}}, \Xi^* \mathcal{L}_V^{\otimes n}).$$

The condition (3.14) can be expressed purely in terms of Fourier coefficients:

$$a_{k,\alpha,\rho k}(g) = \omega'(k^{-1})^{(\exp(\alpha(\det(\rho k)/N)))} a_{k,\alpha,\rho}(g).$$

In particular, the Galois operation on f is determined by

$$a_{k,\alpha^+,1}\left(\frac{{}^{\tau}f}{s}\right) = \omega'\left(\begin{pmatrix} k_\tau & \\ & 1 \end{pmatrix}\right)^{\left(\exp\left(\alpha^+(1/N)\right)\right)} \cdot {}^{\tau}\left(a_{k,\alpha^+,1}\left(\frac{f}{s}\right)\right),$$

where α^+ is one chosen isomorphism $\mathbb{Z} \to \mathbb{Z}(1)$.

We have proven:

3.3.6. Proposition. *If* $\det\colon K \to \widehat{\mathbb{Z}}^*$ *is surjective and* (V, ω) *is a* τ*-semilinear representation, the association* $g \mapsto f := gs$ *determines a bijection*

$$\mathrm{ModForm}(\Gamma, V, k) \to H^0(\mathrm{M}({}^{K}\mathbf{H}_1)_{\mathbb{C}}, (\Xi^* \mathcal{L})_V^{\otimes k}),$$

where $\Gamma = K \cap \mathrm{SL}_2(\mathbb{Z})$*. A section is determined by the Fourier coefficients*

$$a_k(g) = a_{k,\alpha^+,1}\left(\frac{f}{s}\right).$$

Here α^+ *corresponds to the connected component* $\mathbb{H}^+$ *chosen as domain for the classical modular form* g*.* f *is rational, if we have:*

$$a_k(g) = \omega'\left(\begin{pmatrix} k_\tau & \\ & 1 \end{pmatrix}\right)^{\left(\exp\left(\alpha^+(1/N)\right)\right)} \cdot {}^{\tau}\big(a_k(g)\big).$$

It is crucial that, if (V, ω) is given by a $\mathbb{Q}(\zeta_N)$-representation of $\mathrm{SL}(\mathbb{Z}/N\mathbb{Z})$, say, for the Proposition to make sense, the representation must extend τ-semilinearly to $\mathrm{GL}(\mathbb{Z}/N\mathbb{Z})$.

3.3.7. Example. Consider the special case $K = K_0(p)$, p an odd prime. Let ω be the τ-semilinear representation $\left(\begin{smallmatrix} a & * \\ 0 & d \end{smallmatrix}\right) \mapsto \chi(d) \cdot \mathrm{rec}(ad)$ defined on $\mathrm{GL}_2(\mathbb{Z}/p\mathbb{Z})$ for the unique nontrivial character $\chi\colon (\mathbb{Z}/p\mathbb{Z})^* \to \{\pm 1\}$ of order 2. The action of the Galois group on the Fourier coefficients is the natural, untwisted one.

We have a decomposition (A.3.5)

$$\mathrm{ind}_{K_0(p)}^{K(1)}(\omega) = V^+ \oplus V^-$$

as τ-semilinear representations, and hence Propositions 3.3.4 and 3.3.6 yield

$$\mathrm{ModForm}(\Gamma_0(p), \chi, k) = H^0(\mathrm{M}({}^{K(1)}\mathbf{H}_1)_{\mathbb{C}}, (\Xi^* \mathcal{L})_{V^+}^{\otimes k}) \oplus H^0(\mathrm{M}({}^{K(1)}\mathbf{H}_1)_{\mathbb{C}}, (\Xi^* \mathcal{L})_{V^-}^{\otimes k}).$$

This decomposition has been described by Hecke. We get here a priori that it is actually defined over $\mathbb{Q}$. See also 3.3.10 below.

Let $L_{\mathbb{Z}}$ be a lattice of *even* dimension with nondegenerate quadratic form.

3.3.8. Lemma. *The Weil representation*[4] $\mathrm{Weil}(L_{\mathbb{Z}}^*/L_{\mathbb{Z}})$ *extends to a* τ*-semilinear representation of* $\mathrm{GL}_2(\widehat{\mathbb{Z}})$ *such that the matrix* $\left(\begin{smallmatrix}\alpha & \\ & 1\end{smallmatrix}\right) \in \mathrm{GL}_2(\widehat{\mathbb{Z}})$ *acts as* $\mathrm{rec}(\alpha)$. *The dual of the Weil representation extends to a* τ*-semilinear representation, too.*

Note that this τ-semilinear representation does not involve any choices anymore and does depend only on $L_{\mathbb{Z}}$.

Proof. That the above defines indeed a τ-semilinear representation boils down to the equation:

$$\omega\left(\begin{pmatrix}\alpha & 0\\ 0 & 1\end{pmatrix} k \begin{pmatrix}\alpha^{-1} & 0\\ 0 & 1\end{pmatrix}\right) = \mathrm{rec}(\alpha)\big(\omega(k)\big) \tag{3.15}$$

for the Weil representation (cf. 3.2.1). This is checked as follows: The operators $\varphi \mapsto \big(x^* \mapsto \varphi(\alpha x^*)\big)$ are Galois invariant. If we act on the operator $\varphi \mapsto \big(x^* \mapsto \exp(2\pi i Q_L(x^*)\beta)\varphi(x^*)\big)$ by τ_α, we get the same as by substituting β by $\alpha\beta$. If we act on the operator $\varphi \mapsto (x^* \mapsto \int_{L_{\mathbb{A}^{(\infty)}}} \varphi(\gamma x)\exp(-2\pi i x^* x)\,\mathrm{d}x)$ by τ_α, we get the same as by substituting γ by $\alpha^{-1}\gamma$, provided $\mathrm{d}x$ is chosen Galois invariant (e.g., such that $L_{\widehat{\mathbb{Z}}}$ has a volume in $\mathbb{Q}$). Now an actual operator $\omega\Big(r''\big(g_l(\alpha)\big)\Big)$ differs by those considered by a $\Upsilon_f(\alpha Q_L)/\Upsilon_f(Q_L)$ in the first case which is a sign because m is even, hence Galois invariant. In the third case $\omega\Big(r''\big(d(\gamma)\big)\Big)$ differs by $c(\gamma) := \Upsilon_f(\gamma Q_L)/|\gamma|^{1/2}$, where $|\gamma|$ is calculated with respect to the chosen measure and its dual. Now consider the equation $\big(u(-\gamma^{-1})d(\gamma)\big)^3 = 1$. Since $c(\gamma)^2 \in \mathbb{Q}$, it shows $c(\alpha^{-1}\gamma) = {}^{\tau_\alpha}c(\gamma)$. □

3.3.9. Consider the Weil representation $\mathrm{Weil}(L_{\mathbb{Z}}^*/L_{\mathbb{Z}})$ as a τ-semilinear representation of $\mathrm{SL}_2(\widehat{\mathbb{Z}})$. $K_0(N)$ acts on $\chi_{L_{\widehat{\mathbb{Z}}}}$ by

$$\begin{pmatrix}a & *\\ & d\end{pmatrix}\chi_{L_{\widehat{\mathbb{Z}}}} = \Upsilon(dQ_L)\Upsilon(Q_L)^{-1}\chi_{L_{\widehat{\mathbb{Z}}}}$$

which is a μ_2-valued character that we will denote by ε in this section.

We have hence maps

$$\alpha\colon \varepsilon \to \mathrm{Weil}(L_{\mathbb{Z}}^*/L_{\mathbb{Z}}),$$

and we have also the evaluation at 0 map:

$$\beta\colon \mathrm{Weil}(L_{\mathbb{Z}}^*/L_{\mathbb{Z}}) \to \varepsilon.$$

Using the adjunctions (A.3.3) we get morphisms of τ-semilinear representations

$$\widetilde{\alpha}\colon \mathrm{ind}_{K_0(N)}^{K(1)}\,\varepsilon \to \mathrm{Weil}(L_{\mathbb{Z}}^*/L_{\mathbb{Z}}),$$

and

$$\widetilde{\beta}\colon \mathrm{Weil}(L_{\mathbb{Z}}^*/L_{\mathbb{Z}}) \to \mathrm{ind}_{K_0(N)}^{K(1)}\,\varepsilon.$$

[4]identifying $\exp(2\pi i/N)$, as chosen in the underlying additive characters, with the distinguished $\zeta_N \in \mathbb{Q}^{\mathrm{cycl}}$

Using the functor of Proposition 3.3.4 we get linear homomorphisms

$$\alpha'\colon H^0\Big(\mathrm{M}({}^{K(1)}\mathbf{H}_1)_{\mathbb{Q}}, \Xi^*\mathcal{L}^{\otimes k}_{\mathrm{ind}_{K_0(N)}^{K(1)}\varepsilon}\Big) \to H^0\Big(\mathrm{M}({}^{K(1)}\mathbf{H}_1)_{\mathbb{Q}}, \Xi^*\mathcal{L}^{\otimes k}_{\mathrm{Weil}(L_{\mathbb{Z}}^*/L_{\mathbb{Z}})}\Big)$$

$$\beta'\colon H^0\Big(\mathrm{M}({}^{K(1)}\mathbf{H}_1)_{\mathbb{Q}}, \Xi^*\mathcal{L}^{\otimes k}_{\mathrm{Weil}(L_{\mathbb{Z}}^*/L_{\mathbb{Z}})}\Big) \to H^0\Big(\mathrm{M}({}^{K(1)}\mathbf{H}_1)_{\mathbb{Q}}, \Xi^*\mathcal{L}^{\otimes k}_{\mathrm{ind}_{K_0(N)}^{K(1)}\varepsilon}\Big).$$

We get (this is true more generally) that the operator $\widetilde{\beta}\circ\widetilde{\alpha}$ is given by convolution with the function

$$\varphi(g) := \beta\circ\omega(g)\circ\alpha,$$

i.e., with a specific matrix coefficient of the Weil representation. As a formula:

$$\widetilde{\beta}(\widetilde{\alpha}(f))(h) = \sum_{a\in K_0(p)\backslash K(1)} \beta\Big(\omega(ha^{-1})\alpha\big(f(a)\big)\Big).$$

3.3.10. Example. The following was inspired by [9].

Resuming Example 3.3.7 above, it follows directly from the formulas of the Weil-representation (3.2.2), that this function φ corresponding to $\widetilde{\beta}\circ\widetilde{\alpha}$ is determined by

$$\varphi(1) = 1 \quad \varphi(S) = \frac{\sqrt{-1}^{2-l/2}}{\sqrt{p}}.$$

Using the trichotomy between $p \bmod 4$, the type of $L_{\mathbb{Z}}^*/L_{\mathbb{Z}}$, and $l/2 \bmod 4$, see [9, §2], we get

$$\varphi(S) = \begin{cases} 1/\sqrt{p^*} & \text{if } L_{\mathbb{Z}}^*/L_{\mathbb{Z}} \text{ represents the squares mod } p, \\ -1/\sqrt{p^*} & \text{if } L_{\mathbb{Z}}^*/L_{\mathbb{Z}} \text{ represents the nonsquares mod } p, \end{cases}$$

where $p^* = p$ if $p \equiv 1 \ (4)$ and $p^* = -p$, otherwise. Here $\sqrt{-1}$ is the distinguished primitive 4th root of unity in $\mathbb{Q}^{\mathrm{cycl}}$.

Therefore $\widetilde{\beta}\circ\widetilde{\alpha}$ is a projector (A.3.5) onto V^+, respectively V^-, and thus $\widetilde{\alpha}$ yields an isomorphism

$$V^{\pm} \cong \mathrm{Weil}(L_{\mathbb{Z}}^*/L_{\mathbb{Z}})^+$$

according to the isomorphism class of $L_{\mathbb{Z}}^*/L_{\mathbb{Z}}$. (Cf. 3.2.2 for the decomposition of the Weil representation).

We therefore see from condition (3.12) and the explicit formulas for the Weil-representation that for $\lambda \notin \mathbb{Z}$, we have

$$(V_{\mathbb{C}}^+)_\lambda \cong \begin{cases} \mathbb{C} & p\lambda \text{ is a square mod } p, \\ 0 & p\lambda \text{ is a nonsquare mod } p, \end{cases}$$

$$(V_{\mathbb{C}}^-)_\lambda \cong \begin{cases} \mathbb{C} & p\lambda \text{ is a nonsquare mod } p, \\ 0 & p\lambda \text{ is a square mod } p. \end{cases}$$

Recall that the Fricke involution W_p is given by

$$W_p(f)(\tau) := p^{k/2}(p\tau)^{-k} f\Big(\begin{pmatrix} & -1 \\ p & \end{pmatrix}\tau\Big).$$

In terms of the corresponding $\tilde{f} \in \mathrm{ModForm}(\mathrm{SL}_2(\mathbb{Z}), \mathrm{ind}_{\Gamma_0(0)}^{\mathrm{SL}_2(\mathbb{Z})}\chi, k)$, we get

$$W_p(f)(\tau) = \Big(p^{k/2}\omega_L\Big(\begin{pmatrix} & -1 \\ 1 & \end{pmatrix}\Big)\tilde{f}(p\tau)\Big)(1).$$

Hence

$$\mathrm{ModForm}(\mathrm{SL}_2(\mathbb{Z}), V_{\mathbb{C}}^{\pm}, k)$$

is isomorphic to the space of those

$$f \in \mathrm{ModForm}(\Gamma_0(p), \chi, k)$$

such that the Fourier coefficients a_λ of $W_p(f)$, for $(\lambda, p) = 1$ are 0 if λ is a nonsquare (respectively square) mod p. This is (for $W_p(f)$) the classical condition due to Hecke.

3.3.11. Proposition. *Let k be a (half-)integral weight and Sp_2' either Sp_2 or Mp_2 according to whether $k \in \mathbb{Z}$ or $k \in \frac{1}{2} + \mathbb{Z}$ and suppose $2k \equiv l \bmod 2$.*

For $f \in \mathrm{ModForm}(\mathrm{Sp}_2', \mathrm{Weil}(L_{\mathbb{Z}}^*/L_{\mathbb{Z}})^*, k)$ with Fourier expansion

$$f = \sum_{k \in \mathbb{Q}} a_k q^k,$$

$a_k \in \mathrm{Weil}(L_{\mathbb{Z}}^*/L_{\mathbb{Z}})_k^*$, we have for $\tau \in \mathrm{Aut}(\mathbb{C})$ and

$${}^\tau f := \sum_{k \in \mathbb{Q}} {}^\tau a_k q^k,$$

that ${}^\tau f \in \mathrm{ModForm}(\mathrm{Sp}_2', \mathrm{Weil}(L_{\mathbb{Z}}^*/L_{\mathbb{Z}})^*, k)$, too.

PROOF. We first reduce to the case $k \in \mathbb{Z}$. For half-integral weight, i.e., l odd, consider the space $L_{\mathbb{Z}}' = L_{\mathbb{Z}} \oplus \langle 1 \rangle$. We have

$$\mathrm{Weil}\big((L_{\mathbb{Z}}')^*/L_{\mathbb{Z}}'\big)^* = \mathrm{Weil}(L_{\mathbb{Z}}^*/L_{\mathbb{Z}})^* \otimes \mathrm{Weil}(\langle 1 \rangle^*/\langle 1 \rangle)^*.$$

We have the classical theta function

$$\theta \in \mathrm{HolModForm}(\mathrm{Mp}_2, \mathrm{Weil}(\langle 1 \rangle^*/\langle 1 \rangle)^*, \tfrac{1}{2})$$

(with integral Fourier coefficients!). Multiplication with it yields a map

$$\mathrm{ModForm}(\mathrm{Mp}_2, \mathrm{Weil}(L_{\mathbb{Z}}^*/L_{\mathbb{Z}})^*, k) \longrightarrow \mathrm{ModForm}(\mathrm{Mp}_2, \mathrm{Weil}\big((L_{\mathbb{Z}}')^*/L_{\mathbb{Z}}'\big)^*, k + \tfrac{1}{2})$$

which commutes with the Galois action on Fourier coefficients (because θ has integral Fourier coefficients). Now, if $f \in \mathrm{ModForm}(\mathrm{Sp}_2', \mathrm{Weil}(L_{\mathbb{Z}}^*/L_{\mathbb{Z}})^*, k)$ is given, we get (assuming the integral case) that

$${}^\tau(f \otimes \theta) = ({}^\tau f) \otimes \theta \in \mathrm{ModForm}(\mathrm{Sp}_2, \mathrm{Weil}(L_{\mathbb{Z}}^*/L_{\mathbb{Z}})^*, k + \tfrac{1}{2}),$$

or, in other words:

$$({}^\tau f)(gz) \otimes \theta(g\tau) = j(g, \tau)^{2k+1} \big(\omega_L(g)({}^\tau f)(g\tau)\big) \otimes \big(\omega_{\langle 1 \rangle}(g)\theta(g\tau)\big),$$

where $j(g, \tau)$ is the fundamental cocycle of weight $\frac{1}{2}$ on $\mathrm{Mp}_2(\mathbb{Z}) \times \mathbb{H}$.

For a dense set of τ, $\theta(\tau)$ and $\theta(g\tau)$ are not zero; hence ${}^\tau f$ must converge, too, and

$$({}^\tau f)(gz) = j(g, \tau)^{2k} \big(\omega_L(g)({}^\tau f)(g\tau)\big),$$

using that θ is in $\mathrm{HolModForm}(\mathrm{Mp}_2, \mathrm{Weil}(\langle 1 \rangle^*/\langle 1 \rangle)^*, \frac{1}{2})$. Hence

$${}^\tau f \in \mathrm{ModForm}(\mathrm{Mp}_2, \mathrm{Weil}(L_{\mathbb{Z}}^*/L_{\mathbb{Z}})^*, k).$$

Now we may assume that $k \in \mathbb{Z}$ and $\mathrm{Sp}_2'(\mathbb{Z}) = \mathrm{Sp}_2(\mathbb{Z})$. Then the statement of the theorem follows from Lemma 3.3.9 and Propositions 3.3.4 and 3.3.6. $\square$

3.3.12. We resume the notation from Section 3.3.2. Assume now that $l \geq 5$, or, $l = 4$ and Witt rank of $L_{\mathbb{Q}} = 1$. With the results above, we are able to construct Borcherds products $\Psi(F)$ whose divisor

$$\operatorname{div}(\Psi(F)^2) = \sum_{m \in \mathbb{Q}_{<0}} \mathrm{Z}(L_{\mathbb{Z}}, \langle -m \rangle, a_m; K) + D_\infty$$

satisfies certain special properties (see Theorems 3.3.14– 3.3.17 below).

3.3.13. Lemma. *Let $L_{\mathbb{Z}}$ be a lattice of signature $(l-2,2)$, $l \geq 4$. Assume the Witt rank of $L_{\mathbb{Q}}$ equals 1 if $l = 4$. Let Weil_0 be a subrepresentation of the Weil representation $\mathrm{Weil}(L_{\hat{\mathbb{Z}}}^*/L_{\hat{\mathbb{Z}}})$ which is, as a subspace, defined over $\mathbb{Q}$. Let $\{M(m)\}_{m \in \mathbb{Q}_{>0}}$ be a collection of vector subspaces $M(m) \subseteq \mathrm{Weil}_{0,m}^*$ (see 3.3.9 for the notation) such that $\sum_m \dim(M(m)^\perp) \to \infty$.*

Assume that any form in $\mathrm{HolModForm}(\mathrm{Sp}_2'(\mathbb{Z}), \mathrm{Weil}_0^*, l/2)$ whose Fourier coefficients satisfy $a_m \in M(m)$ for all $m > 0$, vanishes. Then the following holds true:

(1) *There is an $F \in \mathrm{ModForm}(\mathrm{Sp}_2'(\mathbb{Z}), \mathrm{Weil}_0, 2-l/2)$ with Fourier expansion:*

$$\lambda(F) = \sum_{m \in \mathbb{Q}} a_m q^m$$

with $a_0(0) \neq 0$ and all $a_m \perp M(-m)$, $m < 0$,

(2) *Let $0 < t \in \mathbb{Q}$ and $a \in \mathrm{Weil}_{0,-t}$ with $a \pm M(t)$ be given.*

There is an $F \in \mathrm{ModForm}(\mathrm{Sp}_2'(\mathbb{Z}), \mathrm{Weil}_0, 2-l/2)$ with Fourier expansion:

$$\lambda(F) = \sum_{m \in \mathbb{Q}} a_m q^m$$

with $a_0(0) \neq 0$, all $a_{-m} \perp M(m)$, $m > 0$, $m \neq t$ and $a_{-t} - a \perp M(t)$.

In both cases we may assume $a_m \in \mathbb{Z}[L_{\mathbb{Z}}^*/L_{\mathbb{Z}}]$ for all $m \in \mathbb{Q}$, in (2) up to replacing a with some integral multiple.

Proof (Compare [10, Lemma 4.11]). (1) Let

$$\mathrm{Sing}_M(\mathrm{Sp}_2'(\mathbb{Z}), \mathrm{Weil}_0) \subset \mathrm{Sing}(\mathrm{Sp}_2'(\mathbb{Z}), \mathrm{Weil}_0)$$

be the subspace defined by the conditions $a_{-m} \perp M(m)$, $m > 0$. Obviously

$$\mathrm{Sing}_M(\mathrm{Sp}_2'(\mathbb{Z}), \mathrm{Weil}_0)^\perp = \mathrm{PowSer}^M(\mathrm{Sp}_2'(\mathbb{Z}), \mathrm{Weil}_0^*)$$

where $\mathrm{PowSer}^M(\mathrm{Sp}_2'(\mathbb{Z}), \mathrm{Weil}_0^*)$ is defined by the conditions $a_m^* \in M(m)$ for all $m > 0$ and $a_0^* = 0$. Since the index

$$\left|\mathrm{Sing}(\mathrm{Sp}_2'(\mathbb{Z}), \mathrm{Weil}_0) : \lambda(\mathrm{ModForm}(\mathrm{Sp}_2'(\mathbb{Z}), \mathrm{Weil}_0, 2-l/2)_{\mathbb{Q}}) \otimes \mathbb{C}\right|$$

is finite and

$$\sum_m \dim(M(m)^\perp) \to \infty,$$

we have

$$\lambda\left(\mathrm{ModForm}\left(\mathrm{Sp}_2'(\mathbb{Z}), \mathrm{Weil}_0, 2-\frac{l}{2}\right)_{\mathbb{Q}}\right) \cap \mathrm{Sing}_M(\mathrm{Sp}_2'(\mathbb{Z}), \mathrm{Weil}_0)_{\mathbb{Q}} \neq 0.$$

We want to show that the element

$$[a_0(0)]\colon \mathrm{Sing}(\mathrm{Sp}_2'(\mathbb{Z}), \mathrm{Weil}_0)_{\mathbb{Q}} \to \mathbb{Q}$$

does not vanish on this intersection. Since the duality between PowSer$(\cdots)$ and Sing$(\cdots)$ is nondegenerate, we have

$$\left(\lambda\left(\mathrm{ModForm}\left(\mathrm{Sp}_2'(\mathbb{Z}),\mathrm{Weil}_0,2-\frac{l}{2}\right)_{\mathbb{Q}}\right)\cap \mathrm{Sing}_M(\mathrm{Sp}_2'(\mathbb{Z}),\mathrm{Weil}_0)_{\mathbb{Q}}\right)^{\perp}$$
$$=\lambda\left(\mathrm{HolModForm}\left(\mathrm{Sp}_2'(\mathbb{Z}),\mathrm{Weil}_0^*,\frac{l}{2}\right)\right)^{\mathrm{Aut}(\mathbb{C})}+\mathrm{PowSer}^M(\mathrm{Sp}_2'(\mathbb{Z}),\mathrm{Weil}_0^*)_{\mathbb{Q}}.$$

Here we used Proposition 3.3.11, stating that $\lambda\big(\mathrm{HolModForm}(\mathrm{Sp}_2'(\mathbb{Z}),\mathrm{Weil}_0^*,l/2)\big)$ is a Galois invariant subspace because, by assumption, Weil_0^* is, as a subspace, defined over $\mathbb{Q}$. Let us assume that there is a relation

$$[a_0(0)]=\lambda(f)+\phi,$$

where $f\in\mathrm{HolModForm}(\mathrm{Sp}_2'(\mathbb{Z}),\mathrm{Weil}_0^*,l/2)$ and ϕ is in $\mathrm{PowSer}^M(\mathrm{Sp}_2'(\mathbb{Z}),\mathrm{Weil}_0^*)$. From this follows that f is a modular form, whose coefficients a_m^*, $m>0$, satisfy $a_m^*\in M(m)$. Therefore $f=0$ by assumption, a contradiction.

(2) Since $a\perp M(t)$, we find a $a^*\in M(t)$ such that $a^*(a)\neq 0$. We have to show that the element $[a^*q^t]\colon \mathrm{Sing}(\mathrm{Sp}_2'(\mathbb{Z}),\mathrm{Weil}_0)_{\mathbb{Q}}\to\mathbb{Q}$ does not vanish on the intersection

$$\lambda\left(\mathrm{ModForm}\left(\mathrm{Sp}_2'(\mathbb{Z}),\mathrm{Weil}_0,2-\frac{l}{2}\right)_{\mathbb{Q}}\right)\cap\mathrm{Sing}_{\widetilde{M}}(\mathrm{Sp}_2'(\mathbb{Z}),\mathrm{Weil}_0)_{\mathbb{Q}},$$

where $\widetilde{M}=M$ except that $\widetilde{M}(t)$ has been set to $M(t)\cap a^{\perp}$. For if this is not the case we can add an appropriate element not being in its kernel to an F constructed in (1) to get the result. (We have only to avoid that $a_0(0)$ becomes zero.) Suppose that $[a^*q^t]$ vanishes. Then we get a relation:

$$[a^*q^t]=\lambda(f)+\phi,$$

where $f\in\mathrm{HolModForm}(\mathrm{Sp}_2'(\mathbb{Z}),\mathrm{Weil}_0^*,l/2)$ and $\phi\in\mathrm{PowSer}^{\widetilde{M}}(\mathrm{Sp}_2'(\mathbb{Z}),\mathrm{Weil}_0^*)$. Therefore $f=0$ as above, a contradiction. The statement about integrality follows from the fact that all components of the constructed f are in fact modular forms for $\Gamma(N)$ for some N. For integral weight it is well-known that their Fourier coefficients have a bounded denominator. Multiplication with the common denominator yields a form with integral coefficients. For half-integral weight use the reduction to integral weight method of the proof of Proposition 3.3.11. □

We will now show several theorems, stating the existence of Borcherds products, whose divisor has special properties. This is an essential ingredient in the calculation of arithmetic volumes later. To not interrupt the discussion, we refer to the Appendix A.1 for several elementary lemmas on quadratic forms which are needed and to Appendix A.2 for several facts about vanishing of modular forms with sparse Fourier coefficients.

3.3.14. Theorem. (1) *Let $L_{\mathbb{Z}_{(p)}}$ be an* **isotropic lattice of signature** $(l-2,2)$, $l\geq 4$. *There is a saturated lattice $L_{\mathbb{Z}}\subset L_{\mathbb{Z}_{(p)}}$ and*

$$F\in\mathrm{ModForm}\left(\mathrm{Sp}_2'(\mathbb{Z}),\mathrm{Weil}(L_{\mathbb{Z}}^*/L_{\mathbb{Z}}),2-\frac{l}{2}\right)$$

with integral Fourier coefficients, such that $\Psi(F)$ has nonzero weight, all occurring $\mathrm{Z}(L_{\mathbb{Z}},\langle -m\rangle,a_m;K)$, $m<0$ in $\mathrm{div}\big(\Psi(F)\big)$ are p-integral, i.e., consist of *canonical models of Shimura varieties* $\mathrm{M}({}^K\mathbf{O}(L'_{\mathbb{Z}_{(p)}}))$*, for various lattices* $L'_{\mathbb{Z}_{(p)}}$ *with unimodular (at p) quadratic form.*

(2) *We find an F above, such that* $\operatorname{div}(\Psi(F))$ *contains, in addition, precisely one* $\mathrm{Z}(L_{\mathbb{Z}}, \langle m\rangle, \kappa; K)$*,* $p \mid m$ *with nonzero multiplicity.*

PROOF. (1) Choose any saturated lattice of the form $L_{\mathbb{Z}} = H \oplus L'_{\mathbb{Z}}$ in $L_{\mathbb{Z}_{(p)}}$. Let

$$M(m) = \begin{cases} 0 & |m|_p = 1, \\ \mathrm{Weil}^*_m & |m|_p \neq 1. \end{cases}$$

Then all $\mathrm{Z}(L_{\mathbb{Z}}, \langle -m\rangle, a_m; K)$ with $a_m \perp M(m)$ consist of p-integral canonical models of Shimura varieties $\mathrm{M}({}^K\mathbf{O}(L'_{\mathbb{Z}_{(p)}}))$ because $v^\perp$, for any $v \in L_{\mathbb{Z}_p}$ with $Q_L(v) = m$, $|m|_p = 1$, is unimodular.

In view of Theorem 3.2.14, to construct the required F, by Lemma 3.3.13(1), we have to show that any modular form

$$f \in \mathrm{HolModForm}\left(\mathrm{Sp}'_2(\mathbb{Z}), \mathrm{Weil}^*, \frac{l}{2}\right),$$

whose Fourier coefficients are supported only on M, vanishes. But every component of f is (in particular) a modular form for some $\Gamma(N)$, $p \nmid N$. It vanishes by Lemma A.2.1.

For (2), use Lemma 3.3.13(2). □

3.3.15. Theorem. (1) *Let* $L_{\mathbb{Z}_{(p)}}$ *be a unimodular lattice* **of signature (3, 2) and Witt rank 1**. *Up to multiplication of* Q_L *with a scalar* $\in \mathbb{Z}^*_{(p)}$*, there is a saturated lattice* $L_{\mathbb{Z}} \subset L_{\mathbb{Z}_{(p)}}$ *and*

$$F \in \mathrm{ModForm}(\mathrm{Mp}_2(\mathbb{Z}), \mathrm{Weil}, -\tfrac{1}{2})$$

with integral Fourier coefficients, such that $\Psi(F)$ *has nonzero weight, all occurring* $\mathrm{Z}(L_{\mathbb{Z}}, \langle -m\rangle, a_m; K)$ *in* $\operatorname{div}(\Psi(F))$ *are p-integral, i.e., consist of canonical models of Shimura varieties* $\mathrm{M}({}^K\mathbf{O}(L'_{\mathbb{Z}_{(p)}}))$*, for various lattices* $L'_{\mathbb{Z}_{(p)}}$ *with unimodular (at p) quadratic form (i.e., with K admissible), such that* $L'_{\mathbb{Q}}$ *has signature (2, 2) and Witt rank 1.*

(2) *Up to multiplication of* $Q_{L'}$ *with a scalar* $\in \mathbb{Z}^*_{(p)}$*, for every* $L'_{\mathbb{Z}_{(p)}}$ *unimodular* **of signature (2, 2), Witt rank 0**, *we find a lattice* $L_{\mathbb{Z}_{(p)}}$ *of signature (3, 2), Witt rank 1, and an F as in (1), such that in* $\operatorname{div}(\Psi(F))$ *occurs, in addition to the subvarieties above, precisely one* $\mathrm{Z}(L, \langle l\rangle, \kappa; K)$ *with nonzero coefficient, consisting of canonical models of Shimura varieties* $\mathrm{M}({}^{K_i}\mathbf{O}(L'_{\mathbb{Z}_{(p)}}))$ *(with the given* $L'_{\mathbb{Z}_{(p)}}$*) for various different admissible* K_i*'s.*

PROOF. (1) By Lemma A.1.1, we may multiply Q_L by a scalar in such a way that there is a saturated lattice $L_{\mathbb{Z}} \subset L_{\mathbb{Z}_{(p)}}$ with cyclic $L^*_{\mathbb{Z}}/L_{\mathbb{Z}}$ of order $2D'$, where D' is square-free, of the form

$$L_{\mathbb{Z}} = H \oplus L'_{\mathbb{Z}}.$$

By 3.2.2, we know that $\mathrm{Weil}(L^*_{\mathbb{Z}}/L_{\mathbb{Z}})$ decomposes

$$\mathrm{Weil}(L^*_{\mathbb{Z}}/L_{\mathbb{Z}}) = \bigotimes_{l|D} \mathrm{Weil}(L^*_{\mathbb{Z}_l}/L_{\mathbb{Z}_l}),$$

and for each l we have a decomposition

$$\mathrm{Weil}(L^*_{\mathbb{Z}_l}/L_{\mathbb{Z}_l}) = \mathrm{Weil}(L^*_{\mathbb{Z}_l}/L_{\mathbb{Z}_l})^+ \oplus \mathrm{Weil}(L^*_{\mathbb{Z}_l}/L_{\mathbb{Z}_l})^-,$$

into irreducible representations. Here $L^*_{\mathbb{Z}_l}/L_{\mathbb{Z}_l}$ cyclic of odd prime order (respectively of order 2 or 4) is used. (If the order is 2, $\mathrm{Weil}(L^*_{\mathbb{Z}_2}/L_{\mathbb{Z}_2})^-$ is zero).

We will work with the *irreducible* representation

$$\mathrm{Weil}(L^*_{\hat{\mathbb{Z}}}/L_{\mathbb{Z}})^+ := \bigotimes_l \mathrm{Weil}(L^*_{\mathbb{Z}_l}/L_{\mathbb{Z}_l})^+,$$

with product basis build from the bases $(\chi_\kappa + \chi_{-\kappa}) \in \mathrm{Weil}(L^*_{\mathbb{Z}_l}/L_{\mathbb{Z}_l})^+$ for $\kappa \in (L^*_{\mathbb{Z}_l}/L_{\mathbb{Z}_l})/(\pm 1)$. There will be a basis vector of the form

$$\gamma := \sum \chi_\kappa,$$

where the sum runs only over *primitive* $\kappa \in L^*_{\hat{\mathbb{Z}}}/L_{\mathbb{Z}}$.

Let

$$M(m) = \begin{cases} (\mathbb{C}\gamma)^\perp & |m|_p = 1, \\ (\mathrm{Weil}(L^*_{\hat{\mathbb{Z}}}/L_{\mathbb{Z}})^+)^*_m & |m|_p \neq 1. \end{cases}$$

Now for $v \in L_{\mathbb{Q}} \cap \kappa$ with κ such that χ_κ occurs in γ (in particular κ primitive), and $|Q_L(v)|_p = 1$, $v^\perp$ is unimodular. It is also isotropic because by Lemma A.1.3 m is already represented by $L'_{\hat{\mathbb{Z}}}$.

In view of Theorem 3.2.14 again, to construct the required Borcherds form, by Lemma 3.3.13(2), we have to show that any modular form

$$f \in \mathrm{HolModForm}(\mathrm{Mp}_2(\mathbb{Z}), (\mathrm{Weil}(L^*_{\hat{\mathbb{Z}}}/L_{\mathbb{Z}})^+)^*, \tfrac{5}{2})$$

with Fourier coefficients supported only on M vanishes. Note that $\mathrm{Weil}(L^*_{\hat{\mathbb{Z}}}/L_{\mathbb{Z}})^+$ is defined over $\mathbb{Q}$ as a vector subspace. Now $\gamma \circ f$ vanishes by A.2.1, so $f = 0$ by Lemma A.2.3 because $\mathrm{Weil}(L^*_{\hat{\mathbb{Z}}}/L_{\mathbb{Z}})^+$ is irreducible.

(2) Use A.1.5 to construct the lattice $L_{\mathbb{Z}}$, and $\mathrm{Z}(L_{\mathbb{Z}}, \langle x\rangle, \chi_{L_{\hat{\mathbb{Z}}}}; K)$ consists obviously of the required models. Since $\chi_{L_{\hat{\mathbb{Z}}}} \not\perp \gamma^\perp$ apply Lemma 3.3.13(2). □

3.3.16. Theorem. (1) *Let $L_{\mathbb{Z}_{(p)}}$ be a* **isotropic lattice of signature $(2,2)$ and discriminant q, q a prime $\equiv 1$ (4)** *different from p. Up to multiplication of Q_L with a scalar $\in \mathbb{Z}^*_{(p)}$, there is a lattice $L_{\mathbb{Z}} \subset L_{\mathbb{Z}_{(p)}}$ and*

$$F \in \mathrm{ModForm}(\mathrm{Sp}_2(\mathbb{Z}), \mathrm{Weil}(L^*_{\hat{\mathbb{Z}}}/L_{\mathbb{Z}}), 0)$$

with integral Fourier coefficients, such that $\Psi(F)$ has nonzero weight, all occurring $\mathrm{Z}(L, \langle -m\rangle, a_m; K)$ in $\mathrm{div}\big(\Psi(F)\big)$ are p-integral, i.e., consist of canonical models of Shimura varieties $\mathrm{M}\big({}^K\mathbf{O}(L'_{\mathbb{Z}_{(p)}})\big)$, for various lattices $L'_{\mathbb{Z}_{(p)}}$ with unimodular quadratic form, such that L' has signature $(1,2)$ and Witt rank 1.

(2) *For every $L'_{\mathbb{Z}_{(p)}}$ unimodular* **of signature $(1,2)$, Witt rank 0**, *up to multiplication of $Q_{L'}$ with a scalar $\in \mathbb{Z}^*_{(p)}$, we find a lattice $L_{\mathbb{Z}}$ of signature $(2,2)$, Witt rank 1 and discriminant q, q a prime $\equiv 1$ (4) and an F as in (1) such that in $\mathrm{div}\big(\Psi(F)\big)$ occurs, in addition, precisely one $\mathrm{Z}(L, \langle l\rangle, \varphi; K)$ with nonzero coefficient, consisting of canonical models of Shimura varieties $\mathrm{M}\big({}^{K_i}\mathbf{O}(L'_{\mathbb{Z}_{(p)}})\big)$ (with the given $L'_{\mathbb{Z}_{(p)}}$) for various different admissible K_i's.*

PROOF. (1) Let Weil^+ be as in the last theorem.

By Lemma A.1.3, we may multiply Q_L by a scalar such that there is a lattice $L_{\mathbb{Z}} \subset L_{\mathbb{Z}_{(p)}}$, with cyclic $L_{\mathbb{Z}}^*/L_{\mathbb{Z}}$ of order q, of the form

$$L_{\mathbb{Z}} = H \oplus \left\langle x^2 + xy + \frac{1-q}{4}y^2 \right\rangle.$$

A $\mathrm{Z}(L, \langle m \rangle, \kappa; K)$ (for admissible K) consists of Shimura varieties of the required form, if $v^\perp$ for $v \in \pm\kappa + L_{\hat{\mathbb{Z}}}$, $Q_L(v) = m$ is isotropic. This is the case if and only if m is represented by $\langle x^2 + xy + \big((1-q)/4\big)y^2 \rangle$. Hence define

$$M(m) := \begin{cases} (\mathrm{Weil}^+)^*_m & |m|_p \neq 1, \\ (\mathrm{Weil}^+)^*_m \cap R(m)^\perp & |m|_p = 1, \end{cases}$$

where

$$R(m) = \left\{ f \in \mathrm{Weil}_{0,m} \;\middle|\; \exists v \in \left\langle x^2 + xy + \frac{1-q}{4}y^2 \right\rangle_{\mathbb{A}^{(\infty)}} : f(v) \neq 0, Q_L(v) = m \right\}.$$

By 3.3.10 we know that

$$\chi_q(j) = -1 \implies (\mathrm{Weil}^+)^*_{\frac{j}{q}} = 0.$$

The definition of $M(m)$ and A.1.4 imply that for *primes* $p' \neq p$ with $\chi_q(p') = 1$, $M(p'/q)$ is also zero.

In view of Theorem 3.2.14 again, to construct the required meromorphic modular form, by Lemma 3.3.13, we have to show that any modular form in

$$\mathrm{HolModForm}(\mathrm{Sp}_2(\mathbb{Z}), (\mathrm{Weil}^+)^*, 2)$$

with Fourier coefficients supported only on M vanishes. Let f be such a modular form. From the remarks in 3.3.10 it follows that f' defined by

$$f'(\tau) := \left(\omega_L\left(\begin{pmatrix} 0 & -1 \\ 1 & 0 \end{pmatrix}\right) f(q\tau)\right)(\chi_{L_{\hat{\mathbb{Z}}}})$$

is a scalar multiple of $W_q\Big(f(\chi_{L_{\hat{\mathbb{Z}}}})\Big)$, hence also a modular form for $\Gamma_0(q)$. For primes $p' \neq p$, $M(p'/q)$ is zero, hence the Fourier coefficient $a_{p'}$ of f' is zero. Similarly for $(n, q) = 1$, $\chi_p(n) = -1$, a_n is zero. Therefore the vanishing of f' now follows from Proposition A.2.2 and that of f by Lemma A.2.3.

(2) Use A.1.2 to construct the lattice $L_{\mathbb{Z}}$ and $\mathrm{Z}(L_{\mathbb{Z}}, \langle x \rangle, \chi_{L_{\hat{\mathbb{Z}}}}; K)$ consists obviously of the required models. Since $\chi_{L_{\hat{\mathbb{Z}}}} \not\perp R(m)^\perp$, apply Lemma 3.3.13(2). □

3.3.17. Theorem. *Let $L_{\mathbb{Z}_{(p)}}$ be a unimodular* **lattice of signature $(3, 2)$ of Witt rank 2**. *There is a lattice $L_{\mathbb{Z}} \subset L_{\mathbb{Z}_{(p)}}$, such that up to multiplication of Q_L with a scalar $\in \mathbb{Z}_{(p)}^*$, there is an*

$$F \in \mathrm{ModForm}(\mathrm{Mp}_2(\mathbb{Z}), \mathrm{Weil}(L_{\mathbb{Z}}^*/L_{\mathbb{Z}}), -\tfrac{1}{2})$$

with integral Fourier coefficients, such that $\Psi(F)$ has nonzero weight, $\mathrm{div}\big(\Psi(F)\big)$ consist of exactly one $\mathrm{Z}(L_{\mathbb{Z}}, \langle -t \rangle, a_t; K)$ with nonzero coefficient which itself consists of canonical models of Shimura varieties $\mathrm{M}\big({}^K\mathbf{O}(L'_{\mathbb{Z}_{(p)}})\big)$, for any *(a priori) given unimodular lattice $L'_{\mathbb{Z}_{(p)}}$ of signature $(2, 2)$, Witt rank ≥ 1 and various admissible K.*

PROOF. Define $L_{\mathbb{Z}_{(p)}} = L'_{\mathbb{Z}_{(p)}} \perp \langle x \rangle$, where $-x$ is represented by $L'_{\mathbb{Z}_{(p)}}$ and $|x|_p = 1$. Up to multiplication of Q_L by a scalar, we find a saturated lattice $L_{\mathbb{Z}} = H^2 \perp \langle 1 \rangle \subset L_{\mathbb{Z}_{(p)}}$. Weil $=$ Weil$(L_{\mathbb{Z}}^*/L_{\mathbb{Z}})$ is irreducible in this case. W.l.o.g. x is assumed to be represented by $L_{\widehat{\mathbb{Z}}}$. Define

$$M(m) = \begin{cases} \mathrm{Weil}^*_m & m \neq x, \\ 0 & m = x. \end{cases}$$

In view of Theorem 3.2.14 again, to construct the required meromorphic modular form, by Lemma 3.3.13, we have to show that any modular form

$$f \in \mathrm{HolModForm}(\mathrm{Mp}_2(\mathbb{Z}), \mathrm{Weil}^*, \tfrac{5}{2})$$

with Fourier coefficients supported only on M vanishes. This follows because

$$\mathrm{HolModForm}(\mathrm{Mp}_2(\mathbb{Z}), \mathrm{Weil}^*, \tfrac{5}{2})$$

is generated by the Eisenstein series whose coefficient a_x is nonzero because x is represented by $L_{\widehat{\mathbb{Z}}}$. □

3.4. An arithmetic Siegel–Weil formula

3.4.1. Let $L_{\mathbb{Q}}$ be a finite-dimensional vector space with nondegenerate quadratic form Q_L of signature $(l-2, 2)$ with $l \geq 4$, and assume its Witt rank to be 1 if $l = 4$. Let $L_{\mathbb{Z}} \subset L_{\mathbb{Q}}$ be a lattice of the form $L_{\mathbb{Z}} = H \perp L'_{\mathbb{Z}}$. Assume $L'_{\mathbb{Z}}$, hence also $L_{\mathbb{Z}}$ to be unimodular at p. These assumptions imply, in particular, that associated canonical models $\mathrm{M}({}^K\mathbf{O}(L))$ exist (if K is admissible) and are noncompact.

Take a modular form F as in 3.2.3 with Fourier expansion

$$F(\tau) = \sum_{m \in \mathbb{Q}} a_m q^m,$$

where $a_m \in \mathrm{Weil}(L_{\mathbb{Z}}^*/L_{\mathbb{Z}}) \subset S(L_{\mathbb{A}^{(\infty)}})$ with $a_0(0) \neq 0$.

In this section we prove a relation (Theorem 3.4.2(2)) between arithmetic volumes of different Shimura varieties of orthogonal type which involves the Borcherds lift $\Phi(F)$ of F—in particular its integral (3.2.14(3)). Recall the definition of the function $\widetilde{\mu}(L_{\mathbb{Z}}, \langle -q \rangle, \kappa; s)$ from 3.2.12. The relation will be the main ingredient in the proof of the main result of this book, Main Theorem 3.5.5. As will be explained there the arithmetic formula expresses—in some sense—the special derivative of the orbit equation 3.5.2. We mention also a rather well-known geometric analogue (Theorem 3.4.2(1)) which expresses—in the same way—the special value of the orbit equation. It may be seen as an averaged version of the Siegel–Weil formula for indefinite lattices. Recall (3.2.5) the SO-equivariant line bundle $\overline{\mathcal{E}} = (\mathcal{E}, h_{\mathcal{E}})$ with Hermitian metric on $\mathbb{D}_{\mathbf{O}}$.

3.4.2. Theorem. *Under the conditions above, we have*:

(1) **(Geometric formula)**

$$\sum_{q<0} \widetilde{\mu}(L_{\mathbb{Z}}, \langle -q \rangle, a_q; 0) \operatorname{vol}_{\mathcal{E}_{\mathbb{C}}}\left(\mathrm{M}({}^K_\Delta\mathbf{O}(L))\right) = \sum_{q<0} \operatorname{vol}_{\mathcal{E}_{\mathbb{C}}}(\mathrm{Z}(L_{\mathbb{Z}}, \langle -q \rangle, a_q; K))$$

(2) **(Arithmetic formula)**

$$\sum_{q<0}\widetilde{\mu}'(L_{\mathbb{Z}},\langle -q\rangle,a_q;0)\,\mathrm{vol}_{\mathcal{E}_{\mathbb{C}}}\Big(\mathrm{M}\big({}^K_\Delta\mathbf{O}(L)\big)\Big) + \sum_{q<0}\widetilde{\mu}(L_{\mathbb{Z}},\langle -q\rangle,a_q;0)\widehat{\mathrm{vol}}_{\overline{\mathcal{E}},p}\Big(\mathrm{M}\big({}^K_\Delta\mathbf{O}(L)\big)\Big) \equiv \sum_{q<0}\widehat{\mathrm{vol}}_{\overline{\mathcal{E}},p}(\mathrm{Z}(L_{\mathbb{Z}},\langle -q\rangle,a_q;K))$$

in $\mathbb{R}^{(p)}$.

PROOF. Let $f_0 := \Psi(F)$ be the Borcherds product associated with F—cf. Section 3.2.3 ff. and especially Theorem 3.2.14.

We sketch the proof of (1). The assumption on l and the Witt rank imply that the Eisenstein series associated with $L_{\mathbb{Z}}$ (see [26, Section 4]) is holomorphic, i.e., there is a

$$G \in \mathrm{HolModForm}\left(\mathrm{SL}_2'(\mathbb{Z}),\mathrm{Weil}(L_{\mathbb{Z}}^*/L_{\mathbb{Z}})^*,\frac{l}{2}\right)$$

with $a_n(G)(\varphi) = -\widetilde{\mu}(L_{\mathbb{Z}},\langle n\rangle,\varphi;0)$ for $n>0, n\in\mathbb{Q}$ and $a_0(G)(\varphi)=\varphi(0)$. Hence relation (3.13) yields:

$$\sum_q \widetilde{\mu}(L_{\mathbb{Z}},\langle -q\rangle,a_q;0) = a_0(0). \tag{3.16}$$

In addition, we have the relation

$$\begin{aligned} a_0(0)\int_M \mathrm{c}_1(\Xi^*\mathcal{E},\Xi^*h)^{l-2} &= 2\int_M \mathrm{d}\,\mathrm{d}^c\log h_{\mathcal{E}}(f_0)\wedge \mathrm{c}_1(\Xi^*\mathcal{E},\Xi^*h)^{l-3} \\ &= 2\int_{\mathrm{div}(f_0)\cap M}\mathrm{c}_1(\Xi^*\mathcal{E},\Xi^*h)^{l-3}. \end{aligned} \tag{3.17}$$

The justification that here everything is integrable and that the second equality is true without contributions from the boundary, can be found in [7]. From (3.16) and (3.17), (1) follows, taking the description of the divisor of f_0 (Theorem 3.2.14) into account.

(2) We may calculate $\widehat{\mathrm{vol}}_{\overline{\mathcal{E}},p}\Big(\mathrm{M}\big({}^K_\Delta\mathbf{O}(L)\big)\Big)$ in the following way:

First assume w.l.o.g. (e.g., by taking a lattice with large discriminant in the construction of F or by just pulling back $\Psi(F)$ afterwards) that K is neat.

We know by 3.2.17 that $\Xi^*(\mathcal{E})$ is ample on $\mathrm{M}\big({}^K\mathbf{O}(L)\big)$ and semiample on $\mathrm{M}\left({}^K_\Delta\mathbf{O}(L)\right)$. Choose some sufficiently fine, smooth with respect to K, complete, and projective Δ (by Lemma 2.6.24 the arithmetic volume does not depend on this choice) such that all models exist and all special cycles involved embed closedly as projective schemes (cf. 3.1.6). $\Xi^*(\mathcal{E})^{\otimes k}$, for some positive k, defines a morphism

$$\alpha\colon \mathrm{M}\big({}^K_\Delta\mathbf{O}(L)\big)\to\mathbb{P}^r(\mathbb{Z}_{(p)}),$$

whose restriction to $\mathrm{M}\big({}^K\mathbf{O}(L)\big)$ is an embedding. Let D be the boundary divisor. By Main Theorem 2.5.9 $\mathrm{M}\big({}^K_\Delta\mathbf{O}(L)\big)$ and D are defined over $\mathbb{Z}_{(p)}$. We have $\dim(\alpha(D)_{\mathbb{C}}) = $ Witt rank of $L_{\mathbb{Q}}-1$, hence $\dim\big(\alpha(D)\big) = $ Witt rank of $L_{\mathbb{Q}}$, because $\Xi^*(\mathcal{E})^{\otimes k}_{\mathbb{C}}$ induces the Baily–Borel compactification. In addition, we may also choose k and F such that $f_0=\Psi(F)\in H^0(\mathrm{M}({}^K_\Delta\mathbf{O}),\Xi^*(\mathcal{E})^{\otimes k}_{\mathrm{horz}})$ (Theorem 3.2.14).

Up to increasing k again, we may find hyperplanes $H_1,\dots,H_{l-2}\subset\mathbb{P}^r(\mathbb{Z}_{(p)})$ which intersect properly with $\alpha\big(\mathrm{div}(f_0)\big)$, and we may avoid that they intersect simultaneously in any closed subset Z of dimension $< l-2$.

Our assumptions on l and the Witt rank imply that we can take Z to be the union of $\alpha(D)$ (dim $=$ Witt rank $< l-2$) and the locus, where $\alpha_{\mathbb{F}_p}$ is not finite (dim $< l-2$).

We may clearly assume that already $H_{1,\mathbb{C}}$ and $H_{2,\mathbb{C}}$ do not intersect in $\alpha(D)_{\mathbb{C}}$. Let $f_1, \dots, f_{l-2}$ be the corresponding sections.

The construction assures that also $\operatorname{div}(f_0), \operatorname{div}(f_1), \dots, \operatorname{div}(f_{l-2})$ intersect properly and not in D simultaneously. Therefore by Theorem 3.2.14(2),

$$\operatorname{div}(f_0) \cdot \operatorname{div}(f_1) \cdots \operatorname{div}(f_{l-2}) = \frac{1}{2} \sum_{q<0} \mathrm{Z}(L_{\mathbb{Z}}, \langle -q \rangle, a_q; K) \cdot \operatorname{div}(f_1) \cdots \operatorname{div}(f_{l-2}).$$

Note (cf. 2.6.17) that the arithmetic volume is the sum of this expression and

$$\frac{1}{2} \int g_0 * g_1 * \cdots * g_{l-2}, \tag{3.18}$$

where

$$g_i = -\log \Xi^* h_{\mathcal{E}}(f_i)$$

(and the star product has to be interpreted in the sense of 2.6.8). In the following we will denote

$$\Omega := \mathrm{c}_1\big(\Xi^*(\mathcal{E}), \Xi^*(h_{\mathcal{E}})\big)$$

to be the first Chern form of the bundle $\Xi^*(\mathcal{E})$ with respect to the (log-singular) Hermitian metric $\Xi^* h_{\mathcal{E}}$.

On any parametrization defined by a point-like boundary component as in 2.4.5 it is given by

$$-\mathrm{d}\,\mathrm{d}^c \log\big(Q_{z'}(Y)\big)$$

where $Z = X + iY \in U_{\mathbb{C}}$.

By Theorem 3.4.3 below, we may write (3.18) as

$$= \frac{1}{2} \int_{\mathrm{M}({}^K\mathbf{O})_{\mathbb{C}}} g_0 k\Omega \wedge \cdots \wedge k\Omega + \frac{1}{2} \int_{\operatorname{div}(f_0) \cap \mathrm{M}({}^K\mathbf{O})_{\mathbb{C}}} g_1 * \cdots * g_{l-2},$$

hence we get the equation ($\in \mathbb{R}^{(p)}$)

$$\begin{aligned} &k^{l-1} \widehat{\mathrm{vol}}_{\overline{\mathcal{E}},p}\Big(\mathrm{M}({}^K_\Delta\mathbf{O}(L))\Big) \\ &\quad = \frac{1}{2} k^{l-2} \int_{\mathrm{M}({}^K\mathbf{O})_{\mathbb{C}}} g_0 \Omega \wedge \cdots \wedge \Omega + \frac{1}{2} k^{l-2} \sum_{q<0} \widehat{\mathrm{vol}}_{\overline{\mathcal{E}},p}\big(\mathrm{Z}(L_{\mathbb{Z}}, \langle -q \rangle, a_q; K)\big), \end{aligned}$$

(cf. also 2.6.17) and therefore the required one, taking into account that

$$\frac{1}{2} \sum_{q<0} \widetilde{\mu}(L_{\mathbb{Z}}, \langle -q \rangle, a_q; 0) = k = \frac{a_0(0)}{2}$$

(geometric formula—part (1)) and by Theorem 3.2.14(3):

$$\int_{\mathrm{M}({}^K_\Delta\mathbf{O})_{\mathbb{C}}} g_0 \Omega \wedge \cdots \wedge \Omega \equiv -\operatorname{vol}_{\mathcal{E}_{\mathbb{C}}}\big(\mathrm{M}({}^K\mathbf{O}(L))_{\mathbb{C}}\big) \sum_{q<0} \widetilde{\mu}'(L_{\mathbb{Z}}, \langle -q \rangle, a_q; 0)$$

in $\mathbb{R}^{(p)}$. □

3.4.3. Theorem. *Let $L_{\mathbb{Q}}$ be as before, i.e., of signature $(l-2,2)$ with $l \geq 5$, or $l = 4$ and $L_{\mathbb{Q}}$ of Witt rank 1.*

Let $f_0, \dots, f_{l-2} \in H^0(\mathrm{M}({}^K_\Delta\mathbf{O}), \Xi^*(\mathcal{E})^{\otimes k})$ be sections, such that $\mathrm{div}(f_0), \dots, \mathrm{div}(f_{l-2})$ intersect properly:

$$\bigcap_i \mathrm{supp}\big(\mathrm{div}(f_i)\big) = \varnothing,$$

and such that

$$\mathrm{supp}\big(\mathrm{div}(f_1)\big) \cap \mathrm{supp}\big(\mathrm{div}(f_2)\big) \cap \mathrm{supp}(D) = \varnothing$$

(Witt rank 2), respectively

$$\mathrm{supp}\big(\mathrm{div}(f_1)\big) \cap \mathrm{supp}(D) = \varnothing$$

(Witt rank 1).

Then we have

$$\int_{\mathrm{M}({}^K_\Delta\mathbf{O})_{\mathbb{C}}} g_0 * g_1 * \cdots * g_{l-2}$$
$$= \int_{\mathrm{M}({}^K\mathbf{O})_{\mathbb{C}}} g_0 k\Omega \wedge \cdots \wedge k\Omega + \int_{\mathrm{div}(f_0) \cap \mathrm{M}({}^K\mathbf{O})_{\mathbb{C}}} g_1 * \cdots * g_{l-2}.$$

All occurring integrals exist.

Proof. According to [10, Theorem 1.14], we have

$$\int_{\mathrm{M}({}^K_\Delta\mathbf{O})_{\mathbb{C}}} g_0 * \cdots * g_{l-2}$$
$$= \lim_{\varepsilon \to 0} \left(\int_{\mathrm{M}({}^K_\Delta\mathbf{O})_{\mathbb{C}} - B_\varepsilon(D)} g_0(k\Omega) \wedge \cdots \wedge (k\Omega) - \int_{\partial B_\varepsilon(D)} g \wedge \mathrm{d}^{\mathrm{c}} g_0 - g_0 \wedge \mathrm{d}^{\mathrm{c}} g \right)$$
$$+ \int_{\mathrm{div}(f_0) \cap \mathrm{M}({}^K\mathbf{O})_{\mathbb{C}}} g,$$

where

$$g = g_1 * \cdots * g_{l-2}.$$

The integral

$$\int_{\mathrm{M}({}^K\mathbf{O})} g_0(k\Omega) \wedge \cdots \wedge (k\Omega)$$

exists by [7, Theorem 2] because we excluded the cases $l = 3$, Witt rank 1 (modular curves) and $l = 4$, Witt rank 2 (product of modular curves). Therefore we are reduced to show that

$$\lim_{\varepsilon \to 0} \int_{\partial B_\varepsilon(D)} g \wedge \mathrm{d}^{\mathrm{c}} g_0 - g_0 \wedge \mathrm{d}^{\mathrm{c}} g = 0.$$

This will occupy the rest of the section.

3.4.4. We will need special neighborhoods of points on the boundary divisor of $\mathrm{M}({}^K_\Delta\mathbf{O})$. By Main Theorem 2.5.9(3) any such point lies on a stratum

$$\left[\mathrm{Stab}_{\Gamma([\sigma])} \backslash \mathrm{M}\left({}^{K_{[\sigma]}}\mathbf{B}_{[\sigma]}\right)\right]_{\mathbb{C}},$$

where $\mathbf{B}$ corresponds (2.4.5) to an isotropic line or an isotropic surface in $L_{\mathbb{Q}}$. (The corresponding boundary stratum in the Baily–Borel compactification is 0-, respectively 1-dimensional.) We will prepare special neighborhoods of these points and call them of **first (respectively second) type** for the rest of this section. By the boundary isomorphism 2.5.9(4), we transfer the neighborhood to the boundary of the mixed Shimura variety (compactfied only along the unipotent fibre):

$$\mathrm{M}({}^{K'}_{(\Delta')^0}\mathbf{B})_{\mathbb{C}}$$

which is a torus embedding constructed from the (relative) torus $\mathrm{M}({}^{K'}\mathbf{B})_{\mathbb{C}}$ over $\mathrm{M}({}^{K''}\mathbf{B}/U_{\mathbf{B}})_{\mathbb{C}}$ (2.5.8).

3.4.5. For a point of the first type, we have

$$\mathbf{B} \cong \mathbf{H}_0[U_{\mathbb{Z}_{(p)}}, 0], \tag{3.19}$$

where $U_{\mathbb{Z}_{(p)}} = I \otimes (I^\perp/I)$ for an isotropic (saturated) line in $L_{\mathbb{Z}_{(p)}}$ (2.4.5). Recall the trivialization s of $\Xi^*\mathcal{E}$ described in 2.8.2. It is of the form

$$s(z_1, \dots, z_{l-2}) = \big(\log(z_1)\lambda_1 + \cdots + \log(z_n)\lambda_n + \psi(z_{n+1}, \dots, z_{l-2})\big)z'$$

for $z_1, \dots, z_{l-2} \in B_\epsilon(0)^{l-2}$, z' is a vector in the fibre of $\mathcal{E}$ above the point corresponding to the splitting (3.19), i.e., $z' \in I^*$. The λ_i span the rational polyhedral cone σ corresponding to our boundary point and form part of a basis of $(U_{\mathbb{Q}} \cap K')(-1)$ (Δ is smooth w.r.t. K), and ψ is a holomorphic embedding $B_\epsilon(0)^{l-2-k} \to U_{\mathbf{B}}(\mathbb{C}) = U_{\mathbb{C}}$.

Let $r_i := |z_i|$. The norm of s is of the form

$$\sum_{i \le n, j \le n} \langle \lambda_i, \lambda_j \rangle_{z'} \log(r_i)\log(r_j) + \sum_{i \le n} \log(r_i)\Psi_i(z_1, \dots, z_{l-2}) + \Psi_0(z_1, \dots, z_{l-2}),$$

where the $\Psi_i, i > 0$ are smooth functions on $B_\epsilon(0)^{l-2}$ satisfying $\mathrm{d}\,\mathrm{d}^c\Psi_i = 0$, and Ψ_0 is smooth.

3.4.6. For a point of the second type, we have

$$\mathbf{B} \cong \mathbf{H}_1(I)[\Lambda^2 I, I \otimes (I^\perp/I)], \tag{3.20}$$

(2.4.5) and the trivialization s of $\Xi^*\mathcal{E}$ described in 2.8.2. It is of the form

$$s(z_1, z_2, \dots, z_{l-3}, z_{l-2}) \longmapsto \big(\lambda \log(z_1)\big) \circ \big(\psi_2(z_2, \dots, z_{l-3})\big)\big(\psi_1(z_{l-2})\big).$$

Here ψ_1 is a local trivialization of $\mathcal{E}_{\mathbb{C}}$ over a small neighborhood over the image of $\mathbb{D}_{\mathbf{H}_1(I)}$ in $\mathbb{D}_{\mathbf{H}_1(I)[\Lambda^2 I, I \otimes (I^\perp/I)]}$ such that $p(\psi_1(0))$ is a representative of the projection of the boundary point. Since $\mathcal{E}$ is the tautological bundle, we may express this by saying that ψ_1 maps to I^* inducing an embedding of $B_\epsilon(0)$ into $\mathbb{P}(I^*)$. ψ_2 is a holomorphic map $B_\epsilon(0)^{l-4} \to W_{\mathbf{B}}(\mathbb{C})$ which is modulo $U_{\mathbf{B}}(\mathbb{C})$ an open embedding into $V_{\mathbf{B}}(\mathbb{C})/\mathrm{Stab}(x)$ for all x in the image of $p(\psi_1(z_{l-2}))$. λ is a generator of the 1-dimensional $(U_{\mathbf{B}}(\mathbb{Q}) \cap K')(-1)$ and corresponds to the rational polyhedral cone σ (here there is only one possible).

Its norm is of the form

$$h_{\mathcal{E}}(s) = \log(r_1)\Psi_1(z_{l-2}) + \Psi_0(z_2, \dots, z_{l-3}, z_{l-2}),$$

where $\mathrm{d}\,\mathrm{d}^c\Psi_1 = 0$, and Ψ_0 is smooth. Note that Ψ_1 does only depend on z_{l-2}.

3.4.7. We will prepare an ε-tube neighborhood of D as follows. We prepared neighborhoods around every point of the boundary above. Take a finite cover $\{U_i\}$ consisting of these. Let τ_i be a partition of unity defined on $\mathrm{M}\big({}^K_\Delta\mathbf{O}(L)\big)_{\mathbb{C}}$ for the chosen cover, consisting of the neighborhoods considered in 3.4.5 and 3.4.6. Consider a neighborhood U_i with coordinates $z_{i,1}, \dots, z_{i,l-2}$. D has the equation $(z_{i,1})^{a_1} \cdots (z_{i,l-2})^{a_{l-2}} = 0$. On intersections $U_i \cap U_j$ we have

$$z_{j,k_j} = z_{i,k}\big(f^j_{i,k,0}(z_{i,1}, \overset{\hat{k}}{\dots}, z_{i,l-2}) + z_{i,k} f^j_{i,k}(z_{i,1}, \dots, z_{i,l-2})\big),$$

where $f^j_{i,k,0}(z_{i,1}, \overset{\hat{k}}{\dots}, z_{i,l-2})$ does not depend on $z_{i,k}$ and is everywhere nonzero on the overlap in question and where k_j is the index of the kth divisor component in the neighborhood U_j. We define

$$r'_{i,k} = \sum_j \tau_j |z^{j,k_j}| \quad \text{on } U_i.$$

If the k-component does not exist in U_j, we omit the term in the sum.

Define $r' = \min_{k | a_k \neq 0} r'_{i,k}$. This does not depend on i anymore and is a well-defined function in a neighborhood of the divisor. A global ε-tube neighborhood around D may therefore be described as

$$B_\varepsilon(D) := \big\{z \in \mathrm{M}\big({}^K_\Delta\mathbf{O}(L)\big)_{\mathbb{C}} \mid r'(z) \le \varepsilon\big\}.$$

In a local neighborhood U_i write

$$r'_{i,k} = r_{i,k}\left(\sum_j \tau_j |f^j_{0,i,k}(z_{i,1}, \overset{\hat{k}}{\dots}, z_{i,l-2})| + r_{i,k} g\right),$$

where g is a bounded C^∞ function. If ε is small enough, $0 < r'_i \le \varepsilon$ will ensure $r_{i,k} = |z_{i,k}| > 0$.

On U_i we may cover $\partial B_\varepsilon(D)$ by (we omit the index i because it will be fixed from now on)

$$\mathrm{St}'_k(\varepsilon) = \{z \in U_i \mid r'_k = \varepsilon, r'_j \ge \varepsilon \text{ if } a_j \neq 0\},$$

for all k, where $a_k \neq 0$. Note that on overlaps the $r'_{i,k}$ and $r'_{j,k}$ are the same up to permutation of the indices k.

We have for small r'_k and some bounded C^∞-function h:

$$r_k = r'_k\left(\left(\sum \tau_j |f^j_{0,k}(z_1, \overset{\hat{k}}{\dots}, z_{l-2})|\right)^{-1} + r'_k h\right),$$

$$\begin{aligned} \mathrm{d}r_k = {} & \mathrm{d}r'_k\left(\sum \tau_j |f^j_{0,k}(z_1, \overset{\hat{k}}{\dots}, z_{l-2})|\right)^{-1} + r'_k\, \mathrm{d}r'_k h \\ & + r'_k\, \mathrm{d}\left(\left(\sum \tau_j |f^j_{0,k}(z_1, \overset{\hat{k}}{\dots}, z_{l-2})|\right)^{-1} + r'_k h\right). \end{aligned} \tag{3.21}$$

3.4.8. In any of the neighborhoods U_i, we will write the quantities in question in the basis

$$\mathrm{d}\varphi_1, \dots, \mathrm{d}\varphi_{l-2}, \mathrm{d}r_1, \overset{\hat{k}}{\dots}, \mathrm{d}r_{l-2}, \mathrm{d}r'_{l-2}.$$

Only the term

$$\omega := \mathrm{d}\varphi_1 \wedge \cdots \wedge \mathrm{d}\varphi_{l-2} \wedge \mathrm{d}r_1 \wedge \overset{\hat{k}}{\cdots} \wedge \mathrm{d}r_{l-2}$$

gives a nonzero contribution in integrals over $\mathrm{St}'_k(\varepsilon)$. In what follows, we understand:

$$|f(z)\omega| := |f(z)|\omega.$$

We order the coordinates in such a way that D is given by the equation $z_1 \cdots z_n = 0$ on U_i.

Recall the notion of log-log-growth of infinite order [14, 2.2]. A log-log-**form** [14, 2.22] is a differential form generated (over smooth forms) by functions of log-log-growth of infinite order and the differentials

$$\frac{\mathrm{d}\varphi_1}{\log(r_1)}, \ldots, \frac{\mathrm{d}\varphi_{l-2}}{\log(r_n)}, \frac{\mathrm{d}r_1}{r_1 \log(r_1)}, \ldots, \frac{\mathrm{d}r_{l-2}}{r_{l-2}\log(r_{l-2})}$$

such that d, d^c and $\mathrm{d}\,\mathrm{d}^c$ of it are again of this type.

Call a form a **(*)-form** for an index k, if it is generated by functions of log-log-growth of infinite order and the differentials

$$\mathrm{d}\varphi_1, \ldots, \mathrm{d}\varphi_{l-2}, \frac{\mathrm{d}r_1}{r_1 \log(r_1)}, \overset{\hat{k}}{\ldots}, \frac{\mathrm{d}r_{l-2}}{r_{l-2}\log(r_{l-2})}$$

and $f\,\mathrm{d}r'_k$ for any f which is smooth outside D.

3.4.9. Lemma. *For every (*)-form σ for the index k:*

$$\int_{\mathrm{St}'_k(\varepsilon)} |\sigma| = O(\log(|\log\varepsilon|)^M)$$

for some M.

PROOF. It suffices to show that on each of the sets

$$\mathrm{St}'_k(\varepsilon, R) = \{z \in B(R) \mid r'_k = \varepsilon, r_j \geq \delta\varepsilon \text{ if } a_j \neq 0\},$$

for some small δ (note that we wrote r_j instead of r'_j), we have the estimate:

$$\int_{\mathrm{St}'_k(\varepsilon,R)} |\sigma| = O(\log(|\log\varepsilon|)^M).$$

This is easy, see, e.g., [7]. The (maybe singular) terms $f\,\mathrm{d}r'_k$ play no role because terms involving $\mathrm{d}r'_k$ do not give any contribution to the integral. □

3.4.10. Corollary. *For every* log-log *growth form ξ:*

$$\lim_{\varepsilon\to 0}\int_{\mathrm{St}'_k(\varepsilon)} |\xi| = 0.$$

PROOF. We have by (3.21)

$$\frac{1}{\log(r_k)}\,\mathrm{d}\varphi_i \prec \frac{1}{\log(r'_k)}\sigma,$$

and

$$\frac{\mathrm{d}r_k}{r_k\log(r_k)} \prec \frac{1}{\log(r'_k)}\sigma,$$

where the σ are (*)-forms w.r.t. k.

If ξ has correct degree, by definition, it involves at least either $\big(1/\log(r_k)\big)\,\mathrm{d}\varphi_k$ or $\mathrm{d}r_k/\big(r_k\log(r_k)\big)$. Hence

$$\xi \prec \frac{1}{\log(r'_k)}\sigma,$$

for some (*)-form σ. The statement follows using 3.4.9 because, of course, log grows faster than any power of log-log. □

3.4.11. First of all $\operatorname{supp}(\operatorname{div}(f_1)) \cap \operatorname{supp}(\operatorname{div}(f_2)) \cap \operatorname{supp}(D) = \varnothing$ by assumption, hence we may represent (cf. 2.6.9):

$$g = \sigma g_{1,2}\Omega^{l-4} + \mathrm{d}\,\mathrm{d}^c\big((1-\sigma)g_{1,2}\big)g',$$

where σ is equal to 1 in a neighborhood of D and $g_{1,2} = g_1 * g_2$. We may represent this by

$$g_{1,2} = \sigma_1 g_1 \Omega + \mathrm{d}\,\mathrm{d}^c(\sigma_2 g_1) g_2,$$

where $\sigma_1 + \sigma_2 = 1$ is a partition of unity of the following form: The intersection of $\operatorname{div}(f_1)$ with D occurs precisely on the preimage under the projection on the Baily–Borel compactification of a set of isolated points in the 1-dimensional boundary stratum isomorphic to some $\mathrm{M}\big({}^K\mathbf{H}(I)\big)$. Choose a C^∞-function $p\sigma$ on $\mathrm{M}\big({}^K\mathbf{H}(I)\big)$ with support on some disc around one of these points which is 1 on a smaller neighborhood of the point.

We may assume that in $B_\varepsilon(D)$ for small ε, σ_1 is just the preimage of $p\sigma$ under the projection $\mathrm{M}({}^{K'}_{(\Delta')^0}\mathbf{B}) \to \mathrm{M}\big({}^K\mathbf{H}(I)\big)$.

We have hence in a neighborhood of D:

$$(3.22)\quad g = \sigma_1 g_1 \Omega^{l-3} + \sigma_2 g_2 \Omega^{l-3} \\ + \big((\mathrm{d}\sigma_2 \wedge \mathrm{d}^c g_1)g_2 - (\mathrm{d}^c\sigma_2 \wedge \mathrm{d} g_1)g_2 + (\mathrm{d}\,\mathrm{d}^c\sigma_2)g_1 g_2\big) \wedge \Omega^{l-4}.$$

In any of the neighborhoods considered in 3.4.5 and 3.4.6 we have

$$(3.23)\qquad g_0 = \sum_{j\le n} a_j \log(r_j) + \log\big(h_{\mathcal{E}}(s)\big) + \psi(z)$$

for some harmonic function ψ, where s is the corresponding trivializing section of $\Xi^*(\mathcal{E})$.

Recall that we want to show that

$$\lim_{\varepsilon\to 0} \int_{\partial B_\varepsilon(D)} g \wedge \mathrm{d}^c g_0 - g_0 \wedge \mathrm{d}^c g = 0.$$

Inserting (3.22) and (3.23) into this, since in any case (by 3.4.10) the limit of integrals of log-log forms is 0, we are reduced to show, one the one hand (first line in (3.22)), that

$$\lim_{\varepsilon\to 0} \int_{\mathrm{St}'_k(\varepsilon)} |\mathrm{d}\varphi_j \sigma_l g_l \Omega^{l-3}| = 0$$

and

$$\lim_{\varepsilon\to 0} \int_{\mathrm{St}'_k(\varepsilon)} |\log(r_j)\,\mathrm{d}^c(\sigma_l g_l)\Omega^{l-3}| = 0,$$

where $l = 1, 2$, in neighborhoods of both types. Every other term in (3.23) inserted into

$$\lim_{\varepsilon\to 0} \int_{\mathrm{St}'_k(\varepsilon)} |g \wedge \mathrm{d}^c g_0 - g_0 \wedge \mathrm{d}^c g|$$

yields a limit over an integral of a form of log-log type which is zero by Lemma 3.4.10.

If $j = k$ we may rewrite r_j by means of r'_k and apply (3.26) of Lemma 3.4.12. In the other case we apply (3.27) of Lemma 3.4.12. (Note that $\mathrm{d}\varphi_j$ is of the form $\log(r_j)$ times a log-log-form.) After this, the vanishing of the limit $\varepsilon \to 0$ follows from Lemma 3.4.9.

On the other hand (second line in (3.22)), we have to show

$$\lim_{\varepsilon\to 0}\int_{\mathrm{St}_1'(\varepsilon)} |\mathrm{d}\varphi_1(\mathrm{d}\sigma_2\wedge \mathrm{d}^c g_1 - \mathrm{d}^c\sigma_2\wedge \mathrm{d}g_1 + (\mathrm{d}\,\mathrm{d}^c\sigma_2)g_1)\wedge g_2\wedge \Omega^{l-4}| = 0 \tag{3.24}$$

(3.25)

$$\lim_{\varepsilon\to 0}\int_{\mathrm{St}_1'(\varepsilon)} |\log(r_1)(\mathrm{d}\sigma_2\wedge \mathrm{d}^c g_1 - \mathrm{d}^c\sigma_2\wedge \mathrm{d}g_1 + (\mathrm{d}\,\mathrm{d}^c\sigma_2)g_1)\wedge \mathrm{d}^c g_2\wedge \Omega^{l-4}| = 0$$

Here $j = k = 1$ because this is nonzero only in neighborhoods of the second type, and there cannot be an intersection of components above $\mathrm{M}({}^K\mathbf{H}(I))$.

The estimate (3.28) of Lemma 3.4.12, we get vanishing of the limit $\varepsilon\to 0$ of (3.24) by Lemma 3.4.9 again.

For (3.25) note that we constructed σ_2 (in $B_\varepsilon(D)$!) as the preimage (under the projection $\mathrm{M}({}^{K_1}_{\Delta_1^0}\mathbf{B}) \to \mathrm{M}({}^K\mathbf{H}(I))$ of a function supported on a small disc around the zeros or poles of f_1 in the boundary of the Baily–Borel compactification. Let X be the preimage of these discs in some neighborhood of the boundary in $\mathrm{M}({}^K_{\Delta_1^0}\mathbf{B})$. Hence the form

$$\mathrm{d}\big(g_0(\mathrm{d}\sigma_2\wedge \mathrm{d}^c g_1 - \mathrm{d}^c\sigma_2\wedge \mathrm{d}g_1 + (\mathrm{d}\,\mathrm{d}^c\sigma_2)g_1)\wedge \mathrm{d}^c g_2\wedge \mathrm{d}^c\log\big(h_{\mathcal{E}}(s)\big)\wedge\Omega^{l-3}\big)$$

is supported only on X. Applying the Theorem of Stokes to this function on $\partial B_\varepsilon(D)$, we get

$$\begin{aligned}
0 = &-\int_{\partial B_\varepsilon(D)} g_0(\mathrm{d}\sigma_2\wedge \mathrm{d}^c g_1 - \mathrm{d}^c\sigma_2\wedge \mathrm{d}g_1 + (\mathrm{d}\,\mathrm{d}^c\sigma_2)g_1)\wedge \mathrm{d}^c g_2\wedge \mathbf{\Omega}\wedge\Omega^{l-5}\\
&+\int_{\partial B_\varepsilon(D)} g_0\big(\mathrm{d}\sigma_2\wedge \mathrm{d}^c g_1 - \mathrm{d}^c\sigma_2\wedge \mathrm{d}g_1 + (\mathrm{d}\,\mathrm{d}^c\sigma_2)g_1\big)\\
&\qquad\qquad \wedge \mathbf{d}\,\mathrm{d}^c g_2\wedge \mathrm{d}^c\log\big(h_{\mathcal{E}}(s)\big)\wedge\Omega^{l-5}\\
&+\int_{\partial B_\varepsilon(D)} g_0\Big(-\mathrm{d}\sigma_2\wedge \mathbf{d}\,\mathrm{d}^c g_1 \underbrace{-\,\mathbf{d}\,\mathrm{d}^c\sigma_2\wedge \mathrm{d}g_1 + \mathrm{d}\,\mathrm{d}^c\sigma_2\wedge \mathbf{d}g_1}_{=0}\Big)\\
&\qquad\qquad \wedge \mathrm{d}^c g_2\wedge \mathrm{d}^c\log\big(h_{\mathcal{E}}(s)\big)\wedge\Omega^{l-5}\\
&+\int_{\partial B_\varepsilon(D)} \mathbf{d}g_0\wedge(\mathrm{d}\sigma_2\wedge \mathrm{d}^c g_1 - \mathrm{d}^c\sigma_2\wedge \mathrm{d}g_1 + (\mathrm{d}\,\mathrm{d}^c\sigma_2)g_1)\\
&\qquad\qquad \wedge \mathrm{d}^c g_2\wedge \mathrm{d}^c\log\big(h_{\mathcal{E}}(s)\big)\wedge\Omega^{l-5}.
\end{aligned}$$

The limit of the first line is the one we have to estimate. We may hence instead consider the remaining three lines. Those we may again restrict to small neighborhoods U_i (all of the second type!) and estimate the absolute value of the integral. Here all problematic terms are of the form

$$\int_{\mathrm{St}_1'(\varepsilon)} |\log(r_1)\xi'\wedge\xi\wedge \mathrm{d}^c\log\big(h_{\mathcal{E}}(s)\big)\wedge\Omega^{l-4}|$$

or

$$\int_{\mathrm{St}_1'(\varepsilon)} \left|\frac{\mathrm{d}r_1}{r_1}\wedge\xi'\wedge\xi\wedge \mathrm{d}^c g_2\wedge \mathrm{d}^c\log\big(h_{\mathcal{E}}(s)\big)\wedge\Omega^{l-5}\right|,$$

where ξ' is a smooth form of degree at least 1, generated by $r_{l-2}\,\mathrm{d}\varphi_{l-2}$ and $\mathrm{d}r_{l-2}$ and ξ is log-log. The latter poses no problem because $\mathrm{d}r_1/r_1$ is, up to terms involving $\mathrm{d}r_1'$, a smooth form (3.21).

Now, by the estimate (3.29) of Lemma 3.4.12, we get vanishing of the limit $\varepsilon\to 0$ by Lemma 3.4.9 again. □

3.4.12. Lemma. *Let U be a neighborhood of any type and let i be the Witt rank of L. Then for any* log-log *form ξ of rank $2(l-2-i)-1$, we have*

$$\xi \wedge \Omega^i \prec \frac{1}{\log(r'_k)^2}\sigma, \tag{3.26}$$

where σ is a (*)-form w.r.t. to k.

Also we have for any $j \neq k$

$$\xi \wedge \Omega^i \prec \frac{1}{\log(r'_k)\log(r_j)}\sigma, \tag{3.27}$$

where σ is another (*)-form w.r.t. k.

If U is a neighborhood of the second type (3.4.6), *for any* log-log *form $\xi' \wedge \xi$ of degree $2(l-2)-5$, where ξ' is a smooth form of degree at least* 1, *involving only $r_{l-2}\,\mathrm{d}\varphi_{l-2}$ and $\mathrm{d}r_{l-2}$, where r_{l-2}, φ_{l-2} are the polar coordinates of the projection to a $\mathrm{M}({}^K\mathbf{H}(I))$ in a neighborhood of the point, we have*

$$\mathrm{d}\varphi_1 \wedge \xi' \wedge \xi \wedge \Omega \prec \frac{1}{\log(r'_1)}\sigma, \tag{3.28}$$

and

$$\xi' \wedge \xi \wedge \mathrm{d}^c \log\big(h_{\mathcal{E}}(s)\big) \wedge \Omega \prec \frac{1}{\log(r'_1)^2}\sigma, \tag{3.29}$$

where the σ are (*)-forms w.r.t. the index 1.

PROOF. We have in any case:

$$\mathrm{d}^c \log(r_i) = \mathrm{d}\varphi_i, \quad \mathrm{d}\log(r_i) = \frac{\mathrm{d}r_i}{r_i},$$

and

$$\mathrm{d}^c \log h_{\mathcal{E}}(s) = \frac{\mathrm{d}^c h_{\mathcal{E}}(s)}{h_{\mathcal{E}}(s)}, \quad \mathrm{d}\log h_{\mathcal{E}}(s) = \frac{\mathrm{d}h_{\mathcal{E}}(s)}{h_{\mathcal{E}}(s)},$$
$$\Omega \sim \mathrm{d}\,\mathrm{d}^c \log h_{\mathcal{E}}(s) = \frac{\mathrm{d}h_{\mathcal{E}}(s)}{h_{\mathcal{E}}(s)}\frac{\mathrm{d}^c h_{\mathcal{E}}(s)}{h_{\mathcal{E}}(s)} + \frac{\mathrm{d}\,\mathrm{d}^c h_{\mathcal{E}}(s)}{h_{\mathcal{E}}(s)}.$$

In a neighborhood *of the first type* (3.4.5), we may write

$$h_{\mathcal{E}}(s) = \sum_{i,j\leq n} \langle\lambda_i, \lambda_j\rangle \log(r_i)\log(r_j) + \sum_{i\leq n} \log(r_i)\psi_i(z) + \psi_0(z),$$

where ψ_i are harmonic functions and ψ_0 is smooth. Here the λ_i are linearly independent and $\langle\lambda_i, \lambda_j\rangle > 0$ if $i \neq j$. Hence for $i \neq j$ always

$$h_{\mathcal{E}}(s) \gg \log(r_i)\log(r_j), \tag{3.30}$$

and if $Q_L(\lambda_i) > 0$

$$h_{\mathcal{E}}(s) \gg \log(r_i)^2. \tag{3.31}$$

Hence

$$\begin{aligned}\mathrm{d}^c h_{\mathcal{E}}(s) = {}& \log(r_k)\Big(\sum_{j\leq n}\langle\lambda_k, \lambda_j\rangle\,\mathrm{d}\varphi_j + \mathrm{d}^c\psi_k\Big) + \sum_{i\leq n}\mathrm{d}\varphi_i\psi_i \\ &+ \sum_{i\leq n, j\leq n, i\neq k}\langle\lambda_i, \lambda_j\rangle\log(r_i)\,\mathrm{d}\varphi_j + \sum_{i\leq n, i\neq k}\log(r_i)\,\mathrm{d}^c\psi_k + \mathrm{d}^c\psi_0\end{aligned} \tag{3.32}$$

$$\begin{aligned} \mathrm{d}h_{\mathcal{E}}(s) = \log(r_k)\left(\sum_{j\le n}\langle\lambda_k,\lambda_j\rangle\frac{\mathrm{d}r_j}{r_j}+\mathrm{d}\psi_k\right)+\sum_{i\le n}\frac{\mathrm{d}r_j}{r_j}\psi_i \\ +\sum_{i\le n,j\le n,i\ne k}\langle\lambda_i,\lambda_j\rangle\log(r_i)\frac{\mathrm{d}r_j}{r_j}+\sum_{i\le n,i\ne k}\log(r_i)\,\mathrm{d}\psi_k+\mathrm{d}\psi_0 \end{aligned} \tag{3.33}$$

$$\begin{aligned} \mathrm{d}\,\mathrm{d}^c h_{\mathcal{E}}(s) = \mathrm{d}\varphi_k\left(\sum_{j\le n}\langle\lambda_k,\lambda_j\rangle\frac{\mathrm{d}r_j}{r_j}+\mathrm{d}\psi_k\right)+\sum_{i\le n}\frac{\mathrm{d}r_j}{r_j}\mathrm{d}^c\psi_i \\ +\sum_{i\le n,j\le n,i\ne k}\langle\lambda_i,\lambda_j\rangle\,\mathrm{d}\varphi_i\frac{\mathrm{d}r_j}{r_j}+\sum_{i\le n,i\ne k}\mathrm{d}\varphi_i\,\mathrm{d}\psi_k+\mathrm{d}\,\mathrm{d}^c\psi_0. \end{aligned} \tag{3.34}$$

We now substitute $\mathrm{d}r_k/r_k$ by a form of shape

$$f\frac{\mathrm{d}r'_k}{r'_k}+\xi,$$

where ξ is smooth. This is possible by means of formula (3.21).

First assume $i=2$. Ω^2 is proportional to

$$\frac{\mathrm{d}\,\mathrm{d}^c h_{\mathcal{E}}(s)}{h_{\mathcal{E}}(s)}\wedge\frac{\mathrm{d}\,\mathrm{d}^c h_{\mathcal{E}}(s)}{h_{\mathcal{E}}(s)} \tag{3.35}$$

$$+\frac{\mathrm{d}h_{\mathcal{E}}(s)}{h_{\mathcal{E}}(s)}\frac{\mathrm{d}^c h_{\mathcal{E}}(s)}{h_{\mathcal{E}}(s)}\wedge\frac{\mathrm{d}\,\mathrm{d}^c h_{\mathcal{E}}(s)}{h_{\mathcal{E}}(s)}. \tag{3.36}$$

In (3.35), multiplying out the expression (3.34) squared, we have for every occurring summand S

$$S\prec\frac{1}{\log(r_k)^2}\sigma\prec\frac{1}{\log(r'_k)^2}\sigma,$$

where σ is a (*)-form, using the estimate (3.30).

In (3.36), multiplying out the product of (3.32)–(3.34), we have

$$S\prec\frac{1}{\log(r_k)^2}\sigma\prec\frac{1}{\log(r'_k)^2}\sigma,$$

using the estimate (3.30) again, except for the summands of the form

$$\log(r_k)\log(r_k)\left(\sum_{j\le n}\langle\lambda_k,\lambda_j\rangle\,\mathrm{d}\varphi_j+\mathrm{d}^c\psi_k\right)\wedge\left(\sum_{j\le n}\langle\lambda_k,\lambda_j\rangle\frac{\mathrm{d}r_j}{r_j}+\mathrm{d}\psi_k\right)\wedge S, \tag{3.37}$$

where S is any summand of (3.34).

But now in the expression $\xi\wedge\Omega^2$ either occurs a $\mathrm{d}\varphi_k/\log(r_k)$ from ξ, hence the estimate (3.26) is true, or a $r_k\,\mathrm{d}\varphi_k$ occurs in S which satisfies a much stronger estimate, or a $\mathrm{d}\varphi_k$ occurs in S. It occurs, however, multiplied with

$$\Big(\sum_{j\le n}\langle\lambda_k,\lambda_j\rangle\frac{\mathrm{d}r_j}{r_j}+\mathrm{d}\psi_k\Big)$$

so (3.37) is zero in that case.

The estimate (3.27) is more easy and left to the reader.

Now assume $i=1$. Then, by the estimate (3.31), already every summand S in (3.32)–(3.34) divided by $h_{\mathcal{E}}(s)$ satisfies

$$S\prec\frac{1}{\log(r_k)}\sigma\prec\frac{1}{\log(r'_k)}\sigma,$$

with a (*)-form σ.

The estimates (3.28), (3.29) do not involve neighborhoods of the first type.

In a neighborhood *of the second type* (3.4.6), we may write

$$h_{\mathcal{E}}(s) = \log(r_1)\psi_1(z_{l-2}) + \psi_0(z),$$

where ψ_1 is harmonic and depends only on z_{l-2} and ψ_0 is smooth.

Clearly in this case:

$$h_{\mathcal{E}}(s) \gg \log(r_1). \tag{3.38}$$

Also

$$\begin{aligned}
\mathrm{d}^c h_{\mathcal{E}}(s) &= \log(r_1)\,\mathrm{d}^c\psi_1(z_{l-2}) + \mathrm{d}\varphi_1\psi_1(z_{l-2}) + \mathrm{d}^c\psi_0(z),\\
\mathrm{d}h_{\mathcal{E}}(s) &= \log(r_1)\,\mathrm{d}\psi_1(z_{l-2}) + \frac{\mathrm{d}r_1}{r_1}\psi_1(z_{l-2}) + \mathrm{d}\psi_0(z),\\
\mathrm{d}\,\mathrm{d}^c h_{\mathcal{E}}(s) &= \mathrm{d}\varphi_1\,\mathrm{d}\psi_1(z_{l-2}) + \frac{\mathrm{d}r_1}{r_1}\,\mathrm{d}^c\psi_1(z_{l-2}) + \mathrm{d}\,\mathrm{d}^c\psi_0(z).
\end{aligned}$$

We now substitute again $\mathrm{d}r_1/r_1$ by a form of shape

$$f\frac{\mathrm{d}r_1'}{r_1'} + \xi,$$

where ξ is smooth. This is possible by means of formula (3.21).

Here, for (3.26), we may argue exactly as before. (3.27) is vacuous in this type of neighborhood. For (3.28): Write once again Ω as

$$\frac{\mathrm{d}\,\mathrm{d}^c h_{\mathcal{E}}(s)}{h_{\mathcal{E}}(s)} \tag{3.39}$$

$$+\frac{\mathrm{d}^c h_{\mathcal{E}}(s)}{h_{\mathcal{E}}(s)} \wedge \frac{\mathrm{d}\,\mathrm{d}^c h_{\mathcal{E}}(s)}{h_{\mathcal{E}}(s)}. \tag{3.40}$$

In (3.39), using (3.38), we have for any summand

$$S \prec \frac{1}{\log(r_1')}\sigma,$$

for a (*)-form σ.

In (3.40), using (3.38) again, we get the same for any summand except possibly for

$$\frac{1}{h_{\mathcal{E}}(s)^2}\log(r_1)^2\,\mathrm{d}^c\psi_1(z_{l-2}) \wedge \mathrm{d}\psi_1(z_{l-2})$$

which cancels, however, with ξ' because $\mathrm{d}^c\psi_1(z_{l-2}) \wedge \mathrm{d}\psi_1(z_{l-2})$ is proportional to $r_{l-2}\,\mathrm{d}\varphi_{l-2} \wedge \mathrm{d}r_{l-2}$.

For (3.29): Consider:

$$\mathrm{d}^c\log\big(h_{\mathcal{E}}(s)\big) \wedge \Omega = \frac{\mathrm{d}^c h_{\mathcal{E}}(s)}{h_{\mathcal{E}}(s)} \wedge \frac{\mathrm{d}\,\mathrm{d}^c h_{\mathcal{E}}(s)}{h_{\mathcal{E}}(s)}. \tag{3.41}$$

Using (3.38) again, we now have for any summand in

$$S \prec \frac{1}{\log(r_1')^2}\sigma,$$

for a (*)-form σ, except possibly for

$$\frac{1}{h_{\mathcal{E}}(s)^2}\log(r_1)\,\mathrm{d}^c\psi_1(z_{l-2})\,\mathrm{d}\varphi_i\,\mathrm{d}\psi_1(z_{l-2})$$

which cancels with ξ' because $\mathrm{d}^c\psi_1(z_{l-2}) \wedge \mathrm{d}\psi_1(z_{l-2})$ is proportional to $r_{l-2}\,\mathrm{d}\varphi_{l-2} \wedge \mathrm{d}r_{l-2}$ and possibly for

$$\frac{1}{h_{\mathcal{E}}(s)^2}\log(r_1)\,\mathrm{d}^c\psi_1(z_{l-2})\,\mathrm{d}\,\mathrm{d}^c\psi_0(z),$$

But then a $\mathrm{d}\varphi_1/\log(r_1)$ must appear somewhere in the expression (3.29). □

3.5. Main results

This section contains the main results of this book.

3.5.1. Recall the definitions of Section 3.2.12 and 2.6.16. The relation between $\widetilde{\mu}$ and $\tilde{\lambda}$ is given by a global orbit equation:

3.5.2. Theorem ([26, Theorem 10.5]). *Assume $l \geq 3$, $l - m \geq 1$. Let D be the discriminant of $L_{\mathbb{Z}}$ and D' be the D-primary part of the discriminant of $M_{\mathbb{Z}}$.*

$$\tilde{\lambda}^{-1}(L_{\mathbb{Z}};0)\widetilde{\mu}(L_{\mathbb{Z}},M_{\mathbb{Z}},\kappa;0) = \sum_{\alpha\,\mathrm{SO}'(L_{\hat{\mathbb{Z}}})\subset \mathrm{I}(L,M)(\mathbb{A}^{(\infty)})\cap\kappa} \tilde{\lambda}^{-1}(\alpha_{\mathbb{Z}}^{\perp};0)$$

and

$$\frac{\mathrm{d}}{\mathrm{d}s}\left(\tilde{\lambda}^{-1}(L_{\mathbb{Z}};s)\widetilde{\mu}(L_{\mathbb{Z}},M_{\mathbb{Z}},\kappa;s)\right)\bigg|_{s=0} \equiv \sum_{\alpha\,\mathrm{SO}'(L_{\hat{\mathbb{Z}}})\subset \mathrm{I}(L,M)(\mathbb{A}^{(\infty)})\cap\kappa} \frac{\mathrm{d}}{\mathrm{d}s}\tilde{\lambda}^{-1}(\alpha_{\mathbb{Z}}^{\perp};s)\bigg|_{s=0}$$

in $\mathbb{R}_{2DD''}$, where D'' is the product of the primes such that $p^2 \nmid D'$.

There is also a relation between different $\widetilde{\mu}$'s in the style of Kitaoka's recursion formula:

3.5.3. Theorem ([26, Theorem 10.6]). *Assume $l \geq 3$, $l - m \geq 1$. Let D be the discriminant of $L_{\mathbb{Z}}$ and $M_{\mathbb{Z}} = M'_{\mathbb{Z}} \perp M''_{\mathbb{Z}}$. Let D' be the D-primary part of the discriminant of $M'_{\mathbb{Z}}$ (not $M_{\mathbb{Z}}$!). Let $\kappa \in (L^*_{\mathbb{Z}}/L_{\mathbb{Z}}) \otimes M^*_{\mathbb{Z}}$ with a corresponding decomposition $\kappa = \kappa' \oplus \kappa''$. We have*

$$\tilde{\lambda}^{-1}(L_{\mathbb{Z}};0)\widetilde{\mu}(L_{\mathbb{Z}},M_{\mathbb{Z}},\kappa;0) = \sum_{\substack{\alpha\,\mathrm{SO}'(L_{\hat{\mathbb{Z}}})\subset \mathrm{I}(L,M')(\mathbb{A}^{(\infty)})\cap\kappa' \\ \kappa''\cap\alpha^{\perp}_{\mathbb{A}^{(\infty)}}\otimes(M''_{\mathbb{A}^{(\infty)}})^*\neq\varnothing}} \tilde{\lambda}^{-1}(\alpha_{\mathbb{Z}}^{\perp};0)\widetilde{\mu}(\alpha_{\mathbb{Z}}^{\perp},M''_{\mathbb{Z}},\kappa'';0)$$

and

$$\frac{\mathrm{d}}{\mathrm{d}s}\left(\tilde{\lambda}^{-1}(L_{\mathbb{Z}};s)\widetilde{\mu}(L_{\mathbb{Z}},M_{\mathbb{Z}},\kappa;s)\right)\bigg|_{s=0}$$
$$= \sum_{\substack{\alpha\,\mathrm{SO}'(L_{\hat{\mathbb{Z}}})\subset \mathrm{I}(L,M')(\mathbb{A}^{(\infty)})\cap\kappa' \\ \kappa''\cap\alpha^{\perp}_{\mathbb{A}^{(\infty)}}\otimes(M''_{\mathbb{A}^{(\infty)}})^*\neq\varnothing}} \frac{\mathrm{d}}{\mathrm{d}s}\left(\tilde{\lambda}^{-1}(\alpha_{\mathbb{Z}}^{\perp};s)\widetilde{\mu}(\alpha_{\mathbb{Z}}^{\perp},M''_{\mathbb{Z}},\kappa'';s)\right)\bigg|_{s=0}$$

in $\mathbb{R}_{2DD'}$. Here $\kappa'' \in (L^*_{\hat{\mathbb{Z}}}/L_{\hat{\mathbb{Z}}})\otimes(M''_{\mathbb{Z}})^*$ is considered as an element of $\left((\alpha_{\hat{\mathbb{Z}}}^{\perp})^*/\alpha_{\hat{\mathbb{Z}}}^{\perp}\right)\otimes(M''_{\hat{\mathbb{Z}}})^*$ via $\kappa'' \mapsto \kappa'' \cap (\alpha^{\perp}_{\mathbb{A}^{(\infty)}} \otimes (M''_{\mathbb{A}^{(\infty)}})^*)$.

3.5.4. Let $L_{\mathbb{Z}}$ be a lattice with quadratic form of discriminant $D \neq 0$ and signature $(l-2,2)$. Let K be the *discriminant kernel* of $L_{\hat{\mathbb{Z}}}$. It is an admissible compact open subgroup for all $p \nmid D$. Let Δ be a complete and smooth K-admissible rational polyhedral cone decomposition, and let $\mathrm{M}^K = \mathrm{M}({}^K_\Delta\mathbf{O}(L_{\mathbb{Z}}))$ the *global* canonical model (2.6.26) defined over $\mathbb{Z}[1/(2D)]$.

At primes $p \neq 2$, with $p \mid D$ but $p^2 \nmid D$, we can find a lattice $L'_{\mathbb{Z}}$ containing $L_{\mathbb{Z}}$ as a saturated sublattice of codimension one such that $L'_{\mathbb{Z}_{(p)}}$ is unimodular. If Δ is chosen appropriate, we may define thus a local model over $\mathbb{Z}_{(p)}$ of $\mathrm{M}({}^K_\Delta\mathbf{O}(L))_{\mathbb{Q}}$ at those p by taking the normalization of the Zariski closure in the model $\mathrm{M}({}^{K'}_{\Delta'}\mathbf{O}(L'))$ over $\mathrm{Spec}(\mathbb{Z}_{(p)})$. We understand by M^K the so glued global model over $\mathbb{Z}[1/(2D')]$, where D' is the product of the primes p such that $p^2|D$. Let $\overline{\mathcal{E}}$ be as before. In the same way, we glue $\Xi^*\overline{\mathcal{E}}$ on M^K from the pullbacks of $\Xi^*\overline{\mathcal{E}}$ on $\mathrm{M}({}^{K'}_{\Delta'}\mathbf{O}(L'))$ over $\mathrm{Spec}(\mathbb{Z}_{(p)})$. Note that it is not clear from the properties of the theory of canonical models outlined in the first sections, whether this depends on the choice of lattices $L'_{\mathbb{Z}}$. The theorem below, however, is true for any of these choices. It is also not clear whether the model obtained is regular. If it is not, we understand by the arithmetic volume $\widehat{\mathrm{vol}}$ below the height computed with respect to this particular embedding (2.6.17).

3.5.5. Main Theorem. *We have*

$$(1)\ \ \mathrm{vol}_{\mathcal{E}_{\mathbb{C}}}(\mathrm{M}^K) = 4\tilde{\lambda}^{-1}(L_{\mathbb{Z}};0),$$
$$(2)\ \ \widehat{\mathrm{vol}}_{\overline{\mathcal{E}}}(\mathrm{M}^K) \equiv \frac{\mathrm{d}}{\mathrm{d}s}4\tilde{\lambda}^{-1}(L_{\mathbb{Z}};s)|_{s=0} \text{ in } \mathbb{R}_{2D'}.$$

Let $M_{\mathbb{Z}}$ be a lattice of dimension m with positive definite $Q_M \in \mathrm{Sym}^2(M^*_{\mathbb{Q}})$. Let D'' be the product of the primes p such that $M_{\mathbb{Z}_p} \not\subseteq M^*_{\mathbb{Z}_p}$ or $M^*_{\mathbb{Z}_p}/M_{\mathbb{Z}_p}$ is not cyclic. Assume

- $l-m \geq 4$, *or*
- $l=4$, $m=1$, *and* $L_{\mathbb{Q}}$ *is has Witt rank* 1.

Then we have for each $\kappa \in \mathbb{Z}[(L^*_{\mathbb{Z}}/L_{\mathbb{Z}}) \otimes M^*_{\mathbb{Z}}]$:

$$(3)\ \ \mathrm{vol}_{\mathcal{E}_{\mathbb{C}}}(\mathrm{Z}(L_{\mathbb{Z}},M_{\mathbb{Z}},\kappa;K)) = 4\tilde{\lambda}^{-1}(L_{\mathbb{Z}};s)\widetilde{\mu}(L_{\mathbb{Z}},M_{\mathbb{Z}},\kappa;0),$$
$$(4)\ \ \widehat{\mathrm{vol}}_{\overline{\mathcal{E}}}(\mathrm{Z}(L_{\mathbb{Z}},M_{\mathbb{Z}},\kappa;K)) \equiv \frac{\mathrm{d}}{\mathrm{d}s}4\big(\tilde{\lambda}^{-1}(L_{\mathbb{Z}};s)\widetilde{\mu}(L_{\mathbb{Z}},M_{\mathbb{Z}},\kappa;s)\big)\big|_{s=0} \text{ in } \mathbb{R}_{2DD''}.$$

(3) is true without the restriction on l and m. Note that for $m=1$ and integral Q_M, we have trivially always $D''=1$.

3.5.6. Corollary. *Let $L_{\mathbb{Z}} \subset L_{\mathbb{Q}}$ be a lattice with cyclic discriminant group $L^*_{\mathbb{Z}}/L_{\mathbb{Z}}$ of square-free order ε. Let $K \subset \mathrm{SO}(L_{\mathbb{A}^{(\infty)}})$ be the discriminant kernel. We* *have*

$$\mathrm{vol}(\mathrm{M}^K) = 2^{-\lfloor (l-4)/2 \rfloor} \prod_{i=1}^{\lfloor (l-1)/2 \rfloor} \zeta(1-2i) \cdot \begin{cases} L\left(1-\frac{l}{2}, \chi\right) & 2 \mid l, \\ (-1)^{(l^2-1)/8} \prod_{p \mid \epsilon/2} \left(p^{(l-1)/2} + \left(\frac{-1}{p}\right)^{(l-1)/2} \Phi_p\right) & 2 \nmid l, \end{cases}$$

$$\widehat{\mathrm{vol}}(\mathrm{M}^K) \equiv \mathrm{vol}(\mathrm{M}^K) \cdot \left(\frac{l-1}{2} C + \sum_{i=1}^{\lfloor (l-1)/2 \rfloor} \left(-2 \frac{\zeta'(1-2i)}{\zeta(1-2i)} - N_{2i} \right) \right.$$
$$\left. + \begin{cases} -\frac{L'(1-l/2,\chi)}{L(1-l/2,\chi)} - \frac{1}{2} N_{l/2} - \frac{1}{2} \log\left|\frac{\varepsilon}{2}\right| & 2 \mid l, \\ \frac{1}{2} \sum_{p \mid \epsilon/2} \frac{p^{(l-1)/2} - \left(\frac{-1}{p}\right)^{(l-1)/2} \Phi_p}{p^{(l-1)/2} + \left(\frac{-1}{p}\right)^{(l-1)/2} \Phi_p} \log(p) + \frac{1}{2} \log(2) & 2 \nmid l, \end{cases} \right) \quad \text{in } \mathbb{R}_2,$$

where $N_n = \sum_{i=1}^{n-1} 1/i$ *and* $C = \log(2\pi) + \gamma$.

PROOF. Follows from the Main Theorem using Propositions B.1.3 and B.1.4. □

PROOF OF MAIN THEOREM 3.5.5. (1) and (3), which are concerned with geometric volumes, are well-known but proofs are included to emphasize the analogies with the arithmetic volume case.

If formula (1) is true for a lattice L then it is true for any lattice L' with $L_{\mathbb{Q}} \cong L'_{\mathbb{Q}}$ respecting the quadratic form up to scalar. This is because in this case the two associated Shimura varieties are related by a correspondence, and for the assertion we only have to show that

$$\frac{[\mathrm{SO}'(L'_{\hat{\mathbb{Z}}}) : K]}{[\mathrm{SO}'(L_{\hat{\mathbb{Z}}}) : K]} = \frac{\lambda^{-1}(L'_{\mathbb{Z}}; 0)}{\lambda^{-1}(L_{\mathbb{Z}}; 0)}, \tag{3.42}$$

where $K \subseteq \mathrm{SO}'(L'_{\hat{\mathbb{Z}}}) \cap \mathrm{SO}'(L_{\hat{\mathbb{Z}}})$ is any finite index subgroup. Observe that the morphism Ξ (2.5.14) is compatible with change of K (see also 2.6.25). Now $\lambda(L; 0)$ and $\lambda(L'; 0)$ are Euler products, and their factors at almost all p are the same. The finite Euler factors are the volumes of $\mathrm{SO}'(L_{\mathbb{Z}_p})$ and $\mathrm{SO}'(L'_{\mathbb{Z}_p})$, respectively, w.r.t. the canonical measures. This shows (3.42).

We continue by showing (1) for lattices $L_{\mathbb{Z}}$ with signature $(0, 2)$. Let $L'_{\mathbb{Z}}$ be the lattice $\{A \in M_2(\mathbb{Z}) \mid {}^tA = A\}$ with quadratic form $Q_L(A) = \det(A)$. By the remark before, we may assume w.l.o.g. that the discriminant of $L_{\mathbb{Z}}$ is fundamental, and there is a saturated embedding $L_{\mathbb{Z}} \hookrightarrow L'_{\mathbb{Z}}$. The formula (1) is then basically just the classical Dirichlet class number formula. We prove it here as follows: Let $E'(\tau, s)$ be the Eisenstein series defined in [26, 12.4] and $E(\tau, s) = \big(Z(2s)/Z(s)\big)E'(\tau, s)$. We let $E(L_{\mathbb{Z}}; s)$ be its trace (in the stack sense) over the embedded Shimura variety. By Kronecker's limit formula, we have (using our normalization of the metric $h_{\mathcal{E}}$, cf. 3.2.5):

$$2E(L_{\mathbb{Z}}, s+1) = \mathrm{vol}_{\mathcal{E}_{\mathbb{C}}}(\mathrm{M}({}^K\mathbf{O}(L_{\mathbb{Z}})) + \widehat{\mathrm{vol}}_{\overline{\mathcal{E}}}(\mathrm{M}({}^K\mathbf{O}(L_{\mathbb{Z}}))s + O(s^2). \tag{3.43}$$

On the other hand, we have [26, Theorem 12.6(5)]:

$$E(L_{\mathbb{Z}}, s+1) = 2\tilde{\lambda}^{-1}(L_{\mathbb{Z}}; s)$$

(up to multiplication with a rational function in 2^{-s} with is 1 at $s = 0$). Taking the value at $s = 0$, respectively the derivative at $s = 0$, we get formula (1)—respectively (2) for this special lattice and embedding—in this case.

In the remaining case $m \geq 3$, we give two different proofs of formula (1):

First proof. Let $M_{\mathbb{Z}} = \langle q \rangle$, $q \in \mathbb{Q}$ and $\kappa \in (L_{\mathbb{Z}}^*/L_{\mathbb{Z}}) \otimes M_{\mathbb{Z}}^* = L_{\mathbb{Z}}^*/L_{\mathbb{Z}}$ is a coset.

We have the following identity:

$$\mu(L_{\mathbb{Z}}, \langle q \rangle, \kappa; 0) = \frac{\mathrm{vol}_{\mathcal{E}_{\mathbb{C}}}\big(\mathrm{Z}(L_{\mathbb{Z}}, \langle q \rangle, \kappa; K)\big)}{\mathrm{vol}_{\mathcal{E}_{\mathbb{C}}}\Big(\mathrm{M}\big({}^K\mathbf{O}(L)\big)\Big)}. \tag{3.44}$$

This is an application of the Siegel–Weil formula, see [35, 4.17, 4.20, 4.21] and also [25, Theorem 7.6.8]. A certain average over this equation, sufficient for proving (1) by induction follows also from Borcherds' theory—see Theorem 3.4.2(1). However, this would exclude some special cases. We mention it because it is precisely the geometric analogue of the proof of (2) that we will give below.

Comparing 3.44 to the value at $s = 0$ of the global orbit equation (3.5.2)

$$\mu(L_{\mathbb{Z}}, \langle q \rangle, \kappa; 0)\lambda^{-1}(L_{\mathbb{Z}}; 0) = \sum_{\alpha\, \mathrm{SO}'(L_{\hat{\mathbb{Z}}}) \subset \mathrm{I}(L, \langle q \rangle)(\mathbb{A}^{(\infty)}) \cap \kappa} \lambda^{-1}(\alpha^{\perp}; 0),$$

we establish that assertion (1) is true for lattices of signature $(l-2, 2)$ if and only if it is true for lattices of signature $(l-3, 2)$, whereby it is proven by induction on l because the right-hand side is $\mathrm{vol}_{\mathcal{E}_{\mathbb{C}}}\big(\mathrm{Z}(L_{\mathbb{Z}}, \langle q \rangle, \kappa; K)\big)$ by the induction hypothesis. □

Second proof. We assume again $l \geq 3$ and use $\tau\big(\mathrm{SO}(L_{\mathbb{Q}})\big) = 2$, where τ denotes the Tamagawa number. We have the following elementary relation:

$$\Lambda\tau\big(\mathrm{SO}(L_{\mathbb{Q}})\big) = \mathrm{vol}(K)\, \mathrm{vol}_{\mathcal{E}_{\mathbb{C}}}\Big(\mathrm{M}\big({}^K\mathbf{O}(L)\big)\Big),$$

where $\Lambda = 2\lambda_{\infty}^{-1}(L; 0)$ is the comparison factor from Lemma 3.2.18 and $\mathrm{vol}(K)$ is computed with respect to the product of the canonical volumes. Recall that Λ involved the canonical volume form on $\mathrm{SO}(L_{\mathbb{R}})$ and the product over all ν of the canonical measures is a Tamagawa measure ([26, Lemma 10.1]). We have $\mathrm{vol}(K) = \prod_p \lambda_p(L; 0)$ by definition of the λ_p. Everything put together yields $\mathrm{vol}_{\mathcal{E}_{\mathbb{C}}}\Big(\mathrm{M}\big({}^K\mathbf{O}(L)\big)\Big) = 4\lambda^{-1}(L_{\mathbb{Z}}; 0)$ (which equals also $4\tilde{\lambda}^{-1}(L_{\mathbb{Z}}; 0)$). □

(3) follows from (1) using the value of the global orbit equation at $s = 0$:

$$\tilde{\lambda}^{-1}(L_{\mathbb{Z}}; 0)\tilde{\mu}(L_{\mathbb{Z}}, M_{\mathbb{Z}}, \kappa; 0) = \sum_{\alpha\, \mathrm{SO}'(L_{\hat{\mathbb{Z}}}) \subset \mathrm{I}(L, M)(\mathbb{A}^{(\infty)}) \cap \kappa} \tilde{\lambda}^{-1}(\alpha_{\mathbb{Z}}^{\perp}; 0).$$

(2), *as an identity* $\mathbb{R}_{2D}$, follows (using (1)) from the following local statement:

3.5.7. Assertion. *Let $p \neq 2$. Let $L_{\mathbb{Z}_{(p)}}$ be a unimodular (at p) lattice and signature $(l-2,2)$. Let K be any admissible compact open subgroup and Δ be complete and smooth w.r.t. K, and let* $\mathrm{M} = \mathrm{M}({}^K_\Delta\mathbf{O}(L))$ *the Shimura variety, considered as variety over* $\mathrm{Spec}(\mathbb{Z}_{(p)})$. *We have*

$$\widehat{\mathrm{vol}}_{\overline{\mathcal{E}},p}(\mathrm{M}) \equiv \mathrm{vol}_{\mathcal{E}_\mathbb{C}}(\mathrm{M}_\mathbb{C}) \left.\frac{(\mathrm{d}/\,\mathrm{d}s)\lambda^{-1}(L_\mathbb{Z};s)}{\lambda^{-1}(L_\mathbb{Z};s)}\right|_{s=0}$$

in $\mathbb{R}^{(p)}$ *and any* $\mathbb{Z}$*-model* $L_\mathbb{Z} \subset L_{\mathbb{Z}_{(p)}}$.

First of all, to prove Assertion 3.5.7 we may w.l.o.g. multiply Q_L by a scalar $\in \mathbb{Z}_{(p)}^*$ because the Shimura datum and $\mathcal{E}$ as an SO-equivariant bundle are not affected by this and the Hermitian metric changes by a factor in $\mathbb{Z}_{(p)}^*$ with does not change the arithmetic volume considered in $\mathbb{R}^{(p)}$. Assertion 3.5.7 may also be shown for any specific *p-admissible* K and will then be true for any such choice (see 2.6.25).

The strategy of induction, similar to the first proof of (1), is to walk through the set of (unimodular at p) lattices in a special way, starting from lattices $L_{\mathbb{Z}_{(p)}}$ with known

$$\widehat{\mathrm{vol}}_{\overline{\mathcal{E}},p}\Big(\mathrm{M}({}^K_\Delta\mathbf{O}(L_{\mathbb{Z}_{(p)}}))\Big)$$

(for admissible K) and then to construct special Borcherds products by the Theorems 3.3.14–3.3.17 in such a way that all quantities in Theorem 3.4.2(2)–except one–are already known. This path through the set of lattices will be mostly according to the dimension and Witt rank of the lattice. It is illustrated by Figure 3.1 where an arrow indicates logical dependence, i.e., the reverse walking direction.

Step 1. We start with the case of a lattice $L_{\mathbb{Z}_{(p)}}$ of signature $(1,2)$ and Witt rank 1. These are up to multiplication of Q_L by scalars in $\mathbb{Z}_{(p)}^*$ of the form $Q_L(z) = z_1z_2 - z_3^2$ which is equivalently the space $\{X \in M_2(\mathbb{Z}_{(p)}) \mid {}^tX = X\}$ with $Q_L(X) = \det(X)$. The arithmetic volume $\widehat{\mathrm{vol}}_{\overline{\mathcal{E}},p}\Big(\mathrm{M}({}^K_\Delta\mathbf{O}(L_{\mathbb{Z}_{(p)}}))\Big)$ has been calculated in Kühn's thesis [45]. Since multiplying the norm (square root of the metric) by a scalar ρ changes the arithmetic volume of an arithmetic variety X of *arithmetic* dimension $\dim(X)$ by

$$-\dim(X)\log(\rho)\,\mathrm{vol},$$

comparing to [45, Theorem 6.1], we see that

$$\widehat{\mathrm{vol}}_{\overline{\mathcal{E}}}\Big(\mathrm{M}\left({}^{K(1)}\mathbf{O}(L)\right)\Big) = \mathrm{vol}_{\mathcal{E}}\Big(\mathrm{M}({}^{K(1)}\mathbf{O}(L))\Big)\left(-2\frac{\zeta'(-1)}{\zeta(-1)} - 1 + \log(2) + \log(2\pi) + \gamma\right).$$

In (1) we obtained that

$$\mathrm{vol}_{\mathcal{E}}\Big(\mathrm{M}({}^{K(1)}\mathbf{O}(L))\Big) = 4\lambda^{-1}(L_\mathbb{Z};0) = -2\zeta(-1) = \tfrac{1}{6}$$

which is of course well-known, taking into account that $\mathcal{E}$ is $\mathcal{O}(-2)$ (identifying $\mathrm{M}^\vee(\mathbf{O}(L))$ with $\mathbb{P}^1$), hence $\Xi^*(\mathcal{E})$ is the line-bundle of classical modular forms of weight 2, and $\mathrm{SO}(L)$ is PGL_2. On the other hand, we have (see Proposition B.1.4 or [26, Example 11.11]):

$$4\tilde{\lambda}^{-1}(L,s) = -2\zeta(-1) + -2\zeta(-1)\left(-2\frac{\zeta'(-1)}{\zeta(-1)} - 1 + \frac{1}{2}\log(2) + \log(2\pi) + \gamma\right)s + O(s^2).$$

Hence formula (2) is true in $\mathbb{R}_2$.

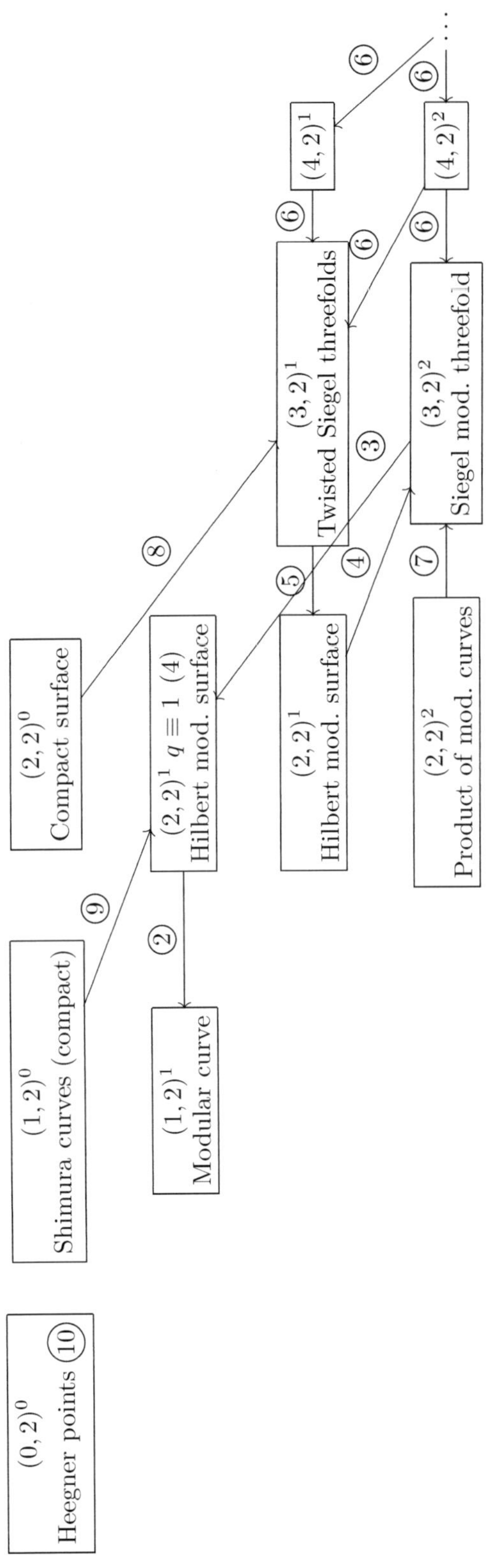

FIGURE 3.1. Strategy of induction on the type of lattice

From now on we proceed by induction along the arrows of the diagram according to the following scheme: Choose a saturated $L_{\mathbb{Z}} \subset L_{\mathbb{Z}_{(p)}}$ such that its discriminant kernel $K \subset \mathrm{SO}(L_{\mathbb{A}^{(\infty)}})$ is neat (it is automatically p-admissible because $L_{\mathbb{Z}_{(p)}}$ was assumed to be unimodular). Assume that there is a modular form F of weight $2 - l/2$ associated with the Weil representation of $L_{\mathbb{Z}}$ with q-expansion

$$F(\tau) = \sum_{\substack{k \in \mathbb{Q} \\ k \gg -\infty}} a_k q^k$$

with the following properties

- $\mathrm{Z}(L, \langle -k \rangle, a_k; K)$ for all $k < 0$ is a union of (images of) canonical models of sub-Shimura varieties (good reduction),
- $\widehat{\mathrm{vol}}_{\overline{\mathcal{E}},p}\left(\mathrm{M}({}^{K'}_{\Delta}\mathbf{O}(L'))\right)$ for all these sub-Shimura varieties is already known,
- $a_0(0) \neq 0$.

Note that the first property is equivalent to $p \mid k \implies a_k = 0$ for all $k < 0$.

Then apply Theorem 3.4.2(2) to get

$$\sum_{q<0} \widetilde{\mu}'(L_{\mathbb{Z}}, \langle -q \rangle, a_q; 0) \, \mathrm{vol}_{\mathcal{E}_{\mathbb{C}}}\left(\mathrm{M}({}^{K}_{\Delta}\mathbf{O}(L))\right)$$
$$+ \sum_{q<0} \widetilde{\mu}(L_{\mathbb{Z}}, \langle -q \rangle, a_q; 0) \widehat{\mathrm{vol}}_{\overline{\mathcal{E}},p}\left(\mathrm{M}({}^{K}_{\Delta}\mathbf{O}(L))\right) = \sum_{q<0} \widehat{\mathrm{vol}}_{\overline{\mathcal{E}},p}(\mathrm{Z}(L_{\mathbb{Z}}, \langle -q \rangle, a_q; K)).$$

Inserting the induction hypothesis and (1), we get:

$$(3.45) \quad \sum_{q<0} \Bigg(\widetilde{\mu}'(L_{\mathbb{Z}}, \langle -q \rangle, a_q; 0) \widetilde{\lambda}^{-1}(L_{\mathbb{Z}}; 0) + \widetilde{\mu}(L_{\mathbb{Z}}, \langle -q \rangle, a_q; 0) \widehat{\mathrm{vol}}_{\overline{\mathcal{E}},p}\left(\mathrm{M}({}^{K}_{\Delta}\mathbf{O}(L))\right)$$
$$- \sum_{K\alpha \subseteq \mathrm{I}(L, \langle -q \rangle)(\mathbb{A}^{(\infty)}) \cap \mathrm{supp}(a_q)} a_q(\alpha) (\widetilde{\lambda}^{-1})'(\alpha_{\mathbb{Z}}^{\perp}; 0) \Bigg) = 0.$$

The derivative of the global orbit equation 3.5.2 multiplied with $a_q(\kappa)$ and summed up over all $\kappa \in L_{\widehat{\mathbb{Z}}}^*/L_{\widehat{\mathbb{Z}}}$ and all $q < 0$ yields:

$$(3.46) \quad \sum_{q<0} \Bigg(\widetilde{\mu}'(L_{\mathbb{Z}}, \langle -q \rangle, a_q; 0) \widetilde{\lambda}^{-1}(L_{\mathbb{Z}}; 0) + \widetilde{\mu}(L_{\mathbb{Z}}, \langle -q \rangle, a_q; 0) (\widetilde{\lambda}^{-1})'(L_{\mathbb{Z}}; 0)$$
$$- \sum_{K\alpha \subseteq \mathrm{I}(L, \langle -q \rangle)(\mathbb{A}^{(\infty)}) \cap \mathrm{supp}(a_q)} a_q(\alpha) (\widetilde{\lambda}^{-1})'(\alpha_{\mathbb{Z}}^{\perp}; 0) \Bigg) = 0.$$

The difference of (3.45) and (3.46) gives

$$\sum_{q<0} \widetilde{\mu}(L_{\mathbb{Z}}, \langle -q \rangle, a_q; 0) \left(\widehat{\mathrm{vol}}_{\overline{\mathcal{E}},p}\left(\mathrm{M}({}^{K}_{\Delta}\mathbf{O}(L))\right) - (\widetilde{\lambda}^{-1})'(L_{\mathbb{Z}}; 0) \right)$$
$$= a_0(0) \left(\widehat{\mathrm{vol}}_{\overline{\mathcal{E}},p}\left(\mathrm{M}({}^{K}_{\Delta}\mathbf{O}(L))\right) - (\widetilde{\lambda}^{-1})'(L_{\mathbb{Z}}; 0) \right) = 0$$

using $\sum_{q<0} \mu(L_{\mathbb{Z}}, \langle -q \rangle, a_q; 0) = a_0(0)$ (see proof of 3.4.2(1)). Since we have $a_0(0) \neq 0$, we get in particular that Assertion 3.5.7 is true for $L_{\mathbb{Z}}$.

This reduces the proof of Assertion 3.5.7, and hence (2), to finding appropriate input forms for Borcherds products. This is done for the individual steps in Figure 3.1 as follows:

Step 2 (Hilbert modular surfaces of prime discriminant $q \equiv 1$ (4)). This is the case considered in [10]. We reproduce their argument here as follows: Theorem 3.3.16(1) shows that a Borcherds lift of nonzero weight can be found such that all occurring Shimura varieties in $\mathrm{Z}(L_{\mathbb{Z}}, \langle -q \rangle, a_q; K)$ for the occurring a_q are of the form already treated in Step 1.

Step 3 (Siegel modular threefold). In Theorem 3.3.17 it is shown that a Borcherds lift of nonzero weight can be found, such that every Shimura variety occurring in $\mathrm{Z}(L_{\mathbb{Z}}, \langle -q \rangle, a_q; K)$ for the nonzero a_q's is of the form considered in Step 2 (even copies of the same).

Step 4 (General Hilbert modular surfaces). In Theorem 3.3.17 it is shown that a Borcherds lift of nonzero weight can be found, such that every Shimura variety occurring in $\mathrm{Z}(L_{\mathbb{Z}}, \langle -q \rangle, a_q; K)$ for the nonzero a_q's is a $\mathrm{M}({}^K_\Delta\mathbf{O}(L_{\mathbb{Z}_{(p)}}))$ for a given lattice $L_{\mathbb{Z}_{(p)}}$ of signature $(2,2)$, Witt rank 1. Now $\widehat{\mathrm{vol}}_{\overline{\mathcal{E}},p}\left(\mathrm{M}({}^K_\Delta\mathbf{O}(L))\right)$ is known by Step 3, and $\widehat{\mathrm{vol}}_{\overline{\mathcal{E}},p}(\mathrm{Z}(L_{\mathbb{Z}}, \langle -q \rangle, a_q; K))$ may be deduced from the comparison of the formula in 3.4.2 with the derivative of the orbit equation by reverting the argument above.

Step 5 (Twisted Siegel threefolds). This is the case L is of signature $(2,3)$ and Witt rank 1. We have only to avoid $\mathrm{Z}(L_{\mathbb{Z}}, \langle -q \rangle, a_q; K)$'s in which Shimura varieties for lattices with are not unimodular at p *or* compact ones occur. This is achieved by Theorem 3.3.15(1).

Step 6 (Dimension 4 and higher). We know all arithmetic volumes for orthogonal Shimura varieties of dimension 3. Hence we may proceed by induction on the dimension and have only to avoid $\mathrm{Z}(L_{\mathbb{Z}}, \langle -q \rangle, a_q; K)$'s containing Shimura varieties for lattices with are not unimodular at p. This is achieved by Theorem 3.3.14.

We are now left with a couple of cases that have been omitted in the above process. They can be treated analogously to Step 4 above, by reverting the argument.

Step 7 (Product of modular curves). This is the case of signature $(2,2)$, Witt rank 2. Use Theorem 3.3.17 again. This can also be treated directly, using Step 1 (Kühn's thesis), see [15, Section 7.8] for a proof.

Step 8 (Shimura surfaces). This is the case of signature $(2,2)$, Witt rank 0. Use Thereom 3.3.15(2).

Step 9 (Shimura curves). This is the case of signature $(1,2)$, Witt rank 0. Use Theorem 3.3.16, 2.

Step 10 (Heegner points). These cases have been treated along the lines of the proof of (1) above.

This completes the proof of Assertion 3.5.7, and hence of (2), as an identity in $\mathbb{R}_{2D}$. We use this, and Theorem 3.4.2 again, to derive the following local assertion, from which (4) follows:

3.5.8. Assertion. *Let $p \neq 2$. Let $L_{\mathbb{Z}_{(p)}}$ be a unimodular (at p) lattice and of signature $(l-2,2)$, Let $M_{\mathbb{Z}_{(p)}}$ be a positive definite lattice of dimension m with cyclic $M^*_{\mathbb{Z}_p}/M_{\mathbb{Z}_p}$. Assume $l-m \geq 4$, or $l=4$, $m=1$ and $L_{\mathbb{Q}}$ is isotropic, or*

$l = 5$, $m = 2$ and $L_{\mathbb{Q}}$ has Witt rank 2. Let $L_{\mathbb{Z}}$, $M_{\mathbb{Z}}$ any saturated lattices in $L_{\mathbb{Z}_{(p)}}$ and $M_{\mathbb{Z}_{(p)}}$, respectively, and let K be any admissible compact open subgroup in the discriminant kernel of $L_{\mathbb{Z}}$. We have:

$$\widetilde{\mu}'(L_{\mathbb{Z}}, M_{\mathbb{Z}}, \kappa; 0)\, \mathrm{vol}_{\mathcal{E}_{\mathbb{C}}}\Big(\mathrm{M}({}^K_\Delta\mathbf{O}(L))\Big) \\ + \deg_E\big(\mathrm{Z}(L_{\mathbb{Z}}, M_{\mathbb{Z}}, \kappa; K)\big)\widehat{\mathrm{vol}}_{\overline{\mathcal{E}},p}\Big(\mathrm{M}({}^K_\Delta\mathbf{O}(L))\Big) \equiv \widehat{\mathrm{vol}}_{\overline{\mathcal{E}},p}\big(\mathrm{Z}(L_{\mathbb{Z}}, M_{\mathbb{Z}}, \kappa; K)\big)$$

in $\mathbb{R}^{(p)}$. Here

$$\deg_E\big(\mathrm{Z}(L_{\mathbb{Z}}, M_{\mathbb{Z}}, \kappa; K)\big) := \frac{\mathrm{vol}_{\mathcal{E}_{\mathbb{C}}}\big(\mathrm{Z}(L_{\mathbb{Z}}, M_{\mathbb{Z}}, \kappa; K)\big)}{\mathrm{vol}_{\mathcal{E}_{\mathbb{C}}}\Big(\mathrm{M}({}^K\mathbf{O}(L))\Big)}$$

is the (relative) geometric degree.

We first prove this statement for $m = 1$. In this case it follows immediately from Theorem 3.4.2(2), using a Borcherds product such that all Z's in the divisor consist of Shimura varieties corresponding to lattices which are unimodular at p, *except* for $\mathrm{Z}(L_{\mathbb{Z}}, \langle q\rangle, \kappa; K)$ which shall occur, too, with nonzero multiplicity. Theorem 3.3.14(2) enables us to construct such a product. Note that the assumptions imply that $\mathrm{M}({}^K\mathbf{O}(L))$ does have cusps.

Proof for $m \geq 2$. Because $M^*_{\mathbb{Z}_{(p)}}/M_{\mathbb{Z}_{(p)}}$ is cyclic, we may find lattices $M_{\mathbb{Z}}$ and $M'_{\mathbb{Z}}$, unimodular at p, and $q \in \mathbb{Z}_{>0}$ such that

$$M_{\mathbb{Z}} = \langle q\rangle \perp M'_{\mathbb{Z}}$$

is a saturated lattice in $M_{\mathbb{Z}_{(p)}}$. Let $\kappa = \kappa_q \oplus \kappa'$ be a corresponding decomposition. Let $\mathrm{SO}'(L_{\hat{\mathbb{Z}}})\alpha_i \subset \mathrm{I}(M'_{\hat{\mathbb{Z}}}, L_{\hat{\mathbb{Z}}}) \cap \kappa'$ be a decomposition into orbits. We may form the cycles $\mathrm{Z}(\alpha_i^\perp, \langle q\rangle, \kappa_q)$ on $\mathrm{M}({}^{K_i}_{\Delta_i}\mathbf{O}(\alpha_i^\perp))$, where the K_i are the respective discriminant kernels (all admissible by construction of $M'_{\mathbb{Z}}$). The latter Shimura varieties are all equipped with morphisms into $\mathrm{M}({}^K_\Delta\mathbf{O}(L_{\mathbb{Z}_{(p)}}))$ (the union of their images is the cycle $\mathrm{Z}(L, M', \kappa', K)$). These Shimura varieties all have the same Shimura datum, and one could see their images as conjugated images of a single Shimura variety with varying K as in the definition of special cycle (3.1.7). In each of these Shimura varieties, we have cycles $\mathrm{Z}(\alpha_i^\perp, \langle q\rangle, \kappa_q; K_i)$. Identifying them with their image in $\mathrm{M}({}^K_\Delta\mathbf{O}(L_{\mathbb{Z}_{(p)}}))$, we get

$$\mathrm{Z}(L, M, \kappa; K) = \bigcup_{\mathrm{SO}'(L_{\hat{\mathbb{Z}}})\alpha \subset \mathrm{I}(L, M')(\mathbb{A}^{(\infty)}) \cap \kappa'} \mathrm{Z}(\alpha_i^\perp, \langle q\rangle, \kappa_q; K_i).$$

Using the $m = 1$ case, we know the height of the right-hand side. For this note that the assumptions imply that $\mathrm{M}({}^{K_i}\mathbf{O}(\alpha_i^\perp))$ has cusps. Comparing this with global Kitaoka (Theorem 3.5.3), we get the general result, and its truth does, of course, not depend on the models $L_{\mathbb{Z}}$, $M_{\mathbb{Z}}$ chosen and not on K (cf. also 2.6.25).

Finally the previous Assertion 3.5.7 at $p \mid D$, where $p \neq 2$ and $p^2 \nmid D$, and therefore (2) as an identity in $\mathbb{R}_{2D'}$ (instead of $\mathbb{R}_{2D}$) now follows from 3.5.8 applied to any of the embeddings used to construct the model M (see 3.5.4). For this we use that the derivative of the global orbit equation 3.5.2 is also valid for $\widetilde{\lambda}$ and $\widetilde{\mu}$ in $\mathbb{R}^{(p)}$ for those p. Recall, that this followed from the fact that the local orbit equation for $\widetilde{\lambda}_p$, and $\widetilde{\mu}_p$ remains true for those p. This is wrong in general if the discriminant at p is not cyclic (cf. also [26, 11.8]). However, then also the q-expansion principle and its application in this form break down because we would have to deal with models of bad reduction in the induction process. □

APPENDIX A

A.1. Lemmas on quadratic forms

A.1.1. Lemma. *Let $p \neq 2$ and let $L_{\mathbb{Z}_{(p)}}$ be a unimodular lattice of signature $(2,2)$ and Witt rank 1, with fundamental*[1] *discriminant $D > 0$ in $\mathbb{Z}$. Up to multiplication of Q_L by a unit in $\mathbb{Z}_{(p)}^*$, there exists a saturated lattice $L_{\mathbb{Z}} \subset L_{\mathbb{Z}_{(p)}}$ of the form*

$$L_{\mathbb{Z}} = H \oplus \left\langle x^2 + xy + \frac{1-D}{4}y^2 \right\rangle$$

if $D \equiv 1 \bmod 4$ or

$$L_{\mathbb{Z}} = H \oplus \left\langle x^2 - \frac{D}{4}y^2 \right\rangle$$

if $4 \mid D$.

PROOF. Up to scaling Q_L we may assume $L_{\mathbb{Z}_{(p)}} = H \oplus L'_{\mathbb{Z}_{(p)}}$ such that $L'_{\mathbb{Z}_{(p)}}$ represents 1. $L'_{\mathbb{Z}_{(p)}}$ is anisotropic by assumption. Then it is well-known and elementary that $L'_{\mathbb{Z}_{(p)}}$ contains a saturated $\mathbb{Z}$-lattice of the required form. $\square$

A.1.2. Lemma. *Let $p \neq 2$ and let $L_{\mathbb{Z}_{(p)}}$ be a unimodular anisotopic lattice of signature $(1,2)$. Up to multiplying Q_L with a unit in $\mathbb{Z}_{(p)}^*$, there exists a unimodular lattice $L'_{\mathbb{Z}_{(p)}} = L_{\mathbb{Z}_{(p)}} \oplus \langle x \rangle_{\mathbb{Z}_{(p)}}$ for some $x \in \mathbb{Z}_{>0}$ which is isotropic and of* prime *fundamental discriminant $q \neq p$ and there exists a saturated lattice $L'_{\mathbb{Z}} \subset L'_{\mathbb{Z}_{(p)}}$ of the form*

$$L'_{\mathbb{Z}} = H \oplus \left\langle x^2 + xy + \frac{1-q}{4}y^2 \right\rangle.$$

PROOF. Write $L_{\mathbb{Z}_{(p)}} = \langle \alpha_1, \alpha_2, \alpha_3 \rangle_{\mathbb{Z}_{(p)}}$, with $\alpha_i \in \mathbb{Z}$ square-free and relatively prime. This is possible without multiplying Q_L by a scalar. Let D be the square-free part of $\prod \alpha_i$. We may find a prime q, different from p, such that $q \equiv 1\ (4)$ and such that for any $l \neq 2$ with $\nu_l(\prod \alpha_i) = 2$ (hence $l \nmid D$), we have $Dq \equiv -\alpha_j\ (l)$, where α_j is the one not divisible by l.

We may also prescribe the residue of q mod 8, such that $L_{\mathbb{Q}_2} \oplus \langle Dq \rangle$ is isotropic. For if the lattices $L_{\mathbb{Q}_2} \oplus \langle Dq \rangle$ were anisotropic for q, q' congruent to $1, 5$ mod 8 respectively, we would get $\langle Dq \rangle_{\mathbb{Q}_2} \cong \langle Dq' \rangle_{\mathbb{Q}_2}$ from Witt's Theorem and the uniqueness of the 4-dimensional anisotropic space which is absurd.

Hence

$$L_{\mathbb{Z}_{(p)}} \oplus \langle Dq \rangle_{\mathbb{Z}_{(p)}}$$

is unimodular, of signature $(2,2)$ with square-free discriminant q. It is isotropic at all $l \neq 2$, too, because of the congruence condition on Dq. Hence the Witt rank

[1] i.e., $D \equiv 1\ (4)$, D square-free, or $D \equiv 0\ (4)$, $D/2$ square-free.

is 1 and we may apply Lemma A.1.1 to it (this changes also the form on $L_{\mathbb{Z}_{(p)}}$ by a scalar) to get the result. □

A.1.3. Lemma. *Let $p \neq 2$ and let $L_{\mathbb{Z}_{(p)}}$ be an unimodular anisotropic lattice of signature $(1,2)$. Up to multiplying Q_L with a unit in $\mathbb{Z}_{(p)}^*$, there is a saturated lattice $L_\mathbb{Z} \subset L_{\mathbb{Z}_{(p)}}$ of discriminant $D = 2D'$, where D' is square-free and such that $L_\mathbb{Z}^*/L_\mathbb{Z}$ is cyclic.*

This lattice has the property that a primitive $\kappa \in L_{\hat{\mathbb{Z}}}^*/L_{\hat{\mathbb{Z}}}$ *represents $m \in \mathbb{Q}$ if and only if $Q_L(\kappa) \equiv m\ (1)$.*

Proof. Write $L_{\mathbb{Z}_{(p)}} = \langle \alpha_1, \alpha_2, \alpha_3 \rangle_{\mathbb{Z}_{(p)}}$, with $\alpha_i \in \mathbb{Z}$ square-free. By multiplying Q_L, if necessary, by a scalar, we may assume that the α_i are pairwise relatively prime.

The lattice $L_\mathbb{Z} = \langle \alpha_1, \alpha_2, \alpha_3 \rangle_\mathbb{Z}$ satisfies the assumption locally for all $l \neq 2$. It then suffices to construct a lattice $L_{\mathbb{Z}_2}$ satisfying the assumption. There are two cases:

(1) The space $L_{\mathbb{Q}_2}$ is isotropic hence there is, up to multiplying the quadratic form by 2, a lattice of the form

$$L_{\mathbb{Z}_2} = H \perp \langle 1 \rangle$$

which has discriminant 2.

(2) The space $L_{\mathbb{Q}_2}$ is anisotropic hence, up to multiplying Q_L by ± 2, it is of the form

$$L_{\mathbb{Q}_2} = \langle 1, 3, 2\epsilon \rangle$$

with $\epsilon \in \{1, 3, -3, -1\}$. It contains the lattice

$$L_{\mathbb{Z}_2} = \langle x^2 + xy + y^2 \rangle \perp \langle 2\epsilon \rangle$$

of discriminant 4, with $L_{\mathbb{Z}_2}^*/L_{\mathbb{Z}_2}$ cyclic of order 4.

The only-if part of the claimed property is clear. For the if part, we have to show that m is represented by κ_l for all l. For $l \nmid D$, the lattice $L_{\mathbb{Z}_l}$ represents every $m \in \mathbb{Z}_l$. For $l \mid D$ and $l \neq 2$, κ_l represents m if the equation

$$\alpha_1 x_1^2 + \alpha_2 x_2^2 + \alpha_3 \left(\left(\frac{\beta}{l} \right)^2 + 2\frac{\beta x_3}{l} + x_3^2 \right) = \frac{n}{l} = m$$

has a solution with $x_i \in \mathbb{Z}_l$, where $\beta \in {\mathbb{Z}_l}^*$ is determined by κ and $n \in {\mathbb{Z}_l}^*$. Here we renumbered the α_i such that $l \mid \alpha_3$.

Now the condition $Q_L(\kappa) \equiv m\ (1)$ implies $(\alpha_3/l)\beta^2 \equiv n\ (l)$. Since $\alpha_1 x_1^2 + \alpha_2 x_2^2$ represents $\mathbb{Z}_l^*$, κ_l represents m (choose x_3 appropriately in the above equation).

For $l = 2$, case (1), κ_2 represents m if

$$x_1 x_2 + \left(\left(\frac{1}{2} \right)^2 + x_3 + x_3^2 \right) = \frac{n}{4} = m$$

has a solution with $x_i \in \mathbb{Z}_l$. This follows from the condition which boils down to $n \equiv 1\ (4)$ in this case.

For $l = 2$, case (2), κ_2 represents m if

$$x_1^2 + x_1 x_2 + x_2^2 + 2\epsilon \left(\left(\frac{\beta}{4} \right)^2 + 2\frac{\beta}{4} x_3 + x_3^2 \right) = \frac{n}{8} = m$$

has a solution with $x_i \in \mathbb{Z}_l$. Here $\beta = \pm 1$ distinguishes the two κ_2. The condition says $\beta^2 \epsilon \equiv n$ (8), hence solubility because $x_1^2 + x_1 x_2 + x_2^2$ represents $\mathbb{Z}_2^*$. □

The lemma, in this strong form, is wrong for lattices of dimension 2. However, we have the following weaker form:

A.1.4. Lemma. *Let $p \neq 2$ and let $L_{\mathbb{Z}_{(p)}}$ be a unimodular anisotropic lattice of signature $(1,1)$ and of discriminant $-q$, where $q \equiv 1$ (4) is prime. Up to multiplying Q_L with a unit $\in \mathbb{Z}_{(p)}^*$, there is a saturated lattice $L_{\mathbb{Z}} \subset L_{\mathbb{Z}_{(p)}}$ of discriminant q of the form*

$$L_{\mathbb{Z}} = \left\langle x^2 + xy + \frac{1-q}{4} y^2 \right\rangle.$$

It has the property that a primitive $\kappa \in L_{\hat{\mathbb{Z}}}^*/L_{\hat{\mathbb{Z}}}$ *represents $m = l'/q \in \mathbb{Q}$, where* $l' \neq 2$ is prime, *if and only if $Q_L(\kappa) \equiv m$ (1).*

PROOF. The existence of the lattice is well-known (see also A.1.1). For the property we have to show that κ_l represents l'/q if $Q_L(\kappa) \equiv l'/q$ (1).

We can write $L_{\mathbb{Z}_l} = \langle 1, -q \rangle$, hence we see (as in the proof of the last lemma) that the condition implies $-x^2 \equiv l'$ (q) has a solution, i.e., $-l'$ is a quadratic residue modulo q. Since $q \equiv 1$ (4) this is equivalent to l' being a quadratic residue modulo q.

For $l \notin \{2, q\}$, the equation

$$x_1^2 - q x_2^2 = \frac{l'}{q}$$

clearly has a solution for $x_1, x_2 \in \mathbb{Z}_l$ if $l \neq l'$, and for $l = l'$ it does if q is a square mod l', and the latter is true by the law of quadratic reciprocity. For $l = q$, the coset κ_q represents m if

$$x_1^2 - q\left(\left(\frac{\beta}{q}\right)^2 + 2\frac{\beta}{q} x_2 + x_2^2\right) = \frac{l'}{q}$$

has a solution with $x_i \in \mathbb{Z}_q$. Since $-\beta^2 \equiv l'$ (q) such a solution clearly exists.

For $l = 2$ we have $L_{\mathbb{Z}_2} = \langle x^2 + xy + z^2 \rangle$ which represents $\mathbb{Z}_2^*$. □

A.1.5. Lemma. *Let $p \neq 2$ and let $L_{\mathbb{Z}_{(p)}}$ be a unimodular anisotopic lattice of signature $(2,2)$. Up to multiplying Q_L with a unit in $\mathbb{Z}_{(p)}^*$, there exists a unimodular lattice $L'_{\mathbb{Z}_{(p)}} = L_{\mathbb{Z}_{(p)}} \oplus \langle x \rangle$ of signature $(3,2)$ and a saturated $\mathbb{Z}$-lattice $L'_{\mathbb{Z}} \subset L'_{\mathbb{Z}_{(p)}}$ of discriminant $D = 2D'$, where D' is square-free, of the form*

$$L'_{\mathbb{Z}} = H \oplus L''_{\mathbb{Z}}$$

and such that $(L'_{\mathbb{Z}})^*/L'_{\mathbb{Z}}$ is cyclic.

PROOF. Take $L'_{\mathbb{Z}_{(p)}} = L_{\mathbb{Z}_{(p)}} \oplus \langle x \rangle$ for an arbitrary positive $x \in \mathbb{Z}_{(p)}^*$. It is automatically isotropic, hence of the form

$$L'_{\mathbb{Z}_{(p)}} = H \oplus L''_{\mathbb{Z}_{(p)}}.$$

Applying Lemma A.1.3 to $L''_{\mathbb{Z}_{(p)}}$, we get the result (we multiply $Q_{L'}$ by the same scalar; Q_L on $L_{\mathbb{Z}_{(p)}}$ gets then also multiplied by the same scalar—note, however, that multiplying the quadratic form on H by a scalar does not affect its isomorphism class). □

A.2. Lacunarity of modular forms

A.2.1. Lemma. *Let $N > 0$ be an integer and p a prime such that $p \nmid N$.*

Let f be a holomorphic modular form of (half-)integral weight $k \neq 0$ for the group $\Gamma(N)$. If f has a Fourier expansion of the form

$$f = \sum_{n \in \mathbb{Q}_{\geq 0}} a_n q^n,$$

where a_n is zero unless n is an integral multiple of p/N, then $f = 0$.

PROOF. The assumption implies that f is periodic with period N/p. It is hence a modular form for the group Γ generated by $\Gamma(N)$ and the matrix $\left(\begin{smallmatrix} 1 & N/p \\ 0 & 1 \end{smallmatrix}\right)$. Γ therefore contains the product

$$C = \begin{pmatrix} 1 & N/p \\ 0 & 1 \end{pmatrix} \begin{pmatrix} 1 & 0 \\ N & 1 \end{pmatrix} = \begin{pmatrix} 1 + N^2/p & N/p \\ N & 1 \end{pmatrix}$$

whose trace is equal to $2 + N^2/p$. Since $p \nmid N$, the p-adic valuation of this number is > 1 so at least one of the eigenvalues of C has p-adic valuation > 1. (We choose some fixed extension of the p-adic valuation to $\overline{\mathbb{Q}}$). It follows that $\|C^i\| \to \infty$ for any chosen p-adic matrix norm $\|\cdot\|$. It follows that $[\Gamma : \Gamma(N)] = \infty$. Indeed, assume that there are finitely many representatives α_i and let ν be the maximum of their p-adic matrix norms. Every element $\gamma \in \Gamma$ is of the form

$$\alpha_i \gamma'$$

for some $\gamma' \in \Gamma(N)$. The matrix norm of γ is hence $\leq \nu$, a contradiction. So Γ cannot be a discrete subgroup, and since $k \neq 0$ we have $f = 0$. □

A.2.2. Proposition. *Let $q \equiv 1$ (4) be a prime and write $\chi_q(x) = \left(\frac{x}{q}\right)$. Let S a finite set of primes and let $k \geq 2$ be an integer.*

If $f \in \mathrm{HolModForm}(\Gamma_0(q), \chi_q, k)$ has a Fourier expansion of the form

$$f = \sum_{n \in \mathbb{Z}_{\geq 0}} a_n q^n,$$

with algebraic a_n such that (i) *$a_n = 0$ whenever $\chi_q(n) = -1$, and* (ii) *$a_p = 0$ whenever $p \notin S$, then $f = 0$.*

PROOF. This follows from an idea of [57], see also [10, Lemma 4.14]. Since $q \equiv 1$ (4), there are no forms of type 'CM' satisfying condition (i).

Since the a_n are algebraic, we can write f as a linear combination

$$f = c_0 E_0 + c_\infty E_\infty + \sum_{i=1}^{n} c_i f_i$$

where the c_i are algebraic, the f_i are cuspidal Hecke eigenforms and E_0, E_∞ are the Eisenstein series.

In [57, Lemma 1] it is shown that for all sufficiently large primes l the image of the mod l representation $\rho := \rho_{1,l} \times \cdots \times \rho_{n,l}$ contains a subgroup G conjugated to

$$\mathrm{SL}_2(\mathbb{F}_1) \times \cdots \times \mathrm{SL}_2(\mathbb{F}_n)$$

and $\mathrm{im}(\rho)/G$ is Abelian, where the $\mathbb{F}_i$'s are defined in [loc. cit.]. Choose l such that all $|c_i|_l = 1$ whenever $c_i \neq 0$.

The mod l representations $\rho_{0,l}$, respectively $\rho_{\infty,l}$ of E_0, respectively E_∞ [59, p. 28] are abelian and $\rho' := (\rho_{0,l} \times \rho_{\infty,l}) \times (\rho_{1,l} \times \cdots \times \rho_{n,l})$ contains a subgroup of the form

$$H \times G$$

because $\mathrm{SL}_2(\mathbb{F}_i)$ has no nontrivial Abelian quotients if $l > 4$. Here

$$H \subset \mathrm{im}(\rho_{0,l} \times \rho_{\infty,l})$$

is a subgroup that, at least, contains a pair of matrices with different traces. Assume first that $c_0 \neq 0$ or $c_\infty \neq 0$ and choose a pair $(M_1, M_2) \in H$ of matrices with different or equal traces, according to whether $c_0 \equiv -c_\infty$ (l) or not.

By Čebotarev's density theorem, there is a positive density of primes p such that the image of Frob_p is conjugated to

$$A = M_1 \times M_2 \times \begin{pmatrix} 0 & 1 \\ -1 & 0 \end{pmatrix} \times \cdots \times \begin{pmatrix} 0 & 1 \\ -1 & 0 \end{pmatrix}.$$

So we have $|a_p(c_0E_0 + c_\infty E_\infty)|_l = 1$ and $|a_p(f_i)|_l < 1$. Hence $a_p(f) \neq 0$ for infinitely many primes, a contradiction. If $c_0 = c_\infty = 0$, and $c_1 \neq 0$ (say) choose

$$A = \begin{pmatrix} 1 & 0 \\ 0 & 1 \end{pmatrix} \times \begin{pmatrix} 1 & 0 \\ 0 & 1 \end{pmatrix} \times \begin{pmatrix} 1 & 0 \\ 0 & 1 \end{pmatrix} \times \begin{pmatrix} 0 & 1 \\ -1 & 0 \end{pmatrix} \times \cdots \times \begin{pmatrix} 0 & 1 \\ -1 & 0 \end{pmatrix}$$

and argue as before. □

A.2.3. Lemma. *Let $\Gamma \subset \mathrm{Mp}_2(\mathbb{R})$ be an arithmetic subgroup and let V_ρ, ρ an irreducible representation of Γ. Let $f \in \mathrm{HolModForm}(\Gamma, \rho, k)$ and $\beta\colon V_\rho \to \mathbb{C}$ be a nonzero linear form. If $\beta \circ f = 0$ then $f = 0$.*

PROOF. This follows immediately from the irreducibility. Indeed choose a basis $\{e_i\}$ of V_ρ such that $\beta = e_0^*$. Since V_ρ is irreducible, for any i there is an operator of the form

$$O_i = \sum_j \alpha_j \rho(\gamma_j)$$

which interchanges e_i and e_0. Consider the form $O_i \circ f$. On the one hand $e_0^* \circ O_i \circ f$ is equal to $e_i^* \circ f$. On the other hand it is equal to $\sum_i \alpha_i (e_0^* \circ f)|_k \gamma_i$ which is zero by assumption. Thus $f = 0$. □

A.3. Semilinear representations

Let K be a pro-finite topological group and let $\tau\colon K \to \mathrm{Gal}(F : \mathbb{Q})$ be a continuous homomorphism, where F is a (possibly infinite) algebraic extension of $\mathbb{Q}$.

A.3.1. Definition (Semilinear representation). By a **τ-semilinear representation** of K on a finite-dimensional F-vector space V, we understand a continuous homomorphism

$$\omega\colon K \to \mathrm{Aut}_{\mathbb{Q}}(V)$$

(where V is equipped with the discrete topology) satisfying

$$\omega(g)(\alpha v) = \tau_g(\alpha)\omega(g)v.$$

A.3.2. Example. For example, a τ-semilinear representation may be given by a $\mathbb{Q}$-vector space V and a continuous homomorphism ω, fitting into the following commutative diagram

$$\begin{array}{ccc} K & \xrightarrow{\omega} & \mathrm{GL}_F(V_F) \rtimes \mathrm{Gal}(F:\mathbb{Q}) \\ & \searrow^{\tau} & \downarrow \\ & & \mathrm{Gal}(F:\mathbb{Q}). \end{array} \tag{A.1}$$

Obvious any such homorphism determines and is determined by its restriction ω' to the first factor which is an element in $H^1\big(K, \mathrm{GL}_F(V_F)\big)$ upon which K acts via τ. Here $\mathrm{GL}_F(V_F)$ is equipped with the discrete topology.

A.3.3. τ-semilinear representations of K form a category, morphisms being linear morphisms of F-vector spaces respecting the representations.

There is a notion of induction for these representations, and we have for $K_1 \subseteq K_2$ (open subgroup) and V_i a τ-semilinear representation of K_i:

$$\mathrm{Hom}_{K_1}(V_2, V_1) \cong \mathrm{Hom}_{K_2}\big(V_2, \mathrm{ind}_{K_1}^{K_2}(V_1)\big)$$

naturally, and since $[K_1 : K_2] < \infty$, we also have

$$\mathrm{Hom}_{K_1}(V_1, V_2) \cong \mathrm{Hom}_{K_2}(\mathrm{ind}_{K_1}^{K_2}(V_1), V_2),$$

naturally. The induction may be defined as

$$\mathrm{ind}_{K_1}^{K_2}(V) = \{f \colon K_2 \to V \mid f(hg) = \omega(h)f(g)\ \forall h \in K_1, g \in K_2\}$$

with the F-linear structure given by

$$(\alpha f)(g) := \tau_g(\alpha) f(g).$$

We have also

$$\begin{aligned} &\mathrm{Hom}_{K_2}\big(\mathrm{ind}_{K_1}^{K_2}(V), \mathrm{ind}_{K_1}^{K_2}(W)\big) \\ &= \left\{ \varphi \colon K_2 \to \mathrm{Hom}_{\mathbb{Q}}(V, W) \;\middle|\; \begin{array}{l} \varphi(h_1 g h_2) = \omega(h_1)\circ\varphi(g)\circ\omega(h_2)\ \forall h_1, h_2 \in K_1, g \in K_2, \\ \varphi(g)\alpha v = \tau_g(\alpha)\varphi(g)v\ \forall g \in K_2, \alpha \in F. \end{array} \right\} \end{aligned}$$

Here a function φ acts by convolution

$$(\varphi * f)(g) := \frac{1}{[K_2 : K_1]} \sum_{x \in K_2 \backslash K_1} \varphi(gx^{-1}) f(x).$$

A.3.4. Example. Any $\mathbb{Q}$-representation ω', V of K gives rise to a τ-semilinear representation $\omega(k) = \omega'(k) \circ \tau_k$ on V_F.

For a τ-semilinear representation V, ω the dual is defined by V^* (the usual dual) acted on by $(\omega^*(g)v^*)(v) := \tau_g\big(v^*(\omega(g^{-1})v)\big)$. It is a τ-semilinear representation, too.

A.3.5. Example. Let p be an odd prime, let χ be the nontrivial character of order 2 of $(\mathbb{Z}/p\mathbb{Z})^*$, and $\tau \colon (\mathbb{Z}/p\mathbb{Z})^* \to \mathrm{Gal}(\mathbb{Q}(\zeta_p)/\mathbb{Q})$ be the natural isomorphism. These data give rise to a semilinear representation

$$\begin{pmatrix} a & * \\ 0 & d \end{pmatrix} \mapsto \chi(a)\tau_{ad}.$$

A function $\varphi \in \operatorname{Hom}(\operatorname{ind}_{K_1}^{K_2} \chi, \operatorname{ind}_{K_1}^{K_2} \chi)$ is determined by its values at

$$\begin{pmatrix} 1 & 0 \\ 0 & 1 \end{pmatrix} \quad \text{and} \quad \begin{pmatrix} 0 & 1 \\ -1 & 0 \end{pmatrix}.$$

The semilinearity forces

$$\tau_\alpha\left(\varphi\left(\begin{pmatrix} 1 & 0 \\ 0 & 1 \end{pmatrix}\right)\right) = \varphi\left(\begin{pmatrix} 1 & 0 \\ 0 & 1 \end{pmatrix}\right)$$

and

$$\tau_\alpha\left(\varphi\left(\begin{pmatrix} 0 & 1 \\ -1 & 0 \end{pmatrix}\right)\right) = \chi(\alpha)\varphi\left(\begin{pmatrix} 0 & 1 \\ -1 & 0 \end{pmatrix}\right),$$

therefore

$$\varphi\left(\begin{pmatrix} 1 & 0 \\ 0 & 1 \end{pmatrix}\right) \in \mathbb{Q} \quad \text{and} \quad \varphi\left(\begin{pmatrix} 0 & 1 \\ -1 & 0 \end{pmatrix}\right) \in \mathbb{Q} \cdot \sqrt{p^*},$$

where $p^* = p$ if $p \equiv 1 \bmod 4$ and $p^* = -p$ otherwise. A calculation shows that $\operatorname{End}_{K_2}\left(\operatorname{ind}_{K_1}^{K_2} \chi\right) = \mathbb{Q}^2$ as algebras hence the irreducible constituents of $\operatorname{ind}_{K_1}^{K_2} \chi$ as $\mathrm{SL}_2(\mathbb{Z}/p\mathbb{Z})$-representations are also irreducible as semilinear representations of $\mathrm{GL}_2(\mathbb{Z}/p\mathbb{Z})$.

APPENDIX B

B.1. Explicit calculation of $\lambda(L_{\mathbb{Z}}; s)$

B.1.1. Let $L_{\mathbb{Z}}$ be a lattice of signature $(l-2,2)$ such that $L_{\mathbb{Z}}^*/L_{\mathbb{Z}}$ is cyclic of square-free order. In this section, we explicitly calculate the quantity $4\tilde{\lambda}^{-1}(L_{\mathbb{Z}}; s)$ whose value, respectively derivative at $s=0$ encode the geometric, respectively arithmetic volume of the associated orthogonal Shimura variety $\mathrm{M}({}^K\mathbf{O}(L))$, where K is the discriminant kernel.

Recall that $\tilde{\lambda}^{-1}(L_{\mathbb{Z}}; s) = |D|^{-s/2}\lambda^{-1}(L_{\mathbb{Z}}; s)$. We will need the following functional equations:

B.1.2. Theorem ([56, Kap. VII]).

$$L(s,\chi_D) = |D|^{-s+1/2}\frac{(2\pi)^s}{2\Gamma(s)\cos(\pi s/2)}L(1-s,\chi_D) \qquad D>0,\ D \textit{ square-free},$$

$$L(s,\chi_D) = |D|^{-s+1/2}\frac{(2\pi)^s}{2\Gamma(s)\sin(\pi s/2)}L(1-s,\chi_D) \qquad D<0,\ D \textit{ square-free},$$

$$\zeta(s) = \frac{(2\pi)^s}{2\Gamma(s)\cos(\pi s/2)}\zeta(1-s).$$

B.1.3. Proposition. *Let $L_{\mathbb{Z}}$ be a quadratic lattice of signature $(l-2,2)$, of even dimension l, and with cyclic discriminant group $L_{\mathbb{Z}}^*/L_{\mathbb{Z}}$ of square-free order $\varepsilon = \det(L_{\mathbb{Z}})$. Note that ε is necessarily odd. Let χ be the character*

$$n \longmapsto \left(\frac{(-1)^{l/2}\varepsilon}{n}\right).$$

Since ε is positive and there exists a unimodular lattice locally at 2, we have $\varepsilon \equiv (-1)^{l/2}$ (4). This means that the character is of conductor ε. The case $\varepsilon = 1$ is included. *Then we get:*

$$\lambda^{-1}(L_{\mathbb{Z}}; s) = 2^{-l/2+s/2}(2\pi)^{s(l-1)/2}$$
$$\cdot \prod_{i=1}^{l/2-1}\left(\frac{\Gamma(s+2i)\zeta(1-2s-2i)}{\Gamma(2s+2i)\cos(\pi(s+i))}\right)\frac{\Gamma(s/2+l/2)L(1-s-l/2,\chi)}{\Gamma(s+l/2)\cos(\pi(s/2+\lfloor l/4\rfloor))}$$

In particular, the value at $s=0$ equals

$$\lambda^{-1}(L_{\mathbb{Z}};0) = 2^{-l/2}\left(\prod_{i=1}^{l/2-1}\zeta(1-2i)\right)L(1-l/2,\chi) = 2^{-l/2}\left(\prod_{i=1}^{l/2-1}\frac{(-B_{2i})}{2i}\right)\frac{(-B_{l/2,\chi})}{l/2}$$

and the derivative at $s = 0$ *is*

$$\frac{\mathrm{d}}{\mathrm{d}s}(\lambda^{-1}(L_{\mathbb{Z}};s))\bigg|_{s=0} = \lambda^{-1}(L_{\mathbb{Z}};0)$$

$$\cdot\left(\frac{l-1}{2}C - \frac{1}{2}N_{l/2} + \sum_{i=1}^{l/2-1}\left(-2\frac{\zeta'(1-2i)}{\zeta(1-2i)} - N_{2i}\right) - \frac{L'(1-l/2,\chi)}{L(1-l/2,\chi)} + \frac{1}{2}\log(2)\right)$$

where $N_n = \sum_{i=1}^{n-1} 1/i$ *and* $C = \log(2\pi) + \gamma$.

PROOF. We have by definition [26, 10.3]

$$\lambda^{-1}(L_{\mathbb{Z}};s) = \lambda_\infty^{-1}(L_{\mathbb{Z}};s)\prod_p \lambda_p^{-1}(L_{\mathbb{Z}};s).$$

We first derive an expression of $\lambda_\infty^{-1}(L_{\mathbb{Z}};s)$ that will be used in the sequel:

$$\begin{aligned}\lambda_\infty^{-1}(L_{\mathbb{Z}};s) &= \Gamma_{l-1,l}(s)^{-1}\\ &= \prod_{k=2}^{l}\frac{\Gamma((s+k)/2)}{2\pi^{(s+k)/2}}\\ &= \frac{\Gamma((l+s)/2)}{2\pi^{(s+l)/2}}\prod_{i=1}^{l/2-1}\left(\frac{\Gamma(s/2+i)}{2\pi^{(s+2i)/2}}\frac{\Gamma(s/2+i+\frac{1}{2})}{2\pi^{(s+2i+1)/2}}\right)\\ &= \frac{\Gamma((l+s)/2)}{2\pi^{(s+l)/2}}\prod_{i=1}^{l/2-1}\left(\frac{\Gamma(s+2i)}{2(2\pi)^{s+2i}}\right).\end{aligned}$$

By [26, Theorem 8.1, 8.2], we have for all p:

$$\lambda_p^{-1}(L_{\mathbb{Z}};s) = |\varepsilon|_p^{-s-(l-1)/2}\left(\prod_{i=1}^{l/2-1}(1-p^{-2i-2s})^{-1}\right)(1-\chi(p)p^{-l/2-s})^{-1}$$

Therefore we have

$$\begin{aligned}\lambda^{-1}(L_{\mathbb{Z}};s) &= \Gamma_{l-1,l}(s)^{-1}|\varepsilon|^{s+(l-1)/2}\left(\prod_{i=1}^{l/2-1}\zeta(2i+2s)\right)L\left(\chi,\frac{l}{2}+s\right)\\ &= \prod_{k=2}^{l}\frac{\Gamma((s+k)/2)}{2\pi^{(s+k)/2}}|\varepsilon|^{s+(l-1)/2}\left(\prod_{i=1}^{l/2-1}\zeta(2i+2s)\right)L\left(\chi,\frac{l}{2}+s\right).\end{aligned}$$

We calculate using the calculation of $\lambda_\infty^{-1}(L_{\mathbb{Z}};s)$ above and the functional equations of Theorem B.1.2 (note that the parity of χ agrees with the parity of $l/2$):

$$= \prod_{k=2}^{l}\left(\frac{\Gamma((s+k)/2)}{2\pi^{(s+k)/2}}\right)|\varepsilon|^{s+(l-1)/2}\prod_{i=1}^{l/2-1}\left(\frac{(2\pi)^{2s+2i}}{2\Gamma(2i+2s)\cos(\pi(i+s))}\zeta(1-2i-2s)\right)$$

$$\cdot|\varepsilon|^{-l/2-s+1/2}\frac{(2\pi)^{s+l/2}}{2\Gamma(s+l/2)\cos(\pi(s/2+\lfloor l/4\rfloor))}L\left(1-\frac{l}{2}-s,\chi\right)$$

$$= \frac{\Gamma((l+s)/2)}{2\pi^{(s+l)/2}}\prod_{i=1}^{l/2-1}\left(\frac{\Gamma(s+2i)}{2(2\pi)^{s+2i}}\frac{(2\pi)^{2s+2i}}{2\Gamma(2i+2s)\cos(\pi(i+s))}\zeta(1-2i-2s)\right)$$

$$\cdot\frac{(2\pi)^{s+l/2}}{2\Gamma(s+l/2)\cos(\pi(s/2+\lfloor l/4\rfloor))}L\left(1-\frac{l}{2}-s,\chi\right)$$

$$= 2^{l(s-1)/2}\pi^{s(l-1)/2}\frac{\Gamma(s/2+l/2)}{\Gamma(s+l/2)}\prod_{i=1}^{l/2-1}\left(\frac{\Gamma(s+2i)}{\Gamma(2i+2s)\cos\bigl(\pi(s+i)\bigr)}\zeta(1-2i-2s)\right)$$
$$\cdot\frac{1}{\cos\bigl(\pi(s/2+\lfloor l/4\rfloor)\bigr)}L\left(1-\frac{l}{2}-s,\chi\right).$$

This yields immediately the claimed value at $s=0$. We now proceed to calculate the logarithmic derivative at $s=0$:

$$d\log\lambda^{-1}(L_{\mathbb{Z}};s)|_{s=0}=\frac{l-1}{2}\log(2\pi)-\frac{1}{2}\frac{\Gamma'(l/2)}{\Gamma(l/2)}+\sum_{i=1}^{l/2-1}\left(-2\frac{\zeta'(1-2i)}{\zeta(1-2i)}-\frac{\Gamma'(2i)}{\Gamma(2i)}\right)$$
$$-\frac{L'(1-l/2,\chi)}{L(1-l/2,\chi)}+\frac{1}{2}\log(2).$$

Since we have

$$\frac{\Gamma'(n)}{\Gamma(n)}=\sum_{i=1}^{n-1}\frac{1}{i}-\gamma=N_n-\gamma$$

we get:

$$d\log\lambda^{-1}(L_{\mathbb{Z}};s)|_{s=0}=\frac{l-1}{2}\log(2\pi)-\frac{1}{2}(N_{l/2}-\gamma)+\sum_{i=1}^{l/2-1}\left(-2\frac{\zeta'(1-2i)}{\zeta(1-2i)}-N_{2i}+\gamma\right)$$
$$-\frac{L'(1-l/2,\chi)}{L(1-l/2,\chi)}+\frac{1}{2}\log(2).$$

This gives indeed the claimed value. □

B.1.4. Proposition. *Let $L_{\mathbb{Z}}$ be a quadratic lattice of signature $(l-2,2)$, of odd dimension l and with cyclic discriminant group $L_{\mathbb{Z}}^*/L_{\mathbb{Z}}$ of square-free order $\varepsilon=\det(L_{\mathbb{Z}})$. Note that ε is necessarily even.*

Then we get

$$\lambda^{-1}(L_{\mathbb{Z}};s)=2^{s-(l-1)/2}(2\pi)^{(l-1)s/2}\prod_{i=1}^{(l-1)/2}\left(\frac{\Gamma(s+2i)\zeta(1-2i-2s)}{\Gamma(2i+2s)\cos\bigl(\pi(i+s)\bigr)}\right)$$
$$\cdot\prod_{p|\epsilon/2}\left(p^{(l-1)/2+s}+\left(\frac{-1}{p}\right)^{(l-1)/2}\Phi_p\right).$$

where Φ_p is the Hasse-invariant of L at p. In particular the value at 0 is

$$\lambda^{-1}(L_{\mathbb{Z}};0)$$
$$=2^{-(l-1)/2}(-1)^{(l^2-1)/8}\left(\prod_{i=1}^{(l-1)/2}\zeta(1-2i)\right)\prod_{p|\epsilon/2}\left(p^{(l-1)/2}+\left(\frac{-1}{p}\right)^{(l-1)/2}\Phi_p\right)$$
$$=2^{-(l-1)/2}(-1)^{(l^2-1)/8}\left(\prod_{i=1}^{(l-1)/2}\frac{-B_{2i}}{2i}\right)\prod_{p|\epsilon/2}\left(p^{(l-1)/2}+\left(\frac{-1}{p}\right)^{(l-1)/2}\Phi_p\right)$$

and the derivative at $s=0$:

$$\frac{\mathrm{d}}{\mathrm{d}s}\left(\lambda^{-1}(L_{\mathbb{Z}};s)\right)\bigg|_{s=0} = \lambda^{-1}(L_{\mathbb{Z}};0)\cdot\Bigg(\frac{l-1}{2}C+\sum_{i=1}^{(l-1)/2}\left(-2\frac{\zeta'(1-2i)}{\zeta(1-2i)}-N_{2i}\right)$$
$$+\sum_{p\mid\epsilon/2}\left(\frac{p^{(l-1)/2}\log(p)}{p^{(l-1)/2}+\left(\frac{-1}{p}\right)^{(l-1)/2}\Phi_p}\right)+\log(2)\Bigg)$$

with $N_n=\sum_{i=1}^{n-1}1/i$ and $C=\log(2\pi)+\gamma$.

PROOF. We have by definition [26, 10.3]

$$\lambda^{-1}(L_{\mathbb{Z}};s)=\lambda_{\infty}^{-1}(L_{\mathbb{Z}};s)\prod_p\lambda_p^{-1}(L_{\mathbb{Z}};s).$$

We first derive an expression of $\lambda_{\infty}^{-1}(L_{\mathbb{Z}};s)$ that will be used in the sequel (analogous to the calculation in the proof of the Proposition B.1.3):

$$\lambda_{\infty}^{-1}(L_{\mathbb{Z}};s)=\Gamma_{l-1,l}(s)^{-1}=\prod_{i=1}^{(l-1)/2}\left(\frac{\Gamma(s+2i)}{2(2\pi)^{s+2i}}\right).$$

By [26, Theorem 8.1, 8.2], we have for all p:

$$\lambda_p^{-1}(L_{\mathbb{Z}};s)=|\varepsilon|_p^{-s-(l-1)/2}\prod_{i=1}^{(l-3)/2}(1-p^{-2i-2s})^{-1}$$
$$\cdot\begin{cases}\left(1-\left(\dfrac{-1}{p}\right)^{(l-1)/2}\Phi_p p^{-(l-1)/2-s}\right)^{-1} & \text{if } p\mid\varepsilon/2,\\ (1-p^{-l+1-2s})^{-1} & \text{otherwise.}\end{cases}$$

where $\Phi_p=\prod_{i<j}(\varepsilon_i,\varepsilon_j)_p$ is the Hasse invariant, where $L_{\mathbb{Q}_p}\cong<\varepsilon_1,\dots,\varepsilon_l>$. Note that for odd p and $\varepsilon_1,\varepsilon_2\in\mathbb{Z}_p^*$ we have $(\varepsilon_1,\varepsilon_2)_p=1$ whereas $(p\varepsilon_1,\varepsilon_2)_p=\left(\frac{\varepsilon_2}{p}\right)$.

We calculate using the calculation of $\lambda_{\infty}^{-1}(L_{\mathbb{Z}};s)$ above and the functional equation for ζ of Theorem B.1.2:

$$\begin{aligned}
&\lambda^{-1}(L_{\mathbb{Z}};s)\\
&=\Gamma_{l-1,l}(s)^{-1}|\varepsilon|^{s+(l-1)/2}\prod_{i=1}^{(l-1)/2}\zeta(2i+2s)\cdot\prod_{p\mid\epsilon/2}\left(1+\left(\frac{-1}{p}\right)^{(l-1)/2}\Phi_p p^{-(l-1)/2-s}\right)\\
&=|\varepsilon|^{s+(l-1)/2}\prod_{i=1}^{(l-1)/2}\left(\frac{\Gamma(s+2i)}{2(2\pi)^{s+2i}}\frac{(2\pi)^{2s+2i}}{2\Gamma(2i+2s)\cos(\pi(i+s))}\zeta(1-2i-2s)\right)\\
&\qquad\cdot\prod_{p\mid\epsilon/2}\left(1+\left(\frac{-1}{p}\right)^{(l-1)/2}\Phi_p p^{-(l-1)/2-s}\right)\\
&=|\varepsilon|^{s+(l-1)/2}\prod_{i=1}^{(l-1)/2}\left(\frac{(2\pi)^s\Gamma(s+2i)}{4\Gamma(2i+2s)\cos(\pi(i+s))}\zeta(1-2i-2s)\right)\\
&\qquad\cdot\prod_{p\mid\epsilon/2}\left(1+\left(\frac{-1}{p}\right)^{(l-1)/2}\Phi_p p^{-(l-1)/2-s}\right)
\end{aligned}$$

$$= 2^{s+(l-1)/2}(2\pi)^{\frac{l-1}{2}s}2^{-l+1}\prod_{i=1}^{(l-1)/2}\left(\frac{\Gamma(s+2i)}{\Gamma(2i+2s)\cos\bigl(\pi(i+s)\bigr)}\zeta(1-2i-2s)\right)$$
$$\cdot\prod_{p|\epsilon/2}\left(p^{(l-1)/2+s}+\left(\frac{-1}{p}\right)^{(l-1)/2}\Phi_p\right)$$

This yields immediately the claimed value at $s=0$. We now proceed to calculate the logarithmic derivative at $s=0$:

$$\mathrm{d}\log\lambda^{-1}(L_{\mathbb{Z}}; s)|_{s=0} = \frac{l-1}{2}\log(2\pi) + \sum_{i=1}^{(l-1)/2}\left(-2\frac{\zeta'(1-2i)}{\zeta(1-2i)} - \frac{\Gamma'(2i)}{\Gamma(2i)}\right)$$
$$+\sum_{p|\epsilon/2}\frac{p^{(l-1)/2}\log(p)}{p^{(l-1)/2}+\left(\frac{-1}{p}\right)^{(l-1)/2}\Phi_p} + \log(2).$$

Since we have

$$\frac{\Gamma'(n)}{\Gamma(n)} = \sum_{i=1}^{n-1}\frac{1}{i} - \gamma = N_n - \gamma.$$

we get the claimed expression. □

Bibliography

1. A. Ash, D. Mumford, M. Rapoport, and Y. Tai, *Smooth compactification of locally symmetric varieties*, Lie Groups: Hist., Frontiers and Appl., vol. IV, Math. Sci. Press, Brookline, MA, 1975.
2. R. E. Borcherds, *Automorphic forms with singularities on Grassmannians*, Invent. Math. **132** (1998), no. 3, 491–562.
3. ______, *The Gross–Kohnen–Zagier theorem in higher dimensions*, Duke Math. J. **97** (1999), no. 2, 219–233.
4. J.-B. Bost, *Potential theory and Lefschetz theorems for arithmetic surfaces*, Ann. Sci. École Norm. Sup. (4) **32** (1999), no. 2, 241–312.
5. N. Bourbaki, *Commutative algebra. Chapters* 1–7, Elem. Math. (Berlin), Springer, Berlin, 1989. Translated from the French; Reprint of the 1972 edition.
6. J. H. Bruinier, *Borcherds products on* $\mathrm{O}(2,l)$ *and Chern classes of Heegner divisors*, Lecture Notes in Math., vol. 1780, Springer, Berlin, 2002.
7. ______, *Two applications of the curve lemma for orthogonal groups*, Math. Nachr. **274/275** (2004), 19–31.
8. ______, *Vector valued formal Fourier–Jacobi series*, Proc. Amer. Math. Soc., to appear.
9. J. H. Bruinier and M. Bundschuh, *On Borcherds products associated with lattices of prime discriminant*, Ramanujan J. **7** (2003), no. 1-3, 49–61.
10. J. H. Bruinier, J. I. Burgos Gil, and U. Kühn, *Borcherds products and arithmetic intersection theory on Hilbert modular surfaces*, Duke Math. J. **139** (2007), no. 1, 1–88.
11. J. H. Bruinier and U. Kühn, *Integrals of automorphic Green's functions associated to Heegner divisors*, Int. Math. Res. Not. **31** (2003), 1687–1729.
12. J. H. Bruinier and T. Yang, *Faltings heights of CM cycles and derivatives of L-functions*, Invent. Math. **177** (2009), no. 3, 631–681.
13. J. I. Burgos, *Green forms and their product*, Duke Math. J. **75** (1994), no. 3, 529–574.
14. J. I. Burgos Gil, J. Kramer, and U. Kühn, *Arithmetic characteristic classes of automorphic vector bundles*, Doc. Math. **10** (2005), 619–716 (electronic).
15. ______, *Cohomological arithmetic Chow rings*, J. Inst. Math. Jussieu **6** (2007), no. 1, 1–172.
16. P. Deligne, *Travaux de Shimura*, Séminaire Bourbaki, 23ème année (1970/71), Exp. No. 389, Lecture Notes in Math., vol. 244, Springer, Berlin, 1971, pp. 123–165.
17. ______, *Variétés de Shimura: interprétation modulaire, et techniques de construction de modèles canoniques*, Automorphic Forms, Representations and L-Functions. Part 2 (Corvallis, OR, 1977) (A. Borel and W. Casselman, eds.), Proc. Sympos. Pure Math., vol. 33, Amer. Math. Soc., Providence, RI, 1979, pp. 247–289.
18. ______, *Espaces hermitiens symétriques*. Transcription available at `http://www.jmilne.org/`.
19. G. Faltings and C.-L. Chai, *Degeneration of abelian varieties*, with an appendix by D. Mumford, Ergeb. Math. Grenzgeb. (3), vol. 22, Springer, Berlin, 1990.
20. H. Gillet, *Arithmetic intersection theory on Deligne–Mumford stacks*, Motives and Algebraic Cycles (R. de Jeu and J. D. Lewis, eds.), Fields Inst. Commun., vol. 56, Amer. Math. Soc., Providence, RI, 2009, pp. 93–109.
21. B. H. Gross and D. B. Zagier, *Heegner points and derivatives of L-series*, Invent. Math. **84** (1986), no. 2, 225–320.
22. A. Grothendieck et al., *Schémas en groupes*. I: *Propriétés générales des schémas en groupes*, Lecture Notes in Math., vol. 151, Springer, Berlin, 1970; II: *Groupes de type multiplicatif, et structure des schémas en groupes généraux*, Lecture Notes in Math., vol. 152, Springer, Berlin, 1970; III: *Structure des schémas en groupes réductifs*, Lecture Notes in Math., vol. 153, Springer, Berlin, 1970.

23. M. Harris and S. Zucker, *Boundary cohomology of Shimura varieties.* I: *Coherent cohomology on toroidal compactifications*, Ann. Sci. École Norm. Sup. (4) **27** (1994), no. 3, 249–344.
24. F. Hirzebruch and D. Zagier, *Intersection numbers of curves on Hilbert modular surfaces and modular forms of Nebentypus*, Invent. Math. **36** (1976), 57–113.
25. F. Hörmann, *The arithmetic volume of Shimura varieties of orthogonal type*, Dissertation, Humboldt-Universität zu Berlin, 2010.
26. ———, *On recursive properties of certain p-adic Whittaker functions*, J. Number Theory **132** (2012), no. 7, 1438–1482.
27. M. Kisin, *Integral canonical models of Shimura varieties*, J. Théor. Nombres Bordeaux **21** (2009), no. 2, 301–312.
28. ———, *Integral models for Shimura varieties of abelian type*, J. Amer. Math. Soc. **23** (2010), no. 4, 967–1012.
29. K. Köhler, *A Hirzebruch proportionality principle in Arakelov geometry*, Number Fields and Function Fields—Two Parallel Worlds (G. van der Geer, B. Moonen, and R. Schoof, eds.), Progr. Math., vol. 239, Birkhäuser Boston, Boston, MA, 2005, pp. 237–268.
30. R. E. Kottwitz, *Points on some Shimura varieties over finite fields*, J. Amer. Math. Soc. **5** (1992), no. 2, 373–444.
31. S. S. Kudla, *Central derivatives of Eisenstein series and height pairings*, Ann. of Math. (2) **146** (1997), no. 3, 545–646.
32. ———, *Algebraic cycles on Shimura varieties of orthogonal type*, Duke Math. J. **86** (1997), no. 1, 39–78.
33. ———, *Derivatives of Eisenstein series and arithmetic geometry*, Proceedings of the International Congress of Mathematicians, Vol. II (Beijing, 2002), Higher Ed. Press, Beijing, 2002, pp. 173–183.
34. ———, *Modular forms and arithmetic geometry*, Current Developments in Mathematics, 2002 (Cambridge, MA, 2002) (D. Jerison, G. Lustig, B. Mazur, T. Mrowka, W. Schmid, R. Stanley, and S.-T. Yau, eds.), Int. Press, Somerville, MA, 2003, pp. 135–179.
35. ———, *Integrals of Borcherds forms*, Compositio Math. **137** (2003), no. 3, 293–349.
36. ———, *Special cycles and derivatives of Eisenstein series*, Heegner Points and Rankin *L*-Series (Berkeley, CA, 2001) (H. Darmon and S.-W. Zhang, eds.), Math. Sci. Res. Inst. Publ., vol. 49, Cambridge Univ. Press, Cambridge, 2004, pp. 243–270.
37. S. S. Kudla and J. J. Millson, *The theta correspondence and harmonic forms.* I, Math. Ann. **274** (1986), no. 3, 353–378; II, Math. Ann. **277** (1987), no. 2, 267–314.
38. ———, *Intersection numbers of cycles on locally symmetric spaces and Fourier coefficients of holomorphic modular forms in several complex variables*, Inst. Hautes Études Sci. Publ. Math. **71** (1990), 121–172.
39. S. S. Kudla and M. Rapoport, *Arithmetic Hirzebruch–Zagier cycles*, J. Reine Angew. Math. **515** (1999), 155–244.
40. ———, *Height pairings on Shimura curves and p-adic uniformization*, Invent. Math. **142** (2000), no. 1, 153–223.
41. ———, *Cycles on Siegel threefolds and derivatives of Eisenstein series*, Ann. Sci. École Norm. Sup. (4) **33** (2000), no. 5, 695–756.
42. S. S. Kudla, M. Rapoport, and T. Yang, *On the derivative of an Eisenstein series of weight one*, Internat. Math. Res. Notices **7** (1999), 347–385.
43. ———, *Derivatives of Eisenstein series and Faltings heights*, Compos. Math. **140** (2004), no. 4, 887–951.
44. ———, *Modular forms and special cycles on Shimura curves*, Ann. of Math. Stud., vol. 161, Princeton Univ. Press, Princeton, NJ, 2006.
45. U. Kühn, *Generalized arithmetic intersection numbers*, J. Reine Angew. Math. **534** (2001), 209–236.
46. K.-W. Lan, *Arithmetic compactifications of PEL-type Shimura varieties*, Ph.D. Thesis, Harvard University, 2008.
47. R. P. Langlands, *The volume of the fundamental domain for some arithmetical subgroups of Chevalley groups*, Algebraic Groups and Discontinuous Subgroups (Boulder, CO, 1965) (A. Borel and G. D. Mostow, eds.), Proc. Sympos. Pure Math., vol. 9, Amer. Math. Soc., Providence, RI, 1966, pp. 143–148.
48. Y. Liu, *Arithmetic theta lifting and L-derivatives for unitary groups.* I, Algebra Number Theory **5** (2011), no. 7, 849–921; II, Algebra Number Theory **5** (2011), no. 7, 923–1000.

49. K. S. Madapusi Sampath, *Toroidal compactifications of integral models of Shimura varieties of Hodge type*, Ph.D. Thesis, The University of Chicago, 2011.
50. J. S. Milne, *Automorphic vector bundles on connected Shimura varieties*, Invent. Math. **92** (1988), no. 1, 91–128.
51. ______, *Canonical models of (mixed) Shimura varieties and automorphic vector bundles*, Automorphic Forms, Shimura Varieties, and L-Functions, Vol. I (Ann Arbor, MI, 1988) (L. Clozel and J. Milne S., eds.), Perspect. Math., vol. 10, Academic Press, Boston, MA, 1990, pp. 283–414.
52. ______, *Introduction to Shimura varieties*, Harmonic Analysis, the Trace Formula, and Shimura Varieties (Toronto, ON, 2003) (J. Arthur, D. Ellwood, and R. Kottwitz, eds.), Clay Math. Proc., vol. 4, Amer. Math. Soc., Providence, RI, 2005, pp. 265–378.
53. J. S. Milne and K-y. Shih, *Shimura varieties: conjugates and the action of complex multiplication*, available at `http://www.jmilne.org/`.
54. B. Moonen, *Models of Shimura varieties in mixed characteristics*, Galois Representations in Arithmetic Algebraic Geometry (Durham, 1996) (A. J. Scholl and R. L. Taylor, eds.), London Math. Soc. Lecture Note Ser., vol. 254, Cambridge Univ. Press, Cambridge, 1998, pp. 267–350.
55. D. Mumford, *Hirzebruch's proportionality theorem in the noncompact case*, Invent. Math. **42** (1977), 239–272.
56. J. Neukirch, *Algebraische Zahlentheorie*, Springer, Berlin, 1992.
57. K. Ono and C. Skinner, *Fourier coefficients of half-integral weight modular forms modulo l*, Ann. of Math. (2) **147** (1998), no. 2, 453–470.
58. R. Pink, *Arithmetical compactification of mixed Shimura varieties*, Dissertation, Rheinische Friedrich-Wilhelms-Universität Bonn, 1989; Bonner Math. Schriften, vol. 209, Universität Bonn Mathematisches Institut, Bonn, 1990.
59. K. A. Ribet, *Galois representations attached to eigenforms with Nebentypus*, Modular Functions of One Variable. V (Bonn, 1976), Lecture Notes in Math., vol. 601, Springer, Berlin, 1977, pp. 17–51.
60. C. L. Siegel, *Indefinite quadratische Formen und Funktionentheorie.* I, Math. Ann. **124** (1951), 17–54; II, Math. Ann. **124** (1952), 364–387.
61. C. Soulé, *Lectures on Arakelov geometry*, with the collaboration of D. Abramovich, J.-F. Burnol and J. Kramer, Cambridge Stud. Adv. Math., vol. 33, Cambridge Univ. Press, Cambridge, 1992.
62. A. Vasiu, *Integral canonical models of Shimura varieties of preabelian type*, Asian J. Math. **3** (1999), no. 2, 401–518.
63. A. Weil, *Sur certains groupes d'opérateurs unitaires*, Acta Math. **111** (1964), 143–211.
64. ______, *Sur la formule de Siegel dans la théorie des groupes classiques*, Acta Math. **113** (1965), 1–87.
65. ______, *Adeles and algebraic groups*, with appendices by M. Demazure and T. Ono, Progr. Math., vol. 23, Birkhäuser Boston, Boston, MA, 1982.
66. T. Yang, *Faltings heights and the derivative of Zagier's Eisenstein series*, Heegner Points and Rankin L-Series (Berkeley, CA, 2001) (H. Darmon and S.-W. Zhang, eds.), Math. Sci. Res. Inst. Publ., vol. 49, Cambridge Univ. Press, Cambridge, 2004, pp. 271–284.
67. D. Zagier, *Nombres de classes et formes modulaires de poids* 3/2, C. R. Acad. Sci. Paris Sér. A-B **281** (1975), no. 21, Ai, A883–A886.

Index

Table of Notation

$\left[\textbf{\textit{S}-1mot}\right]$: category of 1-motives over S; **2.7.3**

$\left[\textbf{\textit{S}-}{}^{K}\textbf{X-\textit{L}-mot}\right]$: category of 1-motives with extra structure, yields a stack defining the canonical model $\mathrm{M}({}^{K}\mathbf{X})$ for $\mathbf{X} \in \{\mathbf{H}_{g_0}, \mathbf{H}_{g_0}[0, L_0 \otimes I], \mathbf{H}_{g_0}[(I \otimes I)^s, L_0 \otimes I]\}$; **2.7.3**

$\left[\textbf{\textit{S}-}{}^{K}\textbf{X-\textit{L}-mot-triv}\right]$: as before with trivialization of de Rham defining a canonical model of the standard principal bundle $\mathrm{P}({}^{K}\mathbf{X})$; **2.7.3**

$[X/G]$: algebraic or analytic quotient stack

$L^{\otimes}$: $= \bigoplus_{i,j} L^{\otimes i} \otimes (L^*)^{\otimes j}$

$\langle v, w \rangle$: denotes the symplectic form on L—or, in the orthogonal case—$Q_L(v+w) - Q_L(v) - Q_L(w)$

$\langle a_1, \dots, a_n \rangle$: the space $\mathbb{Z}^n$, $\mathbb{Z}_p^n$ or $\mathbb{Z}_{(p)}^n$ (according to the context) equipped with quadratic form $x \mapsto \sum_i a_i x_i^2$. It has discriminant $\prod_i 2a_i$

$\langle Q \rangle$: the space R^n equipped with a quadratic form $Q \in \mathrm{Sym}^2\big((R^n)^*\big)$, where R is mostly $\mathbb{Z}$, $\mathbb{Z}_p$, or $\mathbb{Z}_{(p)}$

$M \perp L$: orthogonal direct sum of two quadratic spaces or lattices

$\mathbb{A}$: $= \prod_\nu' \mathbb{Q}_\nu$ restricted product over all places of $\mathbb{Q}$ (adeles of $\mathbb{Q}$)

$\mathbb{A}^{(S)}$: $= \prod_{\nu \notin S}' \mathbb{Q}_\nu$, e.g., $\mathbb{A}^{(\infty)}$ or $\mathbb{A}^{(\infty,p)}$ where p is a prime

$\mathbb{A}^{(\infty)}(1)$: Tate-twist $\widehat{\mathbb{Z}}(1) \otimes \mathbb{Q}$

$a_{\iota,z',k,\alpha,\rho,\sigma}(f)$: Fourier coefficient of a (meromorphic) section of $\Xi^*\mathcal{E}$, w.r.t. some tube domain realization of $\mathbb{D}_{\mathbf{O}}$, valid in the cone σ; **2.8.5**

$\mathbf{B}$: p-integral mixed Shimura data consisting of $P_{\mathbf{B}}$, $\mathbb{D}_{\mathbf{B}}$, $h_{\mathbf{B}}$ mostly a boundary component of another mixed Shimura datum; **2.2.18**

$\iota\colon \mathbf{B} \Longrightarrow \mathbf{X}$: boundary morphism pf p-MSD—leads to a partial order between boundary components. The group $P_{\mathbf{X}}$ acts by conjugation, preserving order ; **2.2.3**

$(\iota, \rho)\colon {}^{K'}_{\Delta'}\mathbf{B} \Longrightarrow {}^{K}_{\Delta}\mathbf{X}$: boundary morphism of p-ECMSD; **2.2.26**

$\mathrm{C}(L)$: Clifford algebra of L; **2.4.1**

$\mathrm{C}^+(L), \mathrm{C}^-(L)$: even and odd part of the Clifford algebra; **2.4.1**

$C_{\mathbf{X}}$: the conical complex of $\mathbf{X}$; **2.2.21**

$d(L_{\mathbb{Z}}), d(L_{\mathbb{Z}_p}), d(L_{\mathbb{Q}})$: the discriminant of a lattice with quadratic form

$d(\beta)$: certain element of $\mathrm{Sp}(\mathfrak{M})$; **3.2.1**

D: boundary divisor on $\mathrm{M}({}^{K}_{\Delta}\mathbf{X})$; **2.5.9**

$D(L)$: $= \underline{\mathrm{Hom}}(X, \mathbb{G}_{\mathrm{m}})$; **2.1.1**

$\mathbb{D}_{\mathbf{X}}$: generalized symmetric space—part of the datum $\mathbf{X}$; **2.2.3**

Δ: a (partial) rational polyhedral cone decomposition of $C_{\mathbf{X}} \times P_{\mathbf{X}}(\mathbb{A}^{(\infty)})$; **2.2.23**

$E_{\mathbf{X}}$: the reflex field $\subset \mathbb{C}$ of $\mathbf{X}$; **2.5.2**

$\mathcal{E}$: a $P_{\mathbf{X}}$-equivariant bundle on $\mathrm{M}^\vee(\mathbf{X})$; **2.6.1**. In Chapter 3, this is the tautological line bundle on $\mathrm{M}^\vee(\mathbf{O}_n) \subset \mathbb{P}^1(L_{\mathbb{Z}_{(p)}})$; **3.2.5**

$\Xi^*\mathcal{E}$: associated automorphic vector bundle on $\mathrm{M}(^K_\Delta\mathbf{X})$; **2.6.4**

$\overline{\mathcal{E}}$: the pair $(\mathcal{E}, h_{\mathcal{E}})$

$\Xi^*\overline{\mathcal{E}}$: associated Hermitian automorphic vector bundle on $\mathrm{M}(^K_\Delta\mathbf{X})$ with Hermitian metric, singular along D; **2.6.4**

$\mathcal{FTYPE}$: classifying scheme of types of cocharacters; **2.1.14**

ftype: type of a cocharacter; **2.1.14**

$g_l(\alpha)$: certain element of $G_l \subset \mathrm{Sp}(\mathfrak{M})$; **3.2.1**

Υ_R: certain character of the Witt group of $R = \mathbb{Q}_\nu, \mathbb{A}^{(\infty)}, \mathbb{A}$ denoted γ in [63]; **3.2.1**

$\widetilde{\Upsilon}_R$: 'correction factor' in the formulas of the Weil representation; **3.2.1**

$\Gamma_{m,l}(s)$: higher-dimensional gamma function; **B.1.3**, **B.1.4**

$\mathrm{GSp}(L)$: group scheme of symplectic similitudes; **2.3.4**

$\mathrm{GSpin}(L)$: general spin group scheme; **2.4.1**

$G_{\mathbf{X}}$: $P_{\mathbf{X}}/W_{\mathbf{X}}$, the reductive part of $P_{\mathbf{X}}$; **2.2.10**

$\mathbb{G}_{\mathrm{a}} = \mathbb{W}(\mathcal{O}_S)$: the additive group scheme over S

$\mathbb{G}_{\mathrm{m}} = D(\mathbb{Z}_S)$: the multiplicative group scheme over S

$h_{\mathbf{X}}$: $P_{\mathbf{X}}(\mathbb{R})U_{\mathbf{X}}(\mathbb{C})$-equivariant morphism $\mathbb{D}_{\mathbf{X}} \to \mathrm{Hom}(\mathbb{S}_{\mathbb{C}}, P_{\mathbf{X},\mathbb{C}})$ part of the datum $\mathbf{X}$; **2.2.3**

$h_{\mathcal{E}}$: a $P_{\mathbf{X}}(\mathbb{R})U_{\mathbf{X}}(\mathbb{C})$-invariant Hermitian metric defined on the image of the Borel embedding; **2.6.1**. In Chapter 3 this is the metric $v, w \mapsto -\frac{1}{2}\mathrm{e}^{-C}\langle v, \overline{w}\rangle$; **3.2.5**

H_0: Deligne/Pink's reference group; **2.2.16**

$\mathbf{H}_g = \mathbf{H}(L_{\mathbb{Z}_{(p)}})$: Shimura datum of symplectic type, where $P_{\mathbf{H}_g} = \mathrm{GSp}(L_{\mathbb{Z}_{(p)}})$ is the symplectic similitude group scheme of a lattice $L_{\mathbb{Z}_{(p)}}$ of rank $2g$ with nondegenerate symplectic form, $\mathbb{D}_{\mathbf{H}_g}$ = conjugacy class of morphisms $\mathbb{S} \to \mathrm{GSp}(L_{\mathbb{R}})$, yielding polarized Hodge structures of type $(-1,0), (0,-1)$ on $L_{\mathbb{R}}$; **2.3.1**

$\mathbf{H}_0$: **2.3.1**

$\mathbf{H}_{g0}[(I \otimes I)^s, I \otimes L_0]$: Shimura datum of general symplectic type I is any space over $\mathbb{Z}_{(p)}$, L_0 is the symplectic space associated with $\mathbf{H}_{g0}$; **2.3.4**

$\mathrm{HolModForm}(\Gamma, V, k)$: classical holomorphic modular forms for Γ of weight k and representation V; **3.3.1**

$\mathfrak{I}$: $= I^* \oplus I$, with natural symplectic form

I: saturated isotropic submodule of L

$\mathrm{I}(L, M)$: variety of isometries from M to L; **3.1**

K: compact open subgroup of $P_{\mathbf{X}}(\mathbb{A}^{(\infty)})]$

$K(N)$: principal level-N subgroup of $P_{\mathbf{X}}(\mathbb{A}^{(\infty)})$ depending on a $\mathbb{Z}$-lattice in a rational representation of $P_{\mathbf{X},\mathbb{Q}}$; **2.2.7**

L: module, mostly over $\mathbb{Z}_{(p)}$, with nondeg. symplectic form (Section 2.3) or nondegenerate quadratic form Q_L (Chapter 3) —more generally—any representation of $P_{\mathbf{X}}$

L_0: saturated submodule of L, considered together with a (noncanonical) splitting $L = I^* \oplus L \oplus I$, with natural symplectic, resp. orthogonal form

$\tilde{\lambda}^{-1}(L_{\mathbb{Z}}; s)$: naturally defined complex analytic function encoding (up to a factor 4) geometric and arithmetic volume of an orthogonal Shimura variety as special value and derivative at $s = 0$; **3.2.12**

$\Lambda^n(L)$: exterior power

$\mathrm{Laur}(\Gamma)$: space of q-expansions of modular forms; **3.3.2**

$\mathrm{Lie}(G)$: Lie algebra functor for a S-group scheme G; **2.1.2**

$\widetilde{\mu}(L_{\mathbb{Z}}, M_{\mathbb{Z}}, \varphi; s)$: strongly related to the Fourier coefficient of the Eisenstein series associated with the Weil representation, for L; here $\varphi \in S\big((L \otimes M^*)_{\mathbb{A}^{(\infty)}}\big)$ is any Schwartz–Bruhat function; **3.2.12**

$\widetilde{\mu}(L_{\mathbb{Z}}, M_{\mathbb{Z}}, \kappa; s)$: The same as before for φ equal to the characteristic function of $\kappa \in (L^*_{\hat{\mathbb{Z}}}/L_{\hat{\mathbb{Z}}}) \otimes M_{\mathbb{Z}}$

$\mathfrak{M}$: $= M^* \oplus M$, with natural symplectic form

$\mathrm{M}(^K_\Delta\mathbf{X})$: canonical model of the toroidal compactification of a Shimura variety, a Deligne–Mumford stack (and a scheme, if K is neat) over $\mathcal{O}_{\mathbf{X}}$; **2.5.9**

$\mathrm{M}^\vee(\mathbf{X})$: canonical model of the 'compact' dual, a right $P_{\mathbf{X},\mathcal{O}_{\mathbf{X}}}$-scheme over $\mathcal{O}_{\mathbf{X}}$; **2.5.12**

$\mathrm{ModForm}(\Gamma, V, k)$: classical weakly holomorphic modular forms for Γ, of weight k and representation V; **3.3.1**

$\mathrm{Mp}(\mathfrak{M}_R)$: metaplectic group; **3.2.1**

$\mathbf{O}(L_{\mathbb{Z}_{(p)}})$: $= \mathbf{S}(L_{\mathbb{Z}_{(p)}})/\mathbb{G}_{\mathrm{m}}$ Shimura datum of orthogonal type, where $P_{\mathbf{O}} = \mathrm{SO}(L_{\mathbb{Z}_{(p)}})$ is the special orthogonal group scheme, $\mathbb{D}_{\mathbf{O}}$ is set of orthogonal Hodge structures of type $(-1,1),(0,0),(1,-1)$ on $L_{\mathbb{R}}$; **2.4.3**

$\mathcal{O}_{\mathbf{X}}$: the reflex ring of $\mathbf{X}$ at p localization of the ring of integers of $E_{\mathbf{X}}$ with respect to $\mathbb{Z} \setminus (p)$; **2.5.2**

φ_ν, φ: Schwartz-Bruhat functions

p: some prime, mostly fixed, main theorems of Chapter 2 require $p \neq 2$

(P): a certain property of group schemes; **2.1.6**

par, qpar: (quasi-)parabolic associated with a cocharacter; **2.1.14**

$P = U \cdot G_l$: certain standard maximal parabolic subgroup of $\mathrm{Sp}(\mathfrak{M})$; **3.2.1**

$\mathcal{PAR}$: scheme of parabolics of a reductive (or type (P)) group scheme; **2.1.14**

$P_{\mathbf{X}}$: group scheme over $\mathbb{Z}_{(p)}$—part of the datum $\mathbf{X}$; **2.2.3**

$\mathrm{PowSer}(\Gamma)$: space of q-expansions of modular forms; **3.3.2**

$\mathrm{PSp}(L_0, I)$: extension of $\mathrm{GSp}(L_0)$ by $\mathbb{W}(I \otimes L_0)$, occurs as group scheme in the unipotent extension $\mathbf{H}_g[0, L_0 \otimes I]$; **2.3.4**

$\mathrm{P}(^K_\Delta\mathbf{X})$: standard principal bundle, a right $P_{\mathbf{X}}$-torsor over $\mathrm{M}(^K_\Delta\mathbf{X})$ equipped with a morphism $\Pi : \mathrm{P}(^K_\Delta\mathbf{X}) \to \mathrm{M}^\vee(\mathbf{X})$; **2.5.15**

Q_L: denotes the quadratic form on L

$\mathbb{Q}(1)$: Tate-twist $2\pi i\mathbb{Q} \subset \mathbb{C}$

$\mathbb{Q}_l(1)$: Tate-twist $\mathbb{Z}_l(1) \otimes \mathbb{Q}$

$\mathcal{QPAR}$: scheme of quasi-parabolics of a group scheme of type (P); **2.1.14**

$\mathbb{R}_{\mathbb{Q}}$: $\mathbb{R}$ modulo rational multiples of $\log(N)$, $N \in \mathbb{N}$; **2.6.16**

$\mathbb{R}_N$: $\mathbb{R}$ modulo rational multiples of $\log(p)$, $p \mid N$; **2.6.16**

$\mathbb{R}^{(p)}$: $\mathbb{R}$ modulo rational multiples of $\log(q)$, $q \neq p$; **2.6.16**

$\mathbb{S}$: Deligne torus; **2.2**

$S(L_{\mathbb{A}}), S(L_{\mathbb{R}}), S(L_{\mathbb{Q}_p}), \dots$: space of Schwartz–Bruhat functions. For $\mathbb{R}$ these are rapidly decaying and smooth. For $\mathbb{Q}_p$ these are locally constant with compact support.

$(L \otimes \cdots \otimes L)^s$: symmetric elements of $\mathrm{T}^n(L)$

$\mathbf{S}(L_{\mathbb{Z}_{(p)}})$: Shimura datum of orthogonal (spin) type, where $P_{\mathbf{S}} = \mathrm{GSpin}(L_{\mathbb{Z}_{(p)}})$ is the general spin group scheme of a lattice $L_{\mathbb{Z}_{(p)}}$ of rank m with nondegenerate quadratic form of signature $(l-2,2)$, $\mathbb{D}_{\mathbf{S}}$ is the set of orthogonal Hodge structures of type $(-2,0),(-1,-1),(0,-2)$ on $L_{\mathbb{R}}$; **2.4.3**

$\mathrm{Sing}(\Gamma)$: space of singular parts of q-expansions of modular forms; **3.3.2**

$\mathrm{SO}(L)$: (special) orthogonal group scheme

$\mathrm{SO}'(L_{\mathbb{Z}})$, $\mathrm{SO}'(L_{\hat{\mathbb{Z}}})$: discriminant kernel

$\mathrm{Sp}(L)$: symplectic group scheme; **2.3.4**

$\mathrm{Sp}'(\mathfrak{M}_R)$: denotes either $\mathrm{Sp}(\mathfrak{M}_R)$ or $\mathrm{Mp}(\mathfrak{M}_R)$ according to the parity of the dimension of L; **3.2.1**

$\mathrm{Spin}(L)$: spin group scheme; **2.4.1**

$\mathrm{Sym}^n(L)$: symmetric power

$\mathrm{T}(L)$: tensor algebra

$\mathcal{TYPE}$: classifying scheme of types of quasi-parabolics; **2.1.14**

type: type of a quasi-parabolic; **2.1.14**

$U_{\mathbf{X}}$: certain central subgroup scheme of $W_{\mathbf{X}}$; **2.2.10**

$u(B)$: certain element of $U \subset \mathrm{Sp}(\mathfrak{M})$; **3.2.1**

$\mathrm{USp}(L_0, I)$: extension of $\mathrm{GSp}(L_0)$ by $\mathrm{WSp}(L_0, I)$, occurs as group scheme in the unipotent extension $\mathbf{H}_g[(I \otimes I)^s, L_0 \otimes I]$; **2.3.4**

$V_{\mathbf{X}}$: $W_{\mathbf{X}}/U_{\mathbf{X}}$; **2.2.10**

$W_{\mathbf{X}}$: unipotent radical of $P_{\mathbf{X}}$; **2.2.10**

w: natural (weight) morphism $\mathbb{G}_{\mathrm{m},\mathbb{R}} \to \mathbb{S}$; **2.2**

$\mathbb{W}(E)$: 'additive group'-scheme for a coherent sheaf E on S, representing $\mathbb{W}(E)(S') = (E \otimes_{\mathcal{O}_S} \mathcal{O}_{S'})(S')$; **2.1.2**

$\mathrm{Weil}(L_{\mathbb{Z}}^*/L_{\mathbb{Z}})$: denotes the Weil representation restricted to $\mathbb{C}[L_{\mathbb{Z}}^*/L_{\mathbb{Z}}] \subset S(L_{\mathbb{A}^{(\infty)}})$ equipped with natural $\mathbb{Q}$ and $\mathbb{Z}$-structures; **3.2.2**

$\mathrm{WSp}(L_0, I)$: a central extension of $\mathbb{W}(I \otimes L_0)$ by $\mathbb{W}((I \otimes I)^s)$; **2.3.4**

$\Xi_{\mathbf{X}}$: morphism of (Artin) stacks $\mathrm{M}({}^K_\Delta\mathbf{X}) \to [\mathrm{M}^\vee(\mathbf{X})/P_{\mathbf{X}}]$ encoding $\mathrm{P}({}^K_\Delta\mathbf{X})$ and Π; **2.5.14**

$(\Xi^*\mathcal{L})_V$: rational bundle of vector valued modular forms associated with a τ-semilinear representation of $\mathrm{GL}_2(\widehat{\mathbb{Z}})$; **3.3.4**

X: p-integral mixed Shimura data consisting of $P_{\mathbf{X}}$, $\mathbb{D}_{\mathbf{X}}$, $h_{\mathbf{X}}$; **2.2.3**

$\mathbf{X}[U, V]$: unipotent extension, where U, V are representations of $P_{\mathbf{X}}$ with symplectic form $V \times V \to U$ (implicit in the notation), satisfying certain axioms, we always have an iso $\mathbf{X} \cong \mathbf{X}/W_{\mathbf{X}}[U_{\mathbf{X}}(\mathbb{Z}_{(p)}), V_{\mathbf{X}}(\mathbb{Z}_{(p)})]$; **2.2.11**

${}^K_\Delta\mathbf{X}$: extended compactified p-integralmixed Shimura datum (p-ECMSD), consisting of a p-integral mixed Shimura datum $\mathbf{X}$, a p-admissible compact open subgroup $K \subset P_{\mathbf{X}}(\mathbb{A}^{(\infty)})$, and a (partial) rational cone decomposition Δ of $C_{\mathbf{X}}$; **2.2.24**

$(\alpha, \rho)\colon {}^{K'}_{\Delta'}\mathbf{Y} \to {}^K_\Delta\mathbf{X}$: a morphism of p-ECMSD, consisting of a morphism α of p-integral mixed Shimura data and $\rho \in P_{\mathbf{X}}(\mathbb{A}^{(\infty,p)})$, satisfying some compatibility conditions; **2.2.24**

Y: p-integral mixed Shimura data consisting of $P_{\mathbf{Y}}, \mathbb{D}_{\mathbf{Y}}, h_{\mathbf{Y}}$; **2.2.3**

$\alpha\colon \mathbf{X} \to \mathbf{Y}$: morphism of p-integral Shimura data; **2.2.3**

$\mathbb{Z}(1)$: Tate-twist $2\pi i\mathbb{Z} \subset \mathbb{C}$

$\mathbb{Z}_l(1)$: Tate-twist $\varprojlim_n \mu_{l^n}$ with its natural Galois-action

$\widehat{\mathbb{Z}}(1)$: Tate-twist $\prod_l \mathbb{Z}_l(1)$ with its natural Galois-action

$\mathrm{Z}(L, M, \varphi; K)$: special cycle on an orthogonal Shimura variety, M a quadratic space over $\mathbb{Q}$, $\varphi \in S(M^*_{\mathbb{A}^{(\infty)}} \otimes L_{\mathbb{A}^{(\infty)}})$ and K a compact open subgroup of $\mathrm{SO}(\mathbb{A}^{(\infty)})$ fixing φ; **3.1.6**

$\mathrm{Z}(L_{\mathbb{Z}}, M_{\mathbb{Z}}, \kappa)$: same as before, with φ equal to the characteristic function of $\kappa \in (L^*_{\widehat{\mathbb{Z}}}/L_{\widehat{\mathbb{Z}}}) \otimes M^*_{\mathbb{Z}}$ and $K = \mathrm{SO}'(L_{\widehat{\mathbb{Z}}})$; **3.1.6**

Selected Published Titles in This Series

35 **Fritz Hörmann,** The Geometric and Arithmetic Volume of Shimura Varieties of Orthogonal Type, 2014

34 **Leonid Polterovich and Daniel Rosen,** Function Theory on Symplectic Manifolds, 2014

33 **Libor Šnobl and Pavel Winternitz,** Classification and Identification of Lie Algebras, 2014

32 **Pavel Bleher and Karl Liechty,** Random Matrices and the Six-Vertex Model, 2014

31 **Jean-Pierre Labesse and Jean-Loup Waldspurger,** La Formule des Traces Tordue d'après le Friday Morning Seminar, 2013

30 **Joseph H. Silverman,** Moduli Spaces and Arithmetic Dynamics, 2012

29 **Marcelo Aguiar and Swapneel Mahajan,** Monoidal Functors, Species and Hopf Algebras, 2010

28 **Saugata Ghosh,** Skew-Orthogonal Polynomials and Random Matrix Theory, 2009

27 **Jean Berstel, Aaron Lauve, Christophe Reutenauer, and Franco V. Saliola,** Combinatorics on Words, 2008

26 **Victor Guillemin and Reyer Sjamaar,** Convexity Properties of Hamiltonian Group Actions, 2005

25 **Andrew J. Majda, Rafail V. Abramov, and Marcus J. Grote,** Information Theory and Stochastics for Multiscale Nonlinear Systems, 2005

24 **Dana Schlomiuk, AndreĭA. Bolibrukh, Sergei Yakovenko, Vadim Kaloshin, and Alexandru Buium,** On Finiteness in Differential Equations and Diophantine Geometry, 2005

23 **J. J. M. M. Rutten, Marta Kwiatkowska, Gethin Norman, and David Parker,** Mathematical Techniques for Analyzing Concurrent and Probabilistic Systems, 2004

22 **Montserrat Alsina and Pilar Bayer,** Quaternion Orders, Quadratic Forms, and Shimura Curves, 2004

21 **Andrei Tyurin,** Quantization, Classical and Quantum Field Theory and Theta Functions, 2003

20 **Joel Feldman, Horst Knörrer, and Eugene Trubowitz,** Riemann Surfaces of Infinite Genus, 2003

19 **L. Lafforgue,** Chirurgie des grassmanniennes, 2003

18 **G. Lusztig,** Hecke Algebras with Unequal Parameters, 2003

17 **Michael Barr,** Acyclic Models, 2002

16 **Joel Feldman, Horst Knörrer, and Eugene Trubowitz,** Fermionic Functional Integrals and the Renormalization Group, 2002

15 **José I. Burgos Gil,** The Regulators of Beilinson and Borel, 2002

14 **Eyal Z. Goren,** Lectures on Hilbert Modular Varieties and Modular Forms, 2002

13 **Michael Baake and Robert V. Moody, Editors,** Directions in Mathematical Quasicrystals, 2000

12 **Masayoshi Miyanishi,** Open Algebraic Surfaces, 2000

11 **Spencer J. Bloch,** Higher Regulators, Algebraic K-Theory, and Zeta Functions of Elliptic Curves, 2000

10 **James D. Lewis and B. Brent Gordon,** A Survey of the Hodge Conjecture, Second Edition, 1999

9 **Yves Meyer,** Wavelets, Vibrations and Scalings, 1998

8 **Ioannis Karatzas,** Lectures on the Mathematics of Finance, 1997

7 **John Milton,** Dynamics of Small Neural Populations, 1996

6 **Eugene B. Dynkin,** An Introduction to Branching Measure-Valued Processes, 1994

For a complete list of titles in this series, visit the AMS Bookstore at **www.ams.org/bookstore/crmmseries/**.